NANO-BIOREMEDIATION FOR WATER AND SOIL TREATMENT

An Eco-Friendly Approach

NANO-BIOREMEDIATION FOR WATER AND SOIL TREATMENT

An Eco-Friendly Approach

Edited by

Vishnu D. Rajput, PhD

Arpna Kumari, PhD

Tatiana M. Minkina, PhD

APPLE ACADEMIC PRESS

First edition published 2024

Apple Academic Press Inc.
1265 Goldenrod Circle, NE,
Palm Bay, FL 32905 USA

760 Laurentian Drive, Unit 19,
Burlington, ON L7N 0A4, CANADA

CRC Press
2385 NW Executive Center Drive,
Suite 320, Boca Raton FL 33431

4 Park Square, Milton Park,
Abingdon, Oxon, OX14 4RN UK

Apple Academic Press exclusively co-publishes with CRC Press, an imprint of Taylor & Francis Group, LLC

Library and Archives Canada Cataloguing in Publication

Title: Nano-bioremediation for water and soil treatment : an eco-friendly approach / edited by Vishnu D. Rajput, PhD, Arpna Kumari, PhD, Tatiana M. Minkina, PhD.
Names: Rajput, Vishnu D., editor. | Kumari, Arpna, editor. | Minkina, Tatiana M., editor.
Description: First edition. | Includes bibliographical references and index.
Identifiers: Canadiana (print) 20230599249 | Canadiana (ebook) 20230599257 | ISBN 9781774914861 (hardcover) | ISBN 9781774914878 (softcover) | ISBN 9781003414964 (ebook)
Subjects: LCSH: Bioremediation. | LCSH: Nanostructured materials—Environmental aspects. | LCSH: Water—Purification—Materials. | LCSH: Soil remediation.
Classification: LCC TD192.5 .N36 2024 | DDC 628.5—dc23

ISBN: 978-1-77491-486-1 (hbk)
ISBN: 978-1-77491-487-8 (pbk)
ISBN: 978-1-00341-496-4 (ebk)

About the Editors

Vishnu D. Rajput, PhD
Associate Professor, Southern Federal University, Russia

Vishnu D. Rajput, PhD, is working as a Leading Researcher (Associate Professor) at Southern Federal University, Rostov-on-Don, Russia. His ongoing research is based on the toxic effects of bulk- and nano-forms of metals and investigating the bioaccumulation, bio/geo-transformations, uptake, translocation, and toxic effects of bulk- and nano-forms of metals on plant physiology, morphology, anatomy, the ultrastructure of cellular and subcellular organelles, cytomorphometric modifications, and DNA damage. He is also discovering possible remediation approaches, such as using biochar/nano-biochar-based sorbents and nanotechnology. Through long experience and experimental work, he comprehensively detailed the state of environmental science research regarding how nanoparticles/heavy metals interact with plants, soil, microbial community, and the larger environment, as well as possible remediation technology using nanoparticles/nano-biochar.

He has published a total of 325 scientific publications, including 188 peer-reviewed full-length articles, 16 books (Springer, Elsevier, CRC Press, Nova USA), 64 chapters (Scopus-indexed), and 41 conference articles, and achieved an h-index: 32 by Scopus with 4,335 citations as of November 2023. He is an internationally recognized reviewer, having reviewed 199 manuscripts, and has received outstanding reviewing certificates from Elsevier, Springer, and Plants-MDPI. He is an editorial board member of various high-impact journals, such as *Eurasian Journal of Soil Science*, *Biochar*, *Journal of Soil Science and Agroclimatology*, and others.

He is a co-PI, including Mega-Grant, or performer of several national and international grants such as the RSF project assessment of the state of contaminated soils and plants using synchrotron methods; RFBR project on "Phytoremediation Potential of Plants under Conditions of Technogenic Soil Pollution;" the RSF project "Ecological and Geochemical Patterns of Formation of Natural and Anthropogenic Fluxes of Substances in the Soils of the Mouth Area of the Don River and the Coast of the Taganrog Bay," granted by the Russian Government; "Application of Microorganism in Agriculture and Allied Sector" (AMAAS) subtheme: 'Bioremediation of Wastewater for Heavy Metals," funded by ICAR; "Evaluation, Selection, and Application of Mycorrhizae as Bio-Hardening Agents in Tissue Culture-Raised Plants," funded by HSCS&T by the Indian govt.

He received the Certificates for Appreciation (2019–2021), Certificate of Honor 2020, and a Diploma Award 2021 from Southern Federal University, Russia, for outstanding contributions to his academic, creative research, and publication activities. He has also received Highly Qualified Specialist status from the Russian government.

Arpna Kumari, PhD
Former Postdoctoral Researcher, Southern Federal University, Russia; JSPS Postdoc Fellow, The University of Tokyo, Japan

Arpna Kumari, PhD, is currently working as a Postdoctoral Researcher at the Academy of Biology and Biotechnology, Southern Federal University, Russia. She formerly as a Senior Researcher at the Academy of Biology and Biotechnology, Southern Federal University, Russia. After that, she joined The University of Tokyo, Japan, as a Postdoctoral Fellow under the JSPS Standard Fellowship program. She has seven years of research experience in plant science with special emphasis on various abiotic stressors; stress mitigation strategies; applications of nanotechnology in sustainable agriculture and contaminated soil restoration; cytotoxicity and genotoxicity of emerging pollutants and endocrine disruptors. Currently, she is dealing with plant nutrition. She has published 71 scientific publications including 51 full-length articles and 20 book chapters, (51 Scopus-indexed; Scopus h-index: 12) and presented her research at many national and international conferences. She has reviewed more than 70 manuscripts for journals of national and international repute. Dr. Kumari has earned BSc and MSc (Botany) degrees with good academic records from Himachal Pradesh University, Shimla, India. She has earned her doctoral degree in Life Sciences with a specialization in Botany from Guru Nanak Dev University (Punjab), India.

Tatiana M. Minkina, PhD

Head, Department of Soil Science and Land Evaluation, Southern Federal University, Russia

Tatiana M. Minkina, PhD, is the Head of the Soil Science and Land Evaluation Department of Southern Federal University. She is also the Head of International Master's Degree Educational Program Management and Estimation of Land Resources (2015–2022, accreditation by ACQUIN). Her areas of scientific interest are soil science, biogeochemistry of trace elements; environmental soil chemistry; and soil monitoring, assessment, modeling, and remediation using physicochemical treatment methods. She was awarded in 2015 with a Diploma from the Ministry of Education and Science of the Russian Federation for many years of long-term work for the development and improvement of the educational process, a significant contribution to the training of highly qualified specialists. Currently, she is handling projects funded by the Russian Scientific Foundation, the Ministry of Education and Science of the Russian Federation, and the Russian Foundation of Basic Research. She is a member of the Expert Group of the Russian Academy of Science; the International Committee on Contamination Land; Eurasian Soil Science Societies; the International Committee on Protection of the Environment; and the International Scientific Committee of the International Conferences on Biogeochemistry of Trace Elements. She has 757 total scientific publications (389 in English). She is also an invited editor of an open access journal by MDPI (impact factor: 2.524) and editorial board member of *Geochemistry: Environment, Exploration, Analysis* and *Eurasian Journal of Soil Science.*

Contents

Contributors

Hazrat Amin
Academy of Biology and Biotechnology, Southern Federal University, Rostov-on-Don, Russia

Pérez-Moreno Andrea
Sustainability of Natural Resources and Energy Program, Cinvestav-Saltillo, Coahuila, Mexico

Jayati Arora
Amity Institute of Environmental Sciences, Amity University, Noida, Uttar Pradesh, India

Muhammad Arslan Ashraf
Department of Botany, Government College University, Faisalabad, Pakistan

Ritu Bala
Department of Chemistry, Guru Nanak Dev University, Amritsar, Punjab, India

Priyanka Balan
Department of Environmental Science, Dr. Y.S. Parmar University of Horticulture and Forestry, Solan, Himachal Pradesh, India

Zaffar Bashir
Center of Research for Development (CORD), University of Kashmir, Srinagar, Jammu and Kashmir, India

Rupesh Kumar Basniwal
Amity Institute of Advanced Research and Studies (M&D), Amity University, Noida, Uttar Pradesh, India

Akshi Bhardwaj
Department of Botanical and Environmental Sciences, Guru Nanak Dev University, Amritsar, Punjab, India

Abhishek Chauhan
Amity Institute of Environmental Toxicology Safety and Management, Amity University, Noida, Uttar Pradesh, India

Lakha V. Chopda
Government Engineering College, Bhuj, Gujarat, India

Pragnesh N. Dave
Department of Chemistry, Sardar Patel University, Vallabh Vidyanagar, Gujarat, India

Valle-García Jessica Denisse
Transdisciplinary Doctoral Program in Scientific and Technological Development for the Society, Cinvestav-Zacatenco, Mexico City, Mexico

Shubham Durgude
School of Agriculture, Graphic Era Hill University, Bhimtal, Uttarakhand, India

Vázquez-Núñez Edgar
El Colegio de la Frontera Sur, Agroecología, Unidad Campeche, Campeche, Mexico

Fernández-Luqueño Fabián
Sustainability of Natural Resources and Energy Program, Cinvestav-Saltillo, Coahuila, Mexico

Sukanya Ghosh
School of Agriculture, Graphic Era Hill University, Dehradun, Uttarakhand, India

Burhan Hamid
Center of Research for Development (CORD), University of Kashmir, Srinagar, Jammu and Kashmir, India

Amara Hassan
Department of Botany, Government College University, Faisalabad, Pakistan

Pérez-Hernández Hermes
El Colegio de la Frontera Sur, Agroecología, Unidad Campeche, Campeche, Mexico

Humaira
Center of Research for Development (CORD), University of Kashmir, Srinagar, Jammu and Kashmir, India

Iqbal Hussain
Department of Botany, Government College University, Faisalabad, Pakistan

Vera-Reyes Ileana
Department of Plastics in Agriculture, Centro de Investigación en Química Aplicada, Saltillo, Coahuila, Mexico

Naeem Iqbal
Department of Botany, Government College University, Faisalabad, Pakistan

Subheet Kumar Jain
Department of Pharmaceutical Sciences, Guru Nanak Dev University, Amritsar, Punjab, India

V. K. Jain
Amity Institute of Advanced Research and Studies (M&D), Amity University, Noida, Uttar Pradesh, India

Akanksha Jasrotia
Department of Botanical and Environmental Sciences, Guru Nanak Dev University, Amritsar, Punjab, India

Tanu Jindal
Amity Institute of Environmental Toxicology Safety and Management, Amity University, Noida, Uttar Pradesh, India

Rajinder Kaur
Department of Botanical and Environmental Sciences, Guru Nanak Dev University, Amritsar, Punjab, India

Muhammad Kamran Khan
Government College University, Faisalabad, Pakistan

Arpna Kumari
Academy of Biology and Biotechnology, Southern Federal University, Stachki, Rostov-on-Don, Russia

Rinky Kumari
Department of Botanical and Environmental Sciences, Guru Nanak Dev University, Amritsar, Punjab, India

Saqib Mahmood
Department of Botany, Government College University, Faisalabad, Pakistan

Saglara S. Mandzhieva
Academy of Biology and Biotechnology, Southern Federal University, Stachki, Rostov-on-Don, Russia

Tatiana M. Minkina
Academy of Biology and Biotechnology, Southern Federal University, Stachki, Rostov-on-Don, Russia

Khair Ul Nisa
Department of Environmental Science, Center of Research for Development (CORD), University of Kashmir, Srinagar, Jammu and Kashmir, India

Qadrul Nisa
Division of Plant Pathology, SKUAST-K, Shalimar, Srinagar, Jammu and Kashmir, India

Abida Parveen
Department of Botany, Government College University, Faisalabad, Pakistan

Shagufta Perveen
Department of Botany, Government College University, Faisalabad, Pakistan

C. Akshaya Prakash
Department of Botany, St. Joseph's College (Autonomous), Devagiri, Kozhikode, Kerala, India

Evgenya V. Prazdnova
Academy of Biology and Biotechnology, Southern Federal University, Stachki, Rostov-on-Don, Russia

Jos T. Puthur
Plant Physiology and Biochemistry Division, Department of Botany, University of Calicut, C.U. Campus P.O., Kerala, India

Preeti Raina
Department of Botanical and Environmental Sciences, Guru Nanak Dev University, Amritsar, Punjab, India

Pushap Raj
Department of Botanical and Environmental Sciences, Guru Nanak Dev University, Amritsar, Punjab, India

Priyadarshani Rajput
Academy of Biology and Biotechnology, Southern Federal University, Rostov-on-Don, Russia

Sneh Rajput
Department of Botanical and Environmental Sciences, Guru Nanak Dev University, Amritsar, Punjab, India

Vishnu D. Rajput
Academy of Biology and Biotechnology, Southern Federal University, Stachki, Rostov-on-Don, Russia

Anuj Ranjan
Academy of Biology and Biotechnology, Southern Federal University, Stachki, Rostov-on-Don, Russia

Rizwan Rasheed
Department of Botany, Government College University, Faisalabad, Pakistan

Prangya Rath
Amity Institute of Environmental Sciences, Amity University, Noida, Uttar Pradesh, India

Sarabia-Castillo Cesar Roberto
Sustainability of Natural Resources and Energy Program, Cinvestav-Saltillo, Coahuila, Mexico

Neha Sahu
Department of Botany, University of Lucknow, Uttar Pradesh, India

Nair G. Sarath
Plant Physiology and Biochemistry Division, Department of Botany, University of Calicut, C.U. Campus P.O., Kerala, India; Department of Botany, Mar Athanasius College (Autonomous), Kothamangalam, Kerala, India

Delse Parekkattil Sebastian
Department of Botany, St. Joseph's College (Autonomous), Devagiri, Kozhikode, Kerala, India

Gauri Sharma
Department of Botanical and Environmental Sciences, Guru Nanak Dev University, Amritsar, Punjab, India

Rashmi Sharma
School of Agriculture, Graphic Era Hill University, Dehradun, Uttarakhand, India

Sudhir S. Shende
Academy of Biology and Biotechnology, Southern Federal University, Rostov-on-Don, Russia

Priya Shrivastava
Amity Institute of Advance Research and Studies (M&D), Amity University, Noida, Uttar Pradesh, India

Shiv Vendra Singh
College of Agriculture, Rani Lakshmi Bai Central Agricultural University, Jhansi, UP, India

Khalid Sultan
Department of Botany, Government College University, Faisalabad, Pakistan

Svetlana N. Sushkova
Academy of Biology and Biotechnology, Southern Federal University, Stachki, Rostov-on-Don, Russia

Najeebul Tarfeen
Center of Research for Development (CORD), University of Kashmir, Srinagar, Jammu and Kashmir, India

Shubham Thakur
Department of Pharmaceutical Sciences, Guru Nanak Dev University, Amritsar, Punjab, India

Saba Wani
Department of Biochemistry, University of Kashmir, Srinagar, Jammu and Kashmir, India

Shabir H. Wani
Mountain Research Center for Field Crops, Khudwani, Sher-e-Kashmir University of Agricultural Sciences and Technology, Srinagar, Jammu and Kashmir, India

Ali Mohd. Yatoo
Center of Research for Development (CORD), University of Kashmir, Srinagar, Jammu and Kashmir, India

Sara Zafar
Department of Botany, Government College University, Faisalabad, Pakistan

Anton Zhumbei
Academy of Biology and Biotechnology, Southern Federal University, Rostov-on-Don, Russia

Abbreviations

2,4-DCP	glyphosate
AA	*Aquilaria agallocha*
AAS	atomic absorption spectroscopy
AC	activated carbon
ADHD	attention-deficit/hyperactivity disorder
AGS	anaerobic granular sludge
Am	americium
AMAAS	application of microorganisms in agriculture and allied sector
AOP	advanced oxidation process
AR	Allura red
As	arsenic
ASV	anodic stripping voltammetry
BAC	biological accumulation coefficient
BAP	benzo(a)pyrene
BB	brilliant blue
BB	bromothymol blue
BC	bacterial cellulose
BC	biochar
BCF	bioconcentration factor
BCF	biological transfer coefficient
BIS	Bureau of Indian Standards
BNPs	biogenic nanoparticles
BPA	bisphenol A
BPB	bromophenol blue
BTB	bromothymol blue
CA	calcium alginate
CB	conduction band
CCNH	carbon-carbon NHs
CE	capillary electrophoresis
CF	chloroform
CIC	carbo–iron colloids

CMC	carboxymethyl cellulose
CMNH	carbon-metal NHs
CNC	cellulose nanocrystals
CNF	cellulose nanofibrils
CNFs	carbonaceous nanofibers
CNM	chitosan nanofiber mats
CNs	carbonaceous nanomaterials
CNTs	carbon nanotubes
CORD	Center of Research for Development
CP	chlorpyrifos
CR	Congo red
CS	chitosan
CS2	carbon disulfide
CS-CA	chitosan and activated carbon
CS-PVA	chitosan/polyvinyl alcohol
CTC	carbon tetrachloride
CTMAB	hexadecyltrimethylammonium bromide
CVD	chemical vapor deposition
DB	direct blue
DBC	dibenzo[def,p]chrysene
DCMD	direct membrane distillation
DDT	dichloro-diphenyl trichloroethane
DNAPLs	dense nonaqueous phase liquids
DNOA	di-n-octylamine
EB	Evans blue
EBT	eriochrome Black T
EC	emerging contaminants
EDTA	ethylene diamine tetra acetic acid
EIA	environmental impact assessment
EK	electric kinetic remediation
ENMs	electrospun nanofiber membranes
ENMs	engineered nanomaterials
Eu	europium
EY	eosin Y
F	fluoride
Fe	iron
Fe-NFM	nanofiber mat

GAC	granulated activated carbon
GEMs	genetically engineered microorganisms
GM	genetically modified
GO	graphene oxide
GO-COOH/CS	carboxylated graphene oxide-chitosan
GO-CSs	graphene-chitosan sponges
GOQDs	graphene oxide quantum dots
GQD	graphene quantum dots
GR	graphene
GR	green remediation
GS	green S
GSDR	Global Sustainable Development Report
h^+	holes
HA	humic acid
HAP	hydroxyapatite
HAp/NaP	hydroxyapatite/zeolite nanocomposite
HCH	hexachlorocyclohexane
HM	heavy metal
HMO	hydrous manganese oxide
HMs	heavy metals
HZO	hydrous zirconia
IAV	influenza A virus
IMI	imidacloprid
IONP	iron oxide NPs
KGM	konjac glucomannan
LCA	life-cycle assessments
LFBC	lignocellulosic fibers reinforced with biodegradable composites
MB	methylene blue
MBT	2-mercaptobenzothiazole
MG	malachite green
MMNH	metal-metal NHs
MNMs	micro/nanomotors
MNPs	magnetic nanoparticles
MO	methyl orange
MOFs	metal-organic frameworks
MOs	microorganisms
MPAC	multipore activated carbon

MPs	microprobes
MR	methyl red
MSA	mercaptosuccinic acid
MTP	metoprolol
MV	methyl violet
MWCNTs	multi-walled carbon nanotubes
MZ	mancozeb
NBR	nanobioremediation
NF	nanofiltration
NH_2-SG	amino-functionalized silica gel
NH_2-SNHS	amino-functionalized silica nano-hollow sphere
NHs	nanohybrids
NMOs	nanocrystalline metal oxides
NMs	nanomaterials
NO_3^-	nitrate
Np	neptunium
NPCs	nanophotocatalysts
NPL	National Priorities List
NPs	nanoparticles
NT	nanotechnology
NZVI	nanoscale zero-valent iron
O_2^-	superoxide radicals
OA	oleic acid
OECD	Organization for Economic Cooperation and Development
OH^-	hydroxyl
OMCNHs	organo-metal-carbon NHs
PAA/GO/Fe_3O_4	polyacrylic acid/graphene oxide/Fe_3O_4
PAC	powdered activated carbon
PAC/UF	PAC-ultrafiltration
PAFE	*Plumeria alba* flower extract
PAHs	polycyclic aromatic hydrocarbons
PAMAM	polyamidoamine
PAN	polyacrylonitrile
PANI	polyaniline
PANI/PPy	polypyrrole/polyacrylonitrile
PCBs	polychlorinated biphenyls
PCDD	dibenzo-p-dioxins

PCDFs	polychlorinated dibenzofurans
PCE	tetrachlorethylene
PCHA-SiO_2	silica-sphere-poly(catechol hexamethylenediamine)
PDA	polydopamine
PEDOT/PSS	poly(3,4-ethylenedioxythiopheneethylene dioxythiophene)/ polystyrene sulfonate
PEG	polyethylene glycol
PEI	poly(ethyleneimine)
PFCA	polyvinyl fluoride cellulose acetate
PFNCs	polymer functionalized nanocomposites
PL	photoluminescence
PLS	polymer/layered silicate
PM	particulate matter
PMMA	poly(methyl methacrylate)
PNCs	polymer nanocomposites
POPs	are persistent organic pollutants
POSS	polyhedral oligomeric silsesquioxane
PR	phenol red
PRB	permeable reactive barrier
PS	polystyrene
PTO	phosphate TiO_2
Pu	plutonium
PVA	polyvinyl alcohol
QDs	quantum dots
RB	reactive black
RB	rhodamine B
RGO	reduced graphene oxides
RL	rhamnolipid
RNZVI	rhamnolipid-stabilized nanoscale zero-valent iron
ROS	reactive oxygen species
RUSA	Rashtriya Uchchatar Shiksha Abhiyan
SAM	self-assembled monolayers
SANPs	salicylic acid NPs
SD	sustainable development
SDC	Samaria-doped ceria
SDC-F	spherical SDC
SDC-I	cluster plate SDC

SDGs	sustainable development goals
SDS	sodium dodecyl sulfate
SER	surfactant enhanced remediation
SMT	sulfamethazine
SPI	soy protein isolate
SPR	plasmon resonance
Sr	strontium
SR	sustainable remediation
SWCNTs	single-walled carbon nanotubes
Tc	technetium
TC	tetracycline
TCE	trichloroethylene
TCP	tetrachlorophenol
TFC	thin-film composite
Th	thorium
TH	thyroid hormone
TOC	total organic carbon
TPHs	total petroleum hydrocarbons
U	uranium
UFAMs	ultrafiltration-adsorption membranes
USEPA	United States Environmental Protection Agency
VB	valence band
VB	Victoria Blue
VOCs	volatile organic compounds
WCA	water contact angle
WCED	World Commission on Environment and Development
WHO	World Health Organization
XAFS	X-ray absorption fine structure
XFS	X-ray fluorescence spectroscopy
ZVI	zero-valent iron
ZVMs	zero-valent metals

Foreword 1

The paucity of innovative and effective technologies for the eradication of persistent environmental contaminants is a major problem for the scientific community worldwide. In this regard, nano-remediation has emerged as a potentially advancing method for combating environmental pollution. Due to the unique features of engineered nanomaterials, nano-remediation is a revolutionary method for the safe and long-term removal of several types of contaminants. This in-depth book titled, *Nano-Bioremediation for Water and Soil Treatment: An Eco-Friendly Approach,* edited by Dr. Vishnu D. Rajput, Dr. Arpna Kumari, and Prof. Dr. Tatiana M. Minkina offers a significant viewpoint on the most recent breakthroughs and future perspectives of nano-remediation technologies utilized for environmental decontamination, such as nano-photocatalysis, nano-bioremediation, nano-phytoremediation, and nano-sensing. Technologies for the mitigation of heavy metals, pesticides, dyes, and other xenobiotics from polluted land and aquatic environments has been summarized in this book. It then discusses the environmental effects of nanotechnology and gives suggestions on how to use nanotechnology to improve the present and the future.

—Prof. Dr. Rer. Nat. Hassan El-Ramady

Kafrelsheikh University, Egypt
Associate Editor in Frontiers in Soil Science,
Environmental Nanotechnology, Monitoring, and Management,
Egyptian Journal of Soil Science,
Journal of Sustainable Agricultural Sciences,
Editor in Chief of Environment, Biodiversity and Soil Security,
Lead Editor of the book "The Soil of Egypt."

Foreword 2

A burgeoning population, urbanization, and industrial development have generated a considerable amount of contaminants in *the soil-water-plant-animal continuum*, constructing a menace to the ecosystem and human health. Taking up this unprecedented challenge and looking at the available feasible smart options, nanotechnology has emerged as an emerging science that could be employed confidently and consistently. This newly developed strategy based on the use of versatile nanoparticles has received significant interest around the world. Thus, in this book, nanoremediation approaches have been thoroughly explored to provide an overview of their decontamination efficiency with regards to environmental pollutants in polluted soils and water bodies and identify research gaps and future prospects has been identified.

Consequently, I appreciate the efforts made by the editors to choose a topic focused on sustainable and safe approaches for the restoration of polluted ecosystems for this book, keeping in mind sustainably managing its natural resources and taking urgent action as per the 2030 agenda. This book has substantial scientific worth and will be a valuable addition to academic libraries and a helpful resource for academics, scientists, students, policymakers, and industry professionals interested in studying eco-friendly and viable remediation approaches. I commend the editors and authors for bringing out such a timely, unique, and important book.

—Prof. Dr. Amitava Rakshit

FSES, FTWAS.Nxt (Italy), FBiovision.Nxt (France), FCWSS, FSBSRD
Review College Member (British Ecological Society, London)
Member of Global Land Program (https://glp.earth/user/3930)
Department of Soil Science and Agricultural Chemistry,
Institute of Agricultural Science,
Banaras Hindu University (www.bhu.ac.in), Varanasi,
Uttar Pradesh, India

Preface

Global key challenges, such as environmental pollution, climate change, food security, soil security, and energy, pose a significant threat to life on planet earth. Frequent emissions of harmful chemical compounds through natural and anthropogenic processes are polluting diverse environmental matrices, with direct and indirect implications on surrounding flora and fauna, including humans. All contaminants, whether persistent or nonpersistent have a significant influence on the structural and functional characteristics of different ecosystems. These pollutants may originate from a variety of sources, including industrial operations, transportation, mining operations, and livestock farming.

The aggravation of environmental pollution not only impedes sustainable land management but also, directly or indirectly, every ecosystem and lifeform. Thus, soil and water pollution have become a worldwide problem, requiring the deployment of efficient remediation procedures. Therefore, international cooperation between nations should be developed to address global pollution. In this context, to solve this international concern, the emerging application of nanotechnology offers immense promise for the cleanup of contaminated sites. At the same time, nanotechnology's ecotoxicity is still a matter of debate. However, nanoremediation is a potential developing method for addressing environmental pollution, particularly resistant pollutants.

The use of nanoremediation has the potential not only to reduce the overall costs of cleaning up large-scale contamination but also to reduce cleaning frequency, eliminate the need for treatment and removal of contaminated components, and reduce toxin absorption. In light of this, a more comprehensive evaluation of polluted sites with various organic and inorganic contaminants treated using nanoremediation has been prioritized in this book.

It is, therefore, important to gather knowledge about the most recent advances in nanoremediation, especially emergent contaminants that require specific and, in several cases, expensive treatments. The objective of this book is to gather valuable information for the decontamination of soil and water resources because they have a direct influence on human life as well as on agricultural activities. We owe a debt of gratitude to all of the contributors whose efforts made this book a reality.

—Editors

CHAPTER 1

Introductory Overview of Nanobioremediation

VISHNU D. RAJPUT, ARPNA KUMARI, TATIANA M. MINKINA, ANUJ RANJAN, SAGLARA S. MANDZHIEVA, HAZRAT AMIN, SUDHIR S. SHENDE, PRIYADARSHANI RAJPUT, and SVETLANA N. SUSHKOVA

Academy of Biology and Biotechnology, Southern Federal University, Rostov-on-Don, Russia

ABSTRACT

Soil degradation and water pollution due to intensive agriculture and industrialization are global issues and major threats to sustainable crop production. Numerous technical innovations have been used to clean up soil or increase the productivity of degraded soils and contaminated waterbodies, but since they are expensive, impractical in real-world settings, or labor-intensive to a greater extent, they fail to do so, leaving the problem unresolved. Recently, the nanotechnological approaches ensure enhanced crop yield as well as improvement in soil and water quality parameters without disturbing the environment. Various approaches of nanotechnology, such as combining nanoparticles with soil amendment, enhancing the growth of hyperaccumulators, increasing soil microbial functionalities to degrade or change the state of soil pollutants, improving plant root system in soil, and helping to uptake by plant unavailable nutrients could restore degraded or polluted soil. The present chapter envisaged the current status of nanotechnology acceptance in decontaminating polluted soils and water and explored future perspectives.

Nano-Bioremediation for Water and Soil Treatment: An Eco-Friendly Approach.
Vishnu D. Rajput, Arpna Kumari, and Tatiana M. Minkina (Eds.)

1.1 INTRODUCTION

Soil, a natural body, is the organic-carbon-mediated realm that contains liquid, solid, and gaseous phases that interact at a range of scales and make abundant ecosystem goods and services. Soil is fundamental to life on Earth; it represents the very base of our food production; in fact, 95% of the food worldwide is produced in soil [1]. Nevertheless, the soil is a vital living ecosystem that sustains plants and animals; thus, soils have diverse functions that help to mitigate and adapt to climate change, filter water, and improve resilience to floods and droughts [2]. However, significant negative environmental externalities are being produced by population increase, excessive use of natural resources, pollution, and waste created during the manufacture of consumer products. Additionally, it poses a serious risk to the long-term viability of these resources. In recent decades, there have been growing concerns about continued population growth and the ill effects on soil's biochemical, physical, and chemical properties [3]. A study carried out showed that by the year 2030, the world population will reach a size of 8.6 billion people. In addition, it has been estimated that by 2050, the demand for food production will increase by 50% [4]. This will cause an expansion in the agricultural frontier, and it could turn the intensification of the use of the territory to obtain food since more and more food is needed. Healthy soils are the key to food security and our sustainable future [5]. Soil acts as a buffer against pollutants such as agricultural chemicals, organic wastes, and industrial chemicals. Nevertheless, soil contamination is the main global threat to soil resources [6, 7]. The soil is contaminated when the presence of certain chemical components from human activity alters its characteristics and may present a risk to human health and the environment. The effects of soil contamination have increased the concern in initiatives that allow remediation of polluted sites that have been affected by a diverse range of contaminants.

Besides the soil pollution relevant to food security concerns, the application of reclaimed water for irrigation has been reported as a major risk as it contaminates crops, especially leafy ones, which are consumed regularly as salads [8, 9]. In usual circumstances, fresh water is used for irrigation, but this is not the case worldwide. In arid or semi-arid regions where water scarcity poses a serious threat to agricultural viability, the reliability of farmers is reported to increase on marginal water sources, including treated wastewater [10]. In addition, irrigation with treated surface water is a widespread approach in many developing nations. In some regions, municipal, industrial, and agricultural runoff water has also been documented to be used for irrigation purposes [11, 12]. Thus, the contamination of agricultural soil and edible products by the usage of reclaimed water is one of the developing world's most serious environmental and public health issues. Plus, Long-term irrigation with polluted water may contribute to the accumulation of harmful contaminants in surface soil above the permissible background level [13]. Environmental components that are essential for agriculture have been polluted to varying degrees with inorganic and organic, which in recent years has resulted in consistent industrial growth as well as numerous other anthropogenic factors [14].

In this context, it has been revealed by a recent report that nearly 10 million tons of unregulated hazardous substances, such as heavy metals, pesticides, azo dyes, and other compounds, are released into the environment [15]. Hence, to eradicate such environmental issues, proper remediation practices must be applied. One such sustainable option is bioremediation, which provides a contemporary and efficient approach for cleaning up contaminants in a variety of ecosystems while also allowing for relatively scaled management that can be applied internationally [16, 17]. However, like other remediation methods, there are numerous advantages and disadvantages to using microbial technology to degrade pollutants; thus, nanotechnological advancements and the incorporation of nanomaterials could indeed provide more viable methods for pressuring bioremediation much beyond its limitations and current boundaries [18, 19]. Thus, in general, nanobioremediation (NBR) could provide a broader range of options for managing pollutants in aquatic and terrestrial resources. With this viewpoint, this chapter has been designed to provide the current progress on the utilization of nanobioremediation for the elimination of inorganic and organic contaminants from polluted water and soil.

1.2 NANOBIOREMEDIATION: FOR SUSTAINABLE REMOVAL OF CONTAMINANTS

Regarding the advantages of NBR, there are several rationales for the integration of nanotechnology and bioremediation. Because NPs have a large surface area per unit mass, more particles can interact with the environment, speeding up the restoration [20]. Furthermore, NBR endeavors to minimize secondary environmental consequences as well as diminish pollutant concentrations to risk-based thresholds [21]. Additionally, this reclamation technique combines the benefits of nanotechnology with bioremediation to provide a remediation process that is faster, more effective, and less harmful to the environment than the separate techniques. Recent pieces of evidence have reported the use of NPs in the bioremediation of heavy metals, pesticides, and dyes with remarkable outcomes [22, 23].

Nanotechnology and its combination with bioremediation have been the subject of several kinds of study at both the laboratory and field levels, notably in North America and Europe [24]. There is an increasing need for novel methods to rapidly and easily expedite the decontamination of polluted regions with various contaminants. For example, sites polluted with different harmful chemicals are being stabilized, transitioning, and dehalogenation persistent organic compounds using elemental or zero-valent metals like iron and palladium [25, 26]. In the case of heavy metal contaminated sites, it has been documented that the utilization of NPs can alter their speciation for the ease of the process. This claim has also been supported by the observation of a recent study. In this study, the researchers found that in comparison to nanoscale zero-valent iron, rhamnolipid (RL)-stabilized nanoscale zero-valent iron was more capable of converting labile Cd to the stable fraction in river sediments [27]. Besides, the metal-oxide frameworks of the Ce (1,3,5-benzenetricarboxylate) $(H_2O)_6$ (Ce-BTC) are used to form CeO_2 nanofibers, which have been thoroughly studied for their potential use in the adsorption of pesticides from water systems [28]. A study confirmed the maximum adsorption capacity for 2,4-D that was 86.16, 95.78, and 84.29 mg/g, respectively [29].

Therefore, the potential applications and challenges for the utilization of NBR have been presented hereunder (Figure 1.1):

- **Advantages of NBR:**
  - remediation assisted by nanotechnology has the potential to significantly enhance soil remediation and environmental conservation because of its unique features and surface functionalization for the removal of pollutants, which could accelerate the process [30].
  - NBR reduces toxins to a low hazardous level and increases the rate of biodegradation, making it an efficient, sustainable, and successful method for decontaminating contaminated areas in different locations [31].
  - When it comes to removing different organic and inorganic contaminants from soil and water, this technology may provide

an effective and environmentally acceptable alternative to existing remediation methods.
- When it comes to the cleaning up of polluted soil, nanoparticles that come from hyperaccumulator plants, bacteria, yeast, and fungus are of the utmost importance [32]. To attain the sustainability of nanoparticles and comprehend the precise mechanism of their mode of action, however, substantial study is needed.

➢ **Challenges Associated with NBR:**
- In spite of the multiple applications and improvements of nanotechnology, there is still apprehension over its presence in environmental domains, its ultimate fate, and the toxicological effects that will result from its presence due to deliberate emission [18].
- Also, according to certain accounts, the increased use of NPs in agriculture has had negative effects. For instance, the intentional application of NPs may cause their accumulation or a rise in the concentration of their components in the soil, changing the characteristics of the soil [33–35].
- The presence of NPs in soils is reported to alter the soil pH, which is one of the most important parameters that influence soil nutrient availability, microbial dynamics, overall soil health, and plant growth and development [36, 37].

1.3 CURRENT OVERVIEW ON INPUTS OF NANOBIOREMEDIATION FROM THE RESTORATION OF CONTAMINATED SITES

As a triphasic system, the soil environment is complicated and depends on both biotic and abiotic interactions to operate well. The remediation procedure may be sped up by successively implementing nanomaterials and bioremediation or by implementing both concurrently. Additionally, certain NPs might be used to remediate concentrated pollutants at source zones, with the residuals afterwards being biodegraded. The use of NPs is more effective when the initial pollutant concentration is high since high concentrations are where the desired reactions are most likely to occur. Additionally, this procedure can change pollutants into easily biodegradable, less dangerous forms. Because certain microorganisms may provide alternate degradation routes and create more benign end products than abiotic treatments alone, the combination of nano- and bioremediation techniques may increase the range of pollutants that can be degraded.

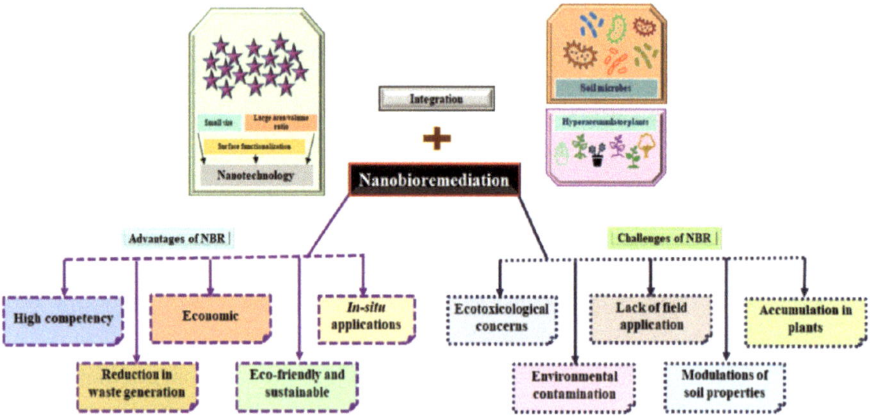

FIGURE 1.1 Schematic representation of advantages and challenges associated with the applications of nanoremediation of contaminated sites.

1.3.1 FROM SOIL

To meet the requirements of a growing population, industrialization and contemporary agricultural methods have contaminated the environment with hazardous heavy metals, pesticides, and other persistent organic pollutants [38]. Therefore, in recent years, nanoremediation has emerged as a dynamic approach for immobilizing heavy metal(oid)s, particularly methods involving the inclusion of nanoscale zero valent iron nanoparticles (nZVI) to stabilize and decrease heavy metal(oid)s. In this context, nZVI and nGoethite were found effective for the immobilization of As even at the lowest dosages tested (0.5 and 0.2%, respectively). The immobilization efficiency of nGoethite was higher than nZVI [39]. In another study, Cu, Pb, and Cd were efficiently immobilized by graphene oxide nanoparticles (nGOx), whereas As and P were mobilized (even at low doses) [40].

Likewise, pesticide removal using nanobioremediation is gaining attention, as reported by Romeh et al. [41]. In this study, the remediation of chlorfenapyr-contaminated soil and water was performed, which involved the applications of green nanoparticles (synthesized from *Ficus*, *Brassica*, and *ipomoea*) solubility-improving agents and phytoremediation using *Plantago major*. This combined approach enables the removal of chlorfenapyr more than 90%, however, the removal was found to be time-dependent [41].

In a study, nZVI (5% w/w) was utilized for the removal of co-pollution of some heavy metals (i.e., Pb, Cd, and Zn). After 120 days, nZVI shown immobilization capability for Pb (20%) but was only marginally effective for Zn (8%) and negligible for Cd [42]. In a study, rhamnolipid (RL)-stabilized nanoscale zero-valent iron (RNZVI) was used to decontaminate river sediments contaminated with Cd. RNZVI altered the speciation of heavy metals and transformed labile Cd to a stable fraction with a maximum residual concentration increasing by 11.37 mg/kg after 42 days of incubation [43]. The more studies associated with NBR of contaminants from soil based on recent literature are presented in Table 1.1.

1.3.2 FROM WATER

Various hazardous compounds have contaminated groundwater and freshwater, posing a major threat to human health and the environment and causing widespread concern among the general public for more than the last 4 decades [44]. A total of 0.127 million polluted sites and 1.17 million potentially polluted locations were found in the European Economic Area in 2011 [45]. With rising population, urbanization, and residential water supply, more wastewater is produced. Thus, a major issue facing the entire world is wastewater treatment and dependable, inexpensive access to clean water. Relying on NBR is a potential strategy that might be used in many wastewater ecosystems to address the issues of wastewater treatment in the future [46]. Therefore, currently, there is an upsurge has been reported in the utilization of nanotechnology in the treatment of water and wastewater [47]. Moreover, NPs moving *via* pore spaces can also enter difficult-to-reach places like fissures and aquifers, negating the need for pricey conventional procedures [48]. Besides, NPs, such as metal ions, polymer films, and zeolites with antimicrobial properties, are capable of disinfecting and controlling microbes that cause disease outbreaks from the water resources or during the treatment of wastewater [49–51].

In a recent study, nZVI and carbo–iron colloids (CIC) were employed as a composite material for the *in-situ* removal of herbicides (atrazine and bromacil) from groundwater. When compared to nZVI, it was shown that CIC reduced bromacil with higher activity. Neither nZVI nor Carbo-Iron significantly degraded atrazine in the conditions employed. Also, findings revealed that CIC can be used as a remedial strategy when an aquifer is bromacil-contaminated [52]. In another study, nZVI was applied to facilitate the *in-situ* remediation of groundwater resources contaminated by organic and inorganic contaminants. The research outcomes established that the contamination mediated by the presence of residual nonaqueous liquid can be effectively treated using nZVI, not only by *in situ* conditions [53]. Thus, we found that the mostly nZVI has been employed mostly for the restoration of aquatic resources and even for the treatment of wastewater,; the other perspective with similar characteristics can be explored for further efficient reclamation of polluted water.

TABLE 1.1 Nanoremediation for the Decontamination of Organic and Inorganic Pollutants from Soil and Water

Contaminated site	Type of Pollutants	Nanoparticles (NPs) Used (size in nm)	Removal/Uptake Efficiency	References
Soil	Cl, Co, Cd, Mg	DTPA functionalized maghemite NPs (47 ± 6.9 nm).	90–100%	[54]
Soil	Pb	Nanoscale zero-valent iron (nZVI) (60 nm).	32%	[55]
Soil	As, Cr, Cu, Pb, Zn	Micro- and nanoscale zero-valent iron (ZVI) and (nZVI) (10–100 nm, average size 50 nm).	10–40%	[56]
Soil	As	α-MnO_2 nanorods (30–40 nm).	60–80%	[57]
Soil	As, Cd, Pb, Zn	nZVI treatment in the rhizosphere.	–	[58]
Soil	Cd, Cu, Ni, Pb	nZVI (10–100 nm) (50 nm average).	50–99.8%	[59]
Soil	Cd	Iron oxide NPs (50–100 nm).	15–20 mg kg^{-1}	[60]
Soil	Phenanthrene / Pentachlorophenol	Natural soil NPs (Inceptisol-NP, Oxisol-NP, Ultisol-NP) (100–131 nm).	70–80%	[61]
Soil	PCB (tri and tetrachlorobiphenyls)	nZVI (50 nm)	98.35%	[62]
Water	Cd, Cu, Ni, Pb	nZVI (10–100 nm) (50 nm)	15–99%	[59]
Water	Cr(VI)	Fe_3O_4@COF(TpPa-1)	245.45 mg g^{-1}	[63]
Water	Pb	Pd NPs (10–20 nm)	12%–90%	[64]
Water	Cr(VI) Cr(VI)	nZVI/Cu (60–120 nm)	94.7%	[65]
Water	Cr(VI)	FeS-coated iron (Fe/FeS) MNPs (70 nm)	69.7 mg g^{-1}	[66]
Water	Tetracycline	Fe/Ni bimetallic NPs	82.3–92.5%	[67]
Water	As(III)	Pd NPs (15 nm)	99.8%	[68]
Water	Selenite	Se NPs (273.8 ± 16.9 nm)	100%	[69]
Water	Cd(II)	Fe_3O_4 NPs (10–30 nm)	6%–56%	[70]

1.4 FUTURE PERSPECTIVES

Due to its remarkable characteristics, nanotechnology has become particularly significant for the restoration of deteriorated sites. Also, it has been reported to play a significant role in addressing many effective and innovative solutions to various ecological challenges [71]. Like other remediation approaches, it has been connected to pros and cons. For the large-scale applications of NBR, there are several ecological challenges like what the fate of NPs is and how to regulate their negative aspects at higher concentrations of soil flora and fauna as these are significant and crucial for the adequate functioning of an ecosystem so cannot be disregarded [18, 23, 72]. Besides, it is known that NPs have no benefit for bioaugmentation since they prevent microbial habitation in polluted environments [73].

When NPs are used for ecological purposes, NPs are deliberately added to the soil or water body. Finally, this has caused growing concern from all across the globe. By causing cellular toxicity and destructive features that are anomalous in small-sized counterparts, the compensation of NPs, such as their minute size, raised activity, and huge capacity, might develop into a potentially fatal feature [36, 37, 74–76]. In stiff water and seawater, NPs tend to congregate, and the sort of natural material or other natural colloids present in freshwater has a significant effect [77]. Several abiotic parameters, such as pH, salinity, and the presence of organic matter, which have an impact on ecotoxicity, including dispersion scenario, have to be rigorously researched [77–79].

Thus, NBR is afflicted by challenges and constraints that jeopardize the commercialization of the method. For the process to be economically viable and scalable for the treatment of various effluents, a number of factors must be studied in depth [80]. Before full-scale implementation, it is necessary to address the large-scale production of biogenic nanoparticles (BNPs) for use in wastewater treatment plants, their long-term stability, technical viability, the disintegration of BNPs from biomass as a result of altered operating parameters, and the reduction of manufacturing costs and labor input [81, 82]. To reduce the total cost, the biosynthesized NPs' reusability or regeneration must be improved. In addition, their destinies, health risks, effects on soil and water, and bioavailability remain unclear. Before verifying their safety, more research on the cytotoxicity of these BNPs is required.

1.5 CONCLUSION

Concerns about environmental contamination with a diverse array of contaminants is not a new issue, but the ever-increasing extent of pollution is gaining a great deal around the globe. For now, with the advancement in science and technology, different technologies have been utilized for the eradication of xenobiotics. In this context, an emerging possibility for sustainable methods is NBR. The combined action of nanoparticles and microorganisms has been established as an effective strategy for hazardous waste removal from polluted places. Despite the numerous advantages of the utilization of NPs for remediation purposes, several concerns need to be addressed before commercialization and field application of this. Therefore, it is urgently necessary to create nanoparticles based on efficient methods for commercial use in the removal of toxins from hazardous places. Future studies using bacteria, fungi, and algae together with nanoparticles could reveal the intricate mechanisms at play as a result of compounding effects. For extensive hazardous pollutant treatment, both native and mixed microbial cultures might be investigated. In addition, to reach any concrete recommendation about NBR's extensive utilization, more studies must be conducted to envisage the fate, ecotoxicological concerns, and re-utilization of NPs.

ACKNOWLEDGMENT

The research was financially supported by the Ministry of Science and Higher Education of the Russian Federation (no. FENW-2023-0008) and the Strategic Academic Leadership Program of the Southern Federal University ("Priority 2030").

KEYWORDS

- **contaminated sites**
- **environmental pollution**
- **nanobioremediation**
- **nanoparticles**
- **remediation**
- **restoration**
- **sustainable agriculture**

REFERENCES

1. Pereira, P., Bogunovic, I., Muñoz-Rojas, M., & Brevik, E. C., (2018). Soil ecosystem services, sustainability, valuation, and management. *Current Opinion in Environmental Science & Health, 5*, 7–13. doi: 10.1016/J.COESH.2017.12.003.
2. Lal, R., (2016). Soil health and carbon management. *Food and Energy Security, 5*(4), 212–222. doi: 10.1002/FES3.96.
3. Yang, L., Wang, Y., Wang, R., et al., (2020). Environmental-social-economic footprints of consumption and trade in the Asia-Pacific region. *Nature Communications, 11*(1), 1–9. doi: 10.1038/s41467-020-18338-3.
4. *World Population Projected to Reach 9.8 Billion in 2050, and 11.2 Billion in 2100* | United Nations. Accessed from: https://www.un.org/en/desa/world-population-projected-reach-98-billion-2050-and-112-billion-2100 (accessed on 23 August 2023).
5. *Soil Matters for Our Future.* Accessed from: https://agriculture.ec.europa.eu/news/soil-matters-our-future-2019-12-05_en (accessed on 23 August 2023).
6. Burauel, P., & Baßmann, F., (2005). Soils as filter and buffer for pesticides—Experimental concepts to understand soil functions. *Environmental Pollution, 133*(1), 11–16. doi: 10.1016/J.ENVPOL.2004.04.011.
7. *Substances in the Soil* | Umweltbundesamt. Accessed from: https://www.umweltbundesamt.de/en/topics/soil-agriculture/land-a-precious-resource/substances-in-the-soil (accessed on 23 August 2023).
8. Lu, S., Wang, J., Pei, L., Bhamidiammarri, R., & Tota-Maharaj, K., (2016). Study on the effects of irrigation with reclaimed water on the content and distribution of heavy metals in soil. *International Journal of Environmental Research and Public Health, 13*(3), 298. doi: 10.3390/IJERPH13030298.
9. Ungureanu, N., Vlăduţ, V., & Voicu, G., (2020). Water scarcity and wastewater reuse in crop irrigation. *Sustainability, 12*(21), 9055. doi: 10.3390/SU12219055.
10. Khanpae, M., Karami, E., Maleksaeidi, H., & Keshavarz, M., (2020). Farmers' attitude towards using treated wastewater for irrigation: The question of sustainability. *Journal of Cleaner Production, 243*, 118541. doi: 10.1016/J.JCLEPRO.2019.118541.
11. Qadir, M., Wichelns, D., Raschid-Sally, L., et al., (2010). The challenges of wastewater irrigation in developing countries. *Agricultural Water Management, 97*(4), 561–568. doi: 10.1016/J.AGWAT.2008.11.004.
12. Valipour, M., Singh, V. P., Valipour, M., & Singh, V. P., (2016). *Global Experiences on Wastewater Irrigation: Challenges and Prospects* (pp. 289–327). doi: 10.1007/978-3-319-28112-4_18.
13. Khalid, S., Shahid, M., Natasha, et al., (2018). A review of environmental contamination and health risk assessment of wastewater use for crop irrigation with a focus on low and high-income countries. *International Journal of Environmental Research and Public Health, 15*(5), 895. doi: 10.3390/IJERPH15050895.
14. Bonelli, M. G., Ferrini, M., & Manni, A., (2017). Artificial neural networks to evaluate organic and inorganic contamination in agricultural soils. *Chemosphere, 186*, 124–131. doi: 10.1016/J.CHEMOSPHERE.2017.07.116.
15. Bhatt, P., Pandey, S. C., Joshi, S., et al., (2022). Nanobioremediation: A sustainable approach for the removal of toxic pollutants from the environment. *Journal of Hazardous Materials, 427*, 128033. doi: 10.1016/J.JHAZMAT.2021.128033.

16. Gul, M. Z., Rupula, K., & Rao, B. S., (2022). Nanobioremediation: A novel application of green-nanotechnology in environmental cleanup. *Microbes and Microbial Biotechnology for Green Remediation* (pp. 823–841). doi: 10.1016/B978-0-323-90452-0.00040-2.
17. De Lorenzo, V., Marlière, P., & Solé, R., (2016). Bioremediation at a global scale: From the test tube to planet Earth. *Microbial Biotechnology*, *9*(5), 618–625. doi: 10.1111/1751-7915.12399.
18. Rajput, V. D., Minkina, T., Kumari, A., et al., (2022). A review on nanobioremediation approaches for restoration of contaminated soil. *Eurasian Journal of Soil Science, 11*(1), 43–60. doi: 10.18393/EJSS.990605.
19. Banerjee, M., Mukherjee, S., & Mukherjee, S., (2022). Facets of nanoparticle-microbe interactions and their roles in nanobioremediation of environmental pollutants: Biochemical, molecular, and technological perspectives. *Nano-Bioremediation: Fundamentals and Applications* (pp. 111–145). doi: 10.1016/B978-0-12-823962-9.00038-6.
20. Khan, I., Saeed, K., & Khan, I. (2019). Nanoparticles: Properties, applications, and toxicities. *Arabian Journal of Chemistry, 12*(7), 908–931. doi: 10.1016/J.ARABJC.2017.05.011.
21. Singh, R., Behera, M., & Kumar, S., (2020). Nano-bioremediation: An innovative remediation technology for treatment and management of contaminated sites. *Bioremediation of Industrial Waste for Environmental Safety* (pp. 165–182). doi: 10.1007/978-981-13-3426-9_7.
22. Alazaiza, M. Y. D., Albahnasawi, A., Ali, G. A. M., et al., (2021). Recent advances of nanoremediation technologies for soil and groundwater remediation: A review. *Water, 13*(16), 2186. doi: 10.3390/W13162186.
23. Rajput, V. D., Minkina, T., Upadhyay, S. K., et al., (2022). Nanotechnology in the restoration of polluted soil. *Nanomaterials, 12*(5), 769. doi: 10.3390/NANO12050769.
24. Cecchin, I., Reddy, K. R., Thomé, A., Tessaro, E. F., & Schnaid, F. (2016). *Nanobioremediation: Integration of Nanoparticles and Bioremediation for Sustainable Remediation of Chlorinated Organic Contaminants in Soils*. doi: 10.1016/j.ibiod.2016.09.027.
25. Fei, L., Bilal, M., Qamar, S. A., et al., (2022). Nano-remediation technologies for the sustainable mitigation of persistent organic pollutants. *Environmental Research, 211*, 113060. doi: 10.1016/J.ENVRES.2022.113060.
26. Sargin, I., Baran, T., & Arslan, G., (2020). Environmental remediation by chitosan-carbon nanotube supported palladium nanoparticles: Conversion of toxic nitroarenes into aromatic amines, degradation of dye pollutants and green synthesis of biaryls. *Separation and Purification Technology, 247*, 116987. doi: 10.1016/J.SEPPUR.2020.116987.
27. Xue, W., Peng, Z., Huang, D., et al., (2018). Nanoremediation of cadmium contaminated river sediments: Microbial response and organic carbon changes. *Journal of Hazardous Materials, 359*, 290–299. doi: 10.1016/J.JHAZMAT.2018.07.062.
28. Roy, A., Roy, M., Alghamdi, S., et al., (2022). Role of microbes and nanomaterials in the removal of pesticides from wastewater. *International Journal of Photoenergy, 2022*. doi: 10.1155/2022/2131583.
29. Abdelillah, A. E. E., Şahin, S., & Bayazit, Ş. S., (2018). Preparation of CeO_2 nanofibers derived from Ce-BTC metal-organic frameworks and their application on pesticide adsorption. *Journal of Molecular Liquids, 255*, 10–17. doi: 10.1016/J.MOLLIQ.2018.01.165.

30. Ningthoujam, R., Singh, Y. D., Babu, P. J., Tirkey, A., Pradhan, S., & Sarma, M., (2022). Nanocatalyst in remediating environmental pollutants. *Chemical Physics Impact, 4*, 100064. doi: 10.1016/J.CHPHI.2022.100064.
31. Pandey, G., (2018). *Oriental Journal of Chemistry Prospects of Nanobioremediation in Environmental Cleanup, 34*(6). doi: 10.13005/ojc/340622.
32. Yadav, K., Singh, J., Gupta, N., & Kumar, V. (2017). A review of nanobioremediation technologies for environmental cleanup: A novel biological approach. *Journal of Materials and Environmental Sciences*. jmaterenvironsci.com. Accessed from http://www.jmaterenvironsci.com/Document/vol8/vol8_N2/78-JMES-2831-Yadav.pdf (accessed on 23 August 2023).
33. Rajput, V., Minkina, T., Mazarji, M., et al., (2020). Accumulation of nanoparticles in the soil-plant systems and their effects on human health. *Annals of Agricultural Sciences, 65*(2), 137–143. doi: 10.1016/J.AOAS.2020.08.001.
34. Mwaanga, P., (2018). Risks, uncertainties, and ethics of nanotechnology in agriculture. *New Visions in Plant Science*. doi: 10.5772/INTECHOPEN.76590.
35. Madanayake, N. H., Perera, N., & Adassooriya, N. M., (2022). Engineered nanomaterials: Threats, releases, and concentrations in the environment. *Emerging Contaminants in the Environment: Challenges and Sustainable Practices* (pp. 225–240). doi: 10.1016/B978-0-323-85160-2.00001-9.
36. Khan, S. T., Adil, S. F., Shaik, M. R., Alkhathlan, H. Z., Khan, M., & Khan, M., (2021). Engineered nanomaterials in soil: Their impact on soil microbiome and plant health. *Plants, 11*(1), 109. doi: 10.3390/PLANTS11010109.
37. Chaudhary, P., Chaudhary, A., Bhatt, P., et al., (2022). Assessment of soil health indicators under the influence of nanocompounds and *Bacillus* spp. in field condition. *Frontiers in Environmental Science, 9*, 648. doi: 10.3389/FENVS.2021.769871/BIBTEX.
38. Briffa, J., Sinagra, E., & Blundell, R. (2020). Heavy metal pollution in the environment and their toxicological effects on humans. *Heliyon, 6*(9), e04691. doi: 10.1016/J.HELIYON.2020.E04691.
39. Baragaño, D., Alonso, J., Gallego, J. R., Lobo, M. C., & Gil-Díaz, M., (2020). Zero valent iron and goethite nanoparticles as new promising remediation techniques for As-polluted soils. *Chemosphere, 238*, 124624. doi: 10.1016/J.CHEMOSPHERE.2019.124624.
40. Baragaño, D., Forján, R., Welte, L., & Gallego, J. L. R., (2020). Nanoremediation of As and metals polluted soils by means of graphene oxide nanoparticles. *Scientific Reports, 10*(1), 1–10. doi: 10.1038/s41598-020-58852-4.
41. Romeh, A. A., & Ibrahim, S. R. A., (2020). Green nano-phytoremediation and solubility-improving agents for the remediation of chlorfenapyr-contaminated soil and water. *Journal of Environmental Management, 260*, 110104. doi: 10.1016/J.JENVMAN.2020.110104.
42. Fajardo, C., Sánchez-Fortún, S., Costa, G., et al., (2020). Evaluation of nanoremediation strategy in Pb, Zn, and Cd contaminated soil. *Science of The Total Environment, 706*, 136041. doi: 10.1016/J.SCITOTENV.2019.136041.
43. Xue, W., Peng, Z., Huang, D., et al., (2018). Nanoremediation of cadmium contaminated river sediments: Microbial response and organic carbon changes. *Journal of Hazardous Materials, 359*, 290–299. doi: 10.1016/J.JHAZMAT.2018.07.062.
44. Phenrat, T., Skácelová, P., Petala, E., et al., (2020). *Nanoscale Zero-Valent Iron Particles for Water Treatment: From Basic Principles to Field-Scale Applications* (pp. 19–52). doi: 10.1007/978-3-030-29840-1_2.

45. *Van Liedekerke: Progress in the Management of Contaminate*. Google Scholar. Accessed https://scholar.google.com/scholar_lookup?&title=Progress%20in%20management%20of%20contaminated%20sites&publication_year=2014 (accessed on 23 August 2023).
46. Prasad, R., & Karchiyappan, T., (2019). *Advanced Research in Nanosciences for Water Technology*. https://link.springer.com/content/pdf/10.1007/978-3-030-02381-2.pdf (accessed on 23 August 2023).
47. Khalid, Z. B., Nasrullah, M., Nayeem, A., Wahid, Z. A., Singh, L., & Krishnan, S., (2020). Application of 2D graphene-based nanomaterials for pollutant removal from advanced water and wastewater treatment processes. *ACS Symposium Series, 1353*, 191–217. doi: 10.1021/BK-2020-1353.CH009.
48. Ingle, A. P., Seabra, A. B., Duran, N., & Rai, M., (2014). Nanoremediation: A new and emerging technology for the removal of toxic contaminants from the environment. *Microbial Biodegradation and Bioremediation* (pp. 233–250). doi: 10.1016/B978-0-12-800021-2.00009-1.
49. Khan, S., Anjum, R., & Muhammad, B., (2021). Revealing chemical speciation behaviors in aqueous solutions for uranium (VI) and europium (III) adsorption on zeolite. *Environmental Technology & Innovation*. Elsevier. https://www.sciencedirect.com/science/article/pii/S2352186421001516?casa_token=YDe1EX7ykzUAAAAA:awJIsbO6KghaoubHa6UNVLMm5cDUM2qPjY6j6c1TrqVdWdgFTMkHU7QG5r2oFO5nOh-ueKGmYQ (accessed on 23 August 2023).
50. Khan, S., Wu, Y., Zhang, X., et al., (2013). Influence of dissolved organic matter from corn straw on Zn and cusorption to Chinese loess. *Toxicological and Environmental Chemistry, 95*(8), 1318-1327. doi: 10.1080/02772248.2013.866238.
51. Khan, S., Sengul, H., & Dan, Z., (2018). *Transport of TiO_2 Nanoparticles and Their Effects on the Mobility of Cu in Soil Media*. https://aperta.ulakbim.gov.tr/record/32705 (accessed on 23 August 2023).
52. Gawel, A., Seiwert, B., Sühnholz, S., Schmitt-Jansen, M., & Mackenzie, K., (2020). In-situ treatment of herbicide-contaminated groundwater–feasibility study for the cases atrazine and bromacil using two novel nanoremediation-type materials. *Journal of Hazardous Materials, 393*, 122470. doi: 10.1016/J.JHAZMAT.2020.122470.
53. Pak, T., De Lima, L. L. F., Tosco, T., Costa, G. S. R., Rosa, P. R. R., & Archilha, N. L., (2020). Pore-scale investigation of the use of reactive nanoparticles for in situ remediation of contaminated groundwater source. *Proc. Natl. Acad. Sci. U S A., 117*(24), 13366–13373. doi: 10.1073/PNAS.1918683117/-/DCSUPPLEMENTAL.
54. Hughes, D. L., Afsar, A., Laventine, D. M., Shaw, E. J., Harwood, L. M., & Hodson, M. E., (2018). Metal removal from soil leachates using DTPA-functionalized maghemite nanoparticles, a potential soil washing technology. *Chemosphere, 209*, 480–488. doi: https://doi.org/10.1016/j.chemosphere.2018.06.121.
55. Gil-Díaz, M., López, L. F., Alonso, J., & Lobo, M. C., (2018). Comparison of nanoscale zero-valent iron, compost, and phosphate for Pb immobilization in acidic soil. *Water, Air, & Soil Pollution, 229*(10), 315. doi: 10.1007/s11270-018-3972-1.
56. Danila, V., Kumpiene, J., Kasiuliene, A., & Vasarevičius, S., (2020). Immobilization of metal(loid)s in two contaminated soils using micro and nano zerovalent iron particles: Evaluating the long-term stability. *Chemosphere, 248*, 126054. doi: https://doi.org/10.1016/j.chemosphere.2020.126054.

57. Li, B., Zhou, S., Wei, D., et al., (2019). Mitigating arsenic accumulation in rice (*Oryza sativa* L.) from typical arsenic-contaminated paddy soil of southern China using nanostructured α-MnO_2: Pot experiment and field application. *Science of The Total Environment, 650*, 546–556. doi: https://doi.org/10.1016/j.scitotenv.2018.08.436.
58. Vítková, M., Puschenreiter, M., & Komárek, M., (2018). Effect of nano zero-valent iron application on As, Cd, Pb, and Zn availability in the rhizosphere of metal(loid) contaminated soils. *Chemosphere, 200*, 217–226. doi: https://doi.org/10.1016/j.chemosphere.2018.02.118.
59. Vasarevičius, S., Danila, V., & Paliulis, D., (2019). Application of stabilized nano zero valent iron particles for immobilization of available Cd^{2+}, Cu^{2+}, Ni^{2+}, and Pb^{2+} ions in the soil. *International Journal of Environmental Research, 13*(3), 465–474. doi: 10.1007/s41742-019-00187-8.
60. Hussain, A., Ali, S., Rizwan, M., et al., (2019). Responses of wheat (*Triticum aestivum*) plants grown in a Cd contaminated soil to the application of iron oxide nanoparticles. *Ecotoxicology and Environmental Safety, 173*, 156–164. doi:https://doi.org/10.1016/j.ecoenv.2019.01.118.
61. Liu, F., Xu, B., He, Y., Brookes, P. C., & Xu, J. (2019). Co-transport of phenanthrene and pentachlorophenol by natural soil nanoparticles through saturated sand columns. *Environmental Pollution, 249*, 406–413. doi: https://doi.org/10.1016/j.envpol.2019.03.052.
62. Liu, J., Chen, T., Qi, Z., Yan, J., Buekens, A., & Li, X., (2014). Thermal desorption of PCBs from contaminated soil using nano zerovalent iron. *Environmental Science and Pollution Research, 21*(22), 12739–12746. doi: 10.1007/s11356-014-3226-8.
63. Zhong, X., Lu, Z., Liang, W., & Hu, B. (2020). The magnetic covalent organic framework as a platform for high-performance extraction of Cr(VI) and bisphenol a from aqueous solution. *Journal of Hazardous Materials, 393*, 122353. doi: https://doi.org/10.1016/j.jhazmat.2020.122353.
64. Sayadi, M. H., Salmani, N., Heidari, A., & Rezaei, M. R., (2018). Bio-synthesis of palladium nanoparticle using Spirulina platensis alga extract and its application as adsorbent. *Surfaces and Interfaces, 10*, 136–143. doi: https://doi.org/10.1016/j.surfin.2018.01.002.
65. Zhu, F., Ma, S., Liu, T., & Deng, X., (2018). Green synthesis of nano zero-valent iron/Cu by green tea to remove hexavalent chromium from groundwater. *Journal of Cleaner Production, 174*, 184–190. doi: https://doi.org/10.1016/j.jclepro.2017.10.302.
66. Gong, Y., Gai, L., Tang, J., Fu, J., Wang, Q., & Zeng, E. Y., (2017). Reduction of Cr(VI) in simulated groundwater by FeS-coated iron magnetic nanoparticles. *Science of the Total Environment, 595*, 743–751. doi:https://doi.org/10.1016/j.scitotenv.2017.03.282.
67. Dong, H., Jiang, Z., Deng, J., et al., (2018). Physicochemical transformation of Fe/Ni bimetallic nanoparticles during aging in simulated groundwater and the consequent effect on contaminant removal. *Water Research, 129*, 51–57. doi: https://doi.org/10.1016/j.watres.2017.11.002.
68. Arsiya, F., Sayadi, M. H., & Sobhani, S., (2017). Arsenic (III) adsorption using palladium nanoparticles from aqueous solution. *Journal of Water and Environmental Nanotechnology, 2*(3), 166–173. doi: 10.22090/JWENT.2017.03.004.
69. Wang, X., Zhang, D., Pan, X., et al., (2017). Aerobic and anaerobic biosynthesis of nano-selenium for remediation of mercury-contaminated soil. *Chemosphere, 170*, 266–273. doi: https://doi.org/10.1016/j.chemosphere.2016.12.020.

70. Kataria, N., & Garg, V. K., (2018). Green synthesis of Fe_3O_4 nanoparticles loaded sawdust carbon for cadmium(II) removal from water: Regeneration and mechanism. *Chemosphere, 208*, 818–828. doi: https://doi.org/10.1016/j.chemosphere.2018.06.022.
71. Babatunde, D. E., Denwigwe, I. H., Babatunde, O. M., Gbadamosi, S. L., Babalola, I. P., & Agboola, O., (2020). Environmental and societal impact of nanotechnology. *IEEE Access, 8*, 4640–4667. doi: 10.1109/ACCESS.2019.2961513.
72. Ray, P. C., Yu, H., & Fu, P. P., (2009). Toxicity and environmental risks of nanomaterials: Challenges and future needs. *J. Environ. Sci. Health C Environ. Carcinog. Ecotoxicol. Rev., 27*(1), 1. doi: 10.1080/10590500802708267.
73. Vázquez-Núñez, E., Molina-Guerrero, C. E., Peña-Castro, J. M., Fernández-Luqueño, F., & De La Rosa-Álvarez, M. G., (2020). Use of nanotechnology for the bioremediation of contaminants: A review. *Processes, 8*(7), 826. doi: 10.3390/PR8070826.
74. Yanga, J., Cao, W., & Rui, Y., (2017). *Interactions Between Nanoparticles and Plants: Phytotoxicity and Defense Mechanisms, 12*(1), 158–169. http://mc.manuscriptcentral.com/tjpi. doi: 10.1080/17429145.2017.1310944.
75. Ruttkay-Nedecky, B., Krystofova, O., Nejdl, L., & Adam, V., (2017). Nanoparticles based on essential metals and their phytotoxicity. *Journal of Nanobiotechnology, 15*(1), 1–19. doi: 10.1186/S12951-017-0268-3/FIGURES/1.
76. Mittal, D., Kaur, G., Singh, P., Yadav, K., & Ali, S. A. (2020). Nanoparticle-based sustainable agriculture and food science: Recent advances and future outlook. *Frontiers in Nanotechnology, 2*, 10. doi: 10.3389/FNANO.2020.579954/BIBTEX.
77. Ghosh, S., Mashayekhi, H., Pan, B., Bhowmik, P., & Xing, B., (2008). Colloidal behavior of aluminum oxide nanoparticles as affected by pH and natural organic matter. *Langmuir, 24*(21), 12385–12391. doi: 10.1021/LA802015F.
78. Turan, N. B., Erkan, H. S., Engin, G. O., & Bilgili, M. S., (2019). Nanoparticles in the aquatic environment: Usage, properties, transformation, and toxicity—A review. *Process Safety and Environmental Protection, 130*, 238–249. doi: 10.1016/J.PSEP.2019.08.014.
79. Cross, R. K., Tyler, C., Galloway, T. S., Cross, R. K., Tyler, C., & Galloway, T. S., (2015). Transformations that affect fate, form, and bioavailability of inorganic nanoparticles in aquatic sediments. *Environmental Chemistry, 12*(6), 627–642. doi: 10.1071/EN14273.
80. Sunanda, S., Misra, M., & Ghosh, S. S., (2022). Nanobioremediation of heavy metals: Perspectives and challenges. *Journal of Basic Microbiology, 62*(3, 4), 428–443. doi: 10.1002/JOBM.202100384.
81. Patil, S., & Chandrasekaran, R., (2020). Biogenic nanoparticles: A comprehensive perspective in synthesis, characterization, application, and its challenges. *Journal of Genetic Engineering and Biotechnology, 18*(1), 1–23. doi: 10.1186/S43141-020-00081-3.
82. Ali, I., Peng, C., Khan, Z. M., et al., (2019). Overview of microbes-based fabricated biogenic nanoparticles for water and wastewater treatment. *Journal of Environmental Management, 230*, 128–150. doi: 10.1016/J.JENVMAN.2018.09.073.

CHAPTER 2

Nano-Phytoremediation: An Emerging Sustainable Reclamation Technique

C. AKSHAYA PRAKASH,[1] NAIR G. SARATH,[2,3]
DELSE PAREKKATTIL SEBASTIAN,[1] and JOS T. PUTHUR[2]

[1]*Department of Botany, St. Joseph's College (Autonomous), Devagiri, Kozhikode, Kerala, India*

[2]*Plant Physiology and Biochemistry Division, Department of Botany, University of Calicut, C.U. Campus P.O., Kerala, India*

[3]*Department of Botany, Mar Athanasius College (Autonomous), Kothamangalam, Kerala, India*

ABSTRACT

Contamination of the environment by pollutants from various sources has become a huge problem worldwide. An environment that is contaminated can be harmful to all life forms. Therefore, efficient methods and technologies are urgently needed to speed up the reclamation process of contaminated areas. Phytoremediation is an upcoming technology that is well appreciated as this technology makes use of the ability of plants in environmental clean-up. It involves the utilization of plants to eliminate, stabilize, or degrade contaminants present in soil or water. Similarly, nanotechnology that includes nanoparticles (NPs) with a large surface area and great absorption capacities is also a favorable technology for the remediation of contaminated areas. Integrating nanotechnology and phytoremediation into a hybrid technology called nano-phytoremediation can lead to more effective restoration of polluted areas. Nanomaterials can facilitate phytoremediation by enhancing the plant growth and bioavailability of pollutants or by eliminating the pollutants directly from

Nano-Bioremediation for Water and Soil Treatment: An Eco-Friendly Approach.
Vishnu D. Rajput, Arpna Kumari, and Tatiana M. Minkina (Eds.)

the medium. Various studies described that nano-phytoremediation can eliminate heavy metals like Cd, Pb, Zn, etc., and organic pollutants like chlorinated hydrocarbons, phenols, etc. Since this technology of nano-phytoremediation is in its infancy, more studies are required to identify the pros and cons and the advancement of knowledge in this field.

2.1 INTRODUCTION

Environmental contamination is a severe problem affecting humankind and is one of the important reasons for diseases and deaths [1]. Even though natural processes like volcanic eruptions and forest fires pollute the environment, air, water, and soil are all polluted by a broad scale of anthropogenic activities, including deforestation, domestic and agricultural waste dumping in aquatic ecosystems, burning of fossil fuels and bushes, mining, increased utilization of chemicals and pesticides and careless discarding of e-waste. The various pollutants include domestic wastes, chemicals, insecticides, medical wastes, inorganic substances like heavy metals, volatile organic compounds, etc. [2]. The increased human population has accelerated and augmented environmental pollution through all the processes mentioned above. The aftereffects of all these processes affect all living organisms, including animals, plants, and microorganisms, which are necessary to maintain the balance of an ecosystem. In humans, pesticides, heavy metals, hydrocarbons, etc., can lead to various deleterious health problems like cancer, liver damage, kidney damage, reproductive disorders, hormonal imbalance, etc. Plants that grow in the contaminated matrix also exhibit reduced growth and crop yields besides transferring the contaminating substances into the food chain [2].

The various negative consequences of pollutants to the environment and organisms call for an urgent and effective removal mechanism of pollutants. For many years, conventional methods have been practiced for removing soil contaminants. Physical remediation techniques like soil capping, soil washing, or soil excavation and chemical remediation methods like immobilization, solidification, electrokinetics, and vitrification can clean the contaminated land in a short period of time [3]. They are found to be efficacious as well. However, these techniques become unattractive due to disadvantages like the generation of additional toxic pollutants, modification of soil properties, and high expense [4]. Recent innovative technologies like phytoremediation and phytoremediation combined with nanotechnology (NT) are great leaps in environmental restoration.

Phytoremediation is a technology that exploits the capability of plants to accumulate, degrade, immobilize, or stabilize compounds present in their growth medium to discard the pollutants that are present in the soil, water, etc. [5]. Since this process is driven by solar energy, the cost of application and maintenance is cheap, eco-friendly, and easy to apply [6]. Phytoremediation occurs through two main strategies: stabilization of contamination in the site and cleaning up the site by removing the contaminants completely from the degraded site [7]. The various strategies, like phytoextraction, phytodegradation, phytostabilization or rhizostabilization, phytovolatilization, etc., may fall into one of the above-mentioned main strategies. Phytoextraction, phytodegradation, and phytovolatilization clean up the contaminated areas while phytostabilization stabilizes the contaminants in the soil, as a result of which their bioavailability is limited [7].

The implementation of phytoremediation can be a promising approach to cleaning polluted areas with less intervention on living organisms and the environment. However, a main disadvantage of the process is its time-consuming nature. The combination of phytoremediation with NT can help in reducing the time period and boosting the efficiency of the technology [8].

NT is a branch of science that deals with the production, depiction or characterization, investigation, and utilization of nanomaterials [NMs]. NMs are those substances whose one dimension, at the minimum, is below 100 nm [9]. The physicochemical characteristics of NMs are quite different from their bulk materials or their atomic-molecular scales, which is responsible for their importance in science and technology. NT deals with science, technology, and engineering, imaging, modeling, measuring, and manipulating matter [10]. The application of NT has increased worldwide in many fields like medicine, food industry, energy, and pollution treatment [11–13].

NT has now been identified as one of the most successful technologies for environmental restoration. A wide scale of materials like nanoscale zeolites, enzymes, carbon nanotubes, metal-oxides, dendrimers, and bimetallic particles useful for remediation purposes are provided by NT. The large surface area provided by the small-sized NPs is accountable for their high efficiency in remediation purposes [11]. The extremely small size of NPs makes it easy to inject these particles into polluted sites. So, excavation of the place is not required for the clean-up process [14]. Traditional methods like disposal of wastes in landfills are expensive and generate additional hazardous waste products. These limitations can be overcome by using NT [15].

NMs can adsorb and degrade pollutants. Plants are capable of absorbing and accumulating pollutants. Besides, several stress response molecules like glutathione, flavonoids, reactive oxygen species, and bioactive molecules present in plants help transform pollutants [16–19]. When plants and NMs are used together, the NM-degraded pollutants, which are less toxic than the initial pollutant, are absorbed by the plants. This integrated technology is called nano-phytoremediation [20]. Using plant-based NPs in environmental restoration is also nano-phytoremediation [21]. Nano-phytoremediation enhances contaminant removal efficiency than nanoremediation or phytoremediation alone could do [20]. The activity of NMs in enhancing or facilitating phytoremediation can be by eliminating the pollutants directly, improving plant growth, and enhancing the bioavailability of pollutants [22]. Nano-phytoremediation is applicable in removing both inorganic and organic contaminants from the environment [23].

2.2 SYNTHESIS OF NPS FOR ENVIRONMENTAL CLEANUP

Physical and chemical methods are used to synthesize NPs by conventional methods. These methods include deposition-precipitation, photocatalytic deposition, chemical vapor decomposition, wet chemical method, chemical solution decomposition, ultrasonic radiation, sol gel, thermal processes, and hydrothermal processes [24]. NPs production by physical methods requires high energy and expense, while chemical methods involve the utilization of toxic reagents. The recent emergence of the biotechnological approach in the synthesis of NPs is considered to be a green alternative to NPs' physical and chemical synthesis. This method uses microbial, algal, or plant biomolecules to produce NPs of desirable size and shape [25]. The oxidation or reduction process is the mechanism for the synthesis of NPs in this case. Reduction and stabilization of metal ions in NPs made from microbes and plants occur as a result of the activity of biomolecules in them [26]. Synthesis of plant-based NPs is convenient, faster, and safer than the production of NPs from microbes. Microbial NPs synthesis is comparatively unsafe for human health and the environment because of the toxic substances in the system [27].

2.2.1 PHYSICAL METHODS OF NP SYNTHESIS

Some of the physical methods used to synthesize NMs are spinning, pyrolysis, laser ablation, and thermal decomposition:

1. **Spinning:** A spinning disc reactor [SDR] is required for NP synthesis by spinning. It consists of a rotating disc inside a chamber. The physical parameters can be regulated inside the reactor. For the elimination of oxygen and to prevent any chemical reactions from occurring inside it, inert gases like nitrogen are fed into the reactor [28]. The precursor material of NM and water is pumped into the reactor, in which the rotating disc rotates at different speeds. The rotation results in the fusion of atoms or molecules, which precipitates. The precipitate is taken and dried to obtain NPs [29].
2. **Pyrolysis:** The most commonly used method for the industrial production of NPs is pyrolysis. Here, at high pressure, the NM precursor is provided into a furnace and burnt with a flame [30]. Then, by air classification, NPs are recovered, and the by-product gases are removed. Pyrolysis is advantageous as it is simple, cheaper, and has higher productivity [31].
3. **Laser Ablation:** This method is generally used to synthesize NPs from different solvents. Here, a laser beam is provided to a metal which is kept in a solution, due to which NPs are generated from a plasma plume [32].
4. **Thermal Decomposition:** The various chemical bonds of a compound are broken by using heat, which results in the formation of NPs. This method is called thermal decomposition [33].

2.2.2 CHEMICAL METHODS OF NP SYNTHESIS

A common and preferred chemical method for NPs synthesis is sol gel process [34]:

- **Sol-Gel:** This method of NPs synthesis is simple and can be utilized for the synthesis of a number of NPs. It involves using a chemical solution, and the precursors of NMs used during this process are metal chlorides and oxides. Here, the precursor is scattered in a liquid by stirring, shaking, etc., resulting in a solid and liquid phase system. Then, the NMs are recovered from this by sedimentation, filtration, and centrifugation. The resultant substance is further dried to eliminate moisture content [35].

2.2.3 BIOLOGICAL METHODS OF NP SYNTHESIS

The biological method of NPs synthesis is eco-friendly and does not involve harmful chemicals. Biological methods involve the synthesis of NPs from microorganisms and plants [36].

Many genera of microorganisms can be used for NPs synthesis. Some genera included in metal NPs synthesis include *Bacillus, Klebsiella, Pseudomonas, Rhodococcus, Weissella*, etc. [37]. Synthesis of NPs from microorganisms can be both intracellular and extracellular [38]:

1. **Extracellular Synthesis of NPs by Microorganisms:** Optimum conditions are provided to the microorganisms in a rotating shaker for one or two days for culturing microorganisms. Then, centrifugation is carried out for biomass removal. The supernatant thus obtained is added to the salt solution of a filter-sterilized metal and is incubated. A change in the color of the medium is an indication of NPs being synthesized. So, color change needs to be noted. After the incubation duration is completed, the mixture is subjected to centrifugation. The NPs are separated, washed, and collected [38].
2. **Intracellular Synthesis of NPs by Microorganisms:** In this method, microorganisms are cultured for a particular period of time. Then, the collection of biomasses is carried out by centrifugation. Thorough washing of this biomass is done using sterile water, which is then fed to the solution of metal salt that is sterilized and incubated. Any color change is visually observed. After incubation, ultrasonication, centrifugation, etc., are carried out to remove the biomass and break the cell wall. Finally, the NPs will be released when the cell walls break. The NPs are collected after centrifugation and washing [38].

2.2.4 NPS FROM PLANTS

Synthesis of NPs from plants is very useful in satisfying the NP requirements of the present in biomedical and environmental areas [39]. Plant extracts can be used for the synthesis of NPs. For this, the plant parts are washed and boiled. The extract thus obtained is subjected to filtration and centrifugation. The plant extract, water, and metal salt solution are taken in different ratios and incubated. Incubation helps in the reduction of metal salt. Any color change is observed. Then, NPs are collected. The production of NPs from plants depends on the plants' species and the various phytochemical components present in them [38].

Extracts of the peel of *Annona squamosa* [40], coir of *Cocos nucifera* [41], leaf extracts of *Nyctanthes arbor-tristis* [42], *Psidium guajava* [43], *Eclipta prostrata* [44] and *Catharanthus roseus* [45] were all successfully used for the preparation of spherical shaped titanium dioxide NPs. Latex of *Calotropis procera* [46], *Physalisalke kengi* [47], *and Aloe vera* [48] were used for the synthesis of zinc oxide NPs. Bran extracts of *Sorghum bicolor* [49] and leaf extracts of *Euphorbia milii*, *Tridax procumbens*, *Tinospora cordifolia*, *Datura innoxia*, *Calotropis procera*, and *Cymbopogon citratus* [50] were used for the synthesis of iron NPs.

2.3 APPLICATION OF NPS IN ENVIRONMENTAL CLEANUP

The efficiency of eliminating toxic substances from soil and water has increased by combining NMs with bioremediation [51]. Nano-bioremediation is the mechanism of using NT to produce NPs from bacteria, algae, and fungi to eliminate organic or inorganic contaminants from the environment [52]. If environmental restoration is carried out using plant-based NPs or simply involving plants along with NMs, the technology is called nano-phytoremediation [21].

NPs can be inorganic, organic, or polymeric. Metal and metal oxide-based NPs, silica, etc., belong to inorganic NPs. Organic NPs consist of carbon NPs. Polyamidoamine dendrimers [PAMAM dendrimers], polymer nanocomposites, etc., are examples of polymer-based NPs (Figure 2.1). Many contaminants can be removed with metal-based NPs, but the elimination of chlorinated organic pollutants and heavy metals are often studied using these categories of NPs [53]. Many authors reported that iron-based NPs remediate water contaminated with chlorinated organic solvents and heavy metals [54–56]. Binary mixed oxide NPs have the capability of cleaning up methylene blue dye-contaminated water, and bimetallic NPs can reclaim water and soil contaminated with chlorinated and brominated compounds [57–61]. The added benefit of using bimetallic NPs compared to monometallic NPs is that the former is more stable in solution, and there is no need to add surfactants for solution stability as in the case of monometallic NPs [62]. The good adsorbing nature of silica NMs makes them excellent materials for gaseous contaminant removal [53]. Amine-modified aluminosilicates and porous silica are reported to remove carbon dioxide, aldehydes, and ketones [63–66]. Organic NMs mainly include carbon-based NMs like fullerene.

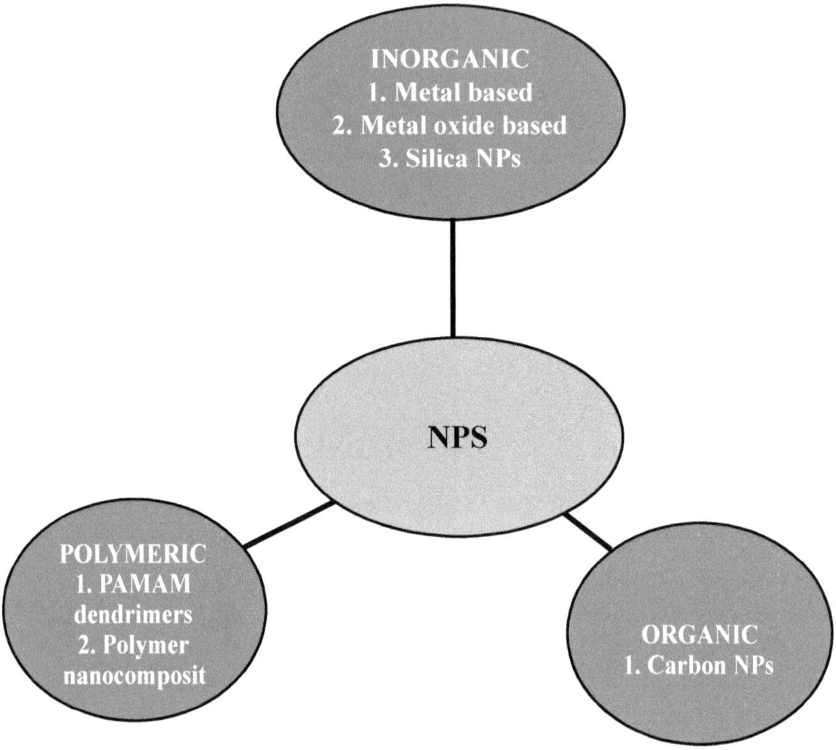

FIGURE 2.1 Different types of NPs are used for the purpose of nano-phytoremediation.

It is possible to remove organic and inorganic pollutants from the environment using these NMs. Here also, the adsorptive nature of the NMs is responsible for their application in contaminant removal [67, 68]. Pristine graphene, a carbon-based nanomaterial, could remediate water contaminated with fluoride [37]. Similarly, CdS graphene NMs were reported to remove heavy metals from water [69]. Polymer-based NMs like amphiphilic polyurethane NPs and PAMAM dendrimers were reported to remove polynuclear aromatic hydrocarbons from soil and heavy metals from wastewater, respectively [70, 71].

2.4 PLANT-NPS INTERACTION

NMs have increasingly been used along with plants for various purposes. NMs help reduce the amount of plant protection products used in agriculture, reduce the loss of nutrients while providing fertilizers, increase abiotic stress tolerance, etc. [72]. The physicochemical properties of NPs, the way of application of NPs to plants [which can be through the soil, as foliar sprays, or in hydroponics], and the quantity provided to plants determine the biological function of NPs on plants. Callus induction, somatic embryogenesis, secondary metabolite production, organogenesis, etc., have been found to be possible *in vitro* studies using NPs [73]. When NPs are provided in soil, many transformations occur to the NPs, which is accountable for NP bioavailability and toxicity. Once the NPs come in direct contact with the roots of plants, NPs are absorbed by the roots, and root-to-shoot translocation occurs. The cell organelles accumulate these NPs [74]. The size of NPs is directly proportional to NP absorption by roots and affects the transport of the NPs into cells or cell organelles. The charge of the NP is another factor that determines the absorption and translocation of NPs. This is because the cell wall has a negative charge, which has a role in its attachment to the NP. The hydrophobicity of plant surface and the structure of NPs are also characteristic features that have a role in NP uptake and translocation [75, 76]. Figure 2.2 is a representation of plant-NP interaction.

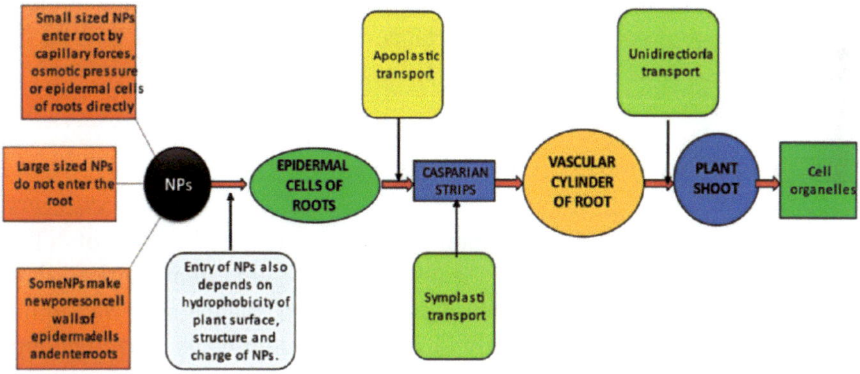

FIGURE 2.2 Plant-NP interaction.

NPs of small size can enter the roots of plants by capillary forces, osmotic pressure, or epidermal cells of roots directly. The entry of large NPs is restricted since the root cell walls of epidermal cells are semipermeable, with only tiny pores. Specific NPs make new pores on the cell walls of epidermal cells and enter into the cells [77, 78]. Once the NPs cross the cell wall, they are carried apoplastically through extracellular spaces to the vascular cylinder. From the vascular cylinder, the NPs are transported unidirectionally. However, for the entry of NPs from the extracellular spaces to the vascular cylinder, NPs have to be transported across the Casparian strips symplastically [79, 80]. The NPs that cannot cross the casparian strips are leftover in aggregates on the Casparian strips. However, the NPs that have entered the xylem are translocated to the shoot and reach the roots via the phloem [79–81]. The NPs that have entered the plant can be found in the cell walls of epidermal cells, the cytoplasm of cortical cells, and nuclei. Those NPs that are not absorbed by the roots remain as aggregates on the surface of the roots and can enhance nutrient absorption [82]. NPs applied as foliar sprays enter inside the plant through cuticle or stomata. NPs having a size lesser than 5 nm enter inside the plant through the cuticle, while NPs of greater size enter through stomata. Apoplastic and symplastic pathways are responsible for the cellular transport of NPs in plants [82]. NPs have been found to increase the length of root and shoot and enhance seed germination, fruit production, and metabolite contents of crop plants [83–85]. The photosynthetic rate and nitrogen use efficiency of some plants have also been increased by the application of NPs [86–88]. Plant consumption of nutrients and plant resistance to diseases and other stresses can be enhanced by applying NPs [89]. Even though NPs have all these beneficial effects on plants, using higher concentrations of NPs can have toxic impacts on the plant [90]. So, it is of utmost importance to analyze the usage of NPs from various dimensions before applying them to plants.

2.5 CONCEPT OF NANO-PHYTOREMEDIATION

Nano-phytoremediation is a technology that blends the benefits of NT and phytoremediation [91]. NPs selected for nano-phytoremediation should satisfy certain criteria like being non-toxic to plants and rhizospheric microbes, possessing the capability to bind to contaminants, and augmenting the bioavailability of contaminants for augmenting the phytoextraction process, etc. It would be good if the NPs could enhance plant growth, resulting in better pollutant removal efficiency by plants [23]. The plants themselves and the soil microbes can produce NPs that have the potential to increase the synthesis of plant growth regulators, thereby increasing the plant biomass and enhancing the phytoextraction efficiency. At present, nano-phytoremediation is confined only to lab and pot culture experiments. More studies are needed to determine if the application of NPs for this purpose becomes a source of secondary pollution [23].

2.5.1 HOW DO THE NMS TAKE PART IN THE PROCESS OF PHYTOREMEDIATION?

NMs mainly carry out their role in phytoremediation through three processes, which are: (i) by the elimination of contaminants directly, (ii) by facilitating the growth of plants, and (iii) by enhancing pollutant bioavailability (Figure 2.3) [91].

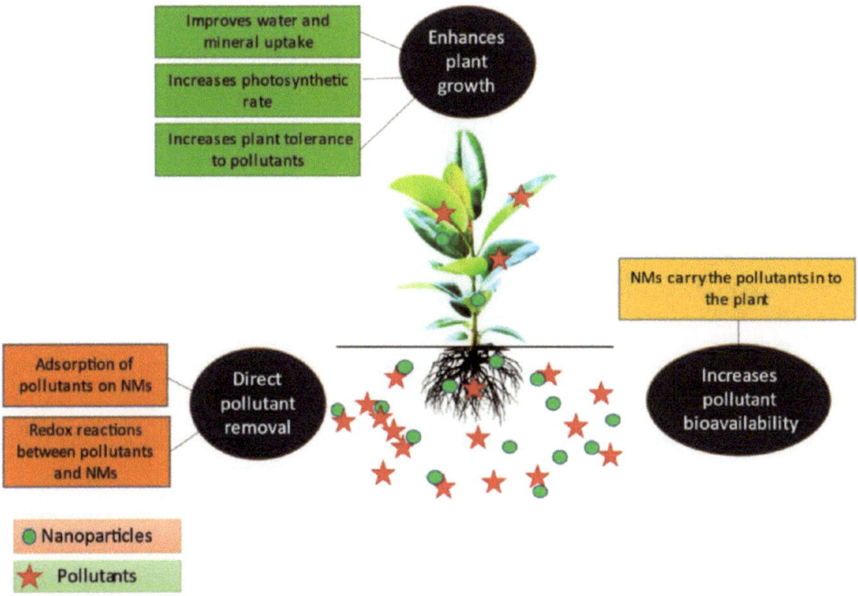

FIGURE 2.3 The role of NMs used in nano-phytoremediation.

1. **Elimination of Contaminants Directly:** The direct elimination of contaminants from the polluted matrix makes the job of phytoremediation easier. This occurs by the adsorption of pollutants on the NMs or by the redox reactions between pollutants and NMs [92]. For instance, carbon nanotubes stabilize organic pollutants by p-p bonding, electrostatic attraction, and hydrophobic interaction. In contrast, heavy metals are stabilized by physical adsorption, surface precipitation, and electrostatic attraction [93].

2. **Facilitation of Growth of Plants:** Certain NMs, like ZnO NPs, silver NPs, etc., are capable of increasing plant growth. This may

be by improving water uptake and mineral absorption from soil, by augmenting the rate of photosynthesis, etc. [94]. Moreover, the plant's tolerance to pollutants is increased by NMs [22].

3. **Enhancement of Pollutant Bioavailability:** An important factor that affects the success of phytoremediation is the bioavailability of pollutants [22]. Lowering the bioavailability of pollutants lowers the phytoremediation efficiency. NMs affect the bioavailability of pollutants in two ways. In one way, NMs increase the bioavailability of contaminants by carrying them after they enter the cell [95, 96]. Another way is that NMs decrease the bioavailability of contaminants by adsorbing the contaminants outside the plant [97]. Therefore, to improve the bioavailability of pollutants using NMs, NMs that are bioavailable and adsorb pollutants have to be used [22].

2.5.2 NANO-PHYTOREMEDIATION OF ORGANIC CONTAMINANTS

Organic chemicals which are man-made and are harmful, called persistent organic pollutants [98]. They are carbon-based compounds that are capable of bioaccumulation. Persistent organic pollutants can be of three types, which include pesticides and their derivatives, industrial chemicals, and by-products of various industrial processes [99]. The majority of persistent organic pollutants are chemicals that are halogenated. The bonds between carbon and halogens in such chemicals are so strong that their degradation in the environment is not possible [100]. As a result, living organisms are exposed to these pollutants, and bioaccumulation occurs. This can lead to a number of health issues like cardiovascular diseases, birth defects, diabetes, cancer, and improper functioning of reproductive and immune systems [101]. Nano-phytoremediation can be used to solve this problem of environmental contamination effectively. Various studies have reported the use of nano-phytoremediation in environmental clean-up. Bootharaju and Pradeep [102] reported that the use of Ag, Au, and Fe NPs in halocarbon-contaminated matrices could remove such contamination by dehalogenating the halocarbon compounds. Pillai and Kottekottil [103] found out that the combined use of zero valent iron NPs and the plant *Alpinia calcarata* enhanced the removal of Endosulfan from the soil by hydrogenolysis, sequential dehalogenation, and phytoextraction. Jesitha and Harikumar [104] reported the enhanced uptake of Endosulfan in the plant *Alpinia calcarata* on treating the soil with nano-zero valent iron NPs. The use of green synthesized iron NPs in combination with solubility-improving agents and phytoremediation was found to be effective in improving soil contaminant degradation and chlorfenapyr removal efficiency by *Plantago major* [105]. Romeh and Saber [105] combined phytoremediation with green NT to remove chlorfenapyr from contaminated soil and water. They synthesized iron NPs from *Ficus* and silver NPs from *Ipomoea* and *Brassica*. They observed that the combined use of the above-mentioned NPs with *Plantago major* resulted in 93.7%, 91.30%, and 92.92% decrease in chlorfenapyr in water after a period of 24 hours, respectively. Similarly, in the soil, *P. major* and the three NPs also reduced the chlorfenapyr content by 71.22%, 57.32%, and 73.10% after 4 days of treatment [105]. Silver NPs synthesized from the leaf extracts of *Lagerstroemia speciosa* were able to degrade the dyes methyl orange and methylene blue within 310 and 290 min [106]. Romeh [107] reported that *Plantago major* and silver NPs could remediate fipronil-contaminated water and soil. Ma and Wang [108] reported that fullerene NPs increased the absorption of trichloroethylene in plants without causing any acute toxicity effects. The absorption of trichloroethylene by plants in the presence of fullerene NPs occurs by the formation of fullerene-trichloroethylene complex, which in turn is formed by the adsorption of trichloroethylene to the surface of fullerene NPs [108]. The application of nano-zero valent iron NPs enhanced the growth and accumulation of polychlorinated biphenyls by the plant *Impatiens balsamina* [109]. Table 2.1 shows the list of NMs involved in the nano-phytoremediation of organic contaminants.

2.5.3 NANO-PHYTOREMEDIATION OF HEAVY METALS

Heavy metals are those metals that have high density and atomic number. They include Cd, Hg, Pb, Cr, Ni, Cu, Zn, etc. [110]. Heavy metals occur naturally in the earth's crust, but anthropogenic activities have increased their concentration in the environment, resulting in contamination. They are persistent pollutants that cannot be degraded biologically [111]. This makes their removal essential. Many authors have reported the use of nano-phytoremediation to reclaim heavy metal-contaminated environments. Some of such studies are mentioned here.

Salicylic acid NPs could increase root and shoot length and biomass of *Isatis cappadocica* under arsenic stress. In addition, the NPs also increased the accumulation of arsenic in the shoot and root of the plant

TABLE 2.1 NPs Used in Nano-Phytoremediation of Organic Pollutants

NPs Used	Size of NP	Applied Concentration of NP	Plant Used	Organic Contaminant	Plant Response	References
Fullerene NPs	100 nm	Up to 15 mg/L	*Populus deltoides*	Trichloroethylene [TCE]	Increased accumulation of TCE.	[108]
Zerovalent iron NPs	–	0.075–0.1%	*Impatiens balsamina*	Polychlorinated biphenyl [PCB]	Improved the growth of plants and accumulation of PCB.	[109]
Zerovalent iron NPs	<100 nm	1,000 mg/kg	*Alpinia calcarata*	Endosulfan	Enhanced removal of Endosulfan	[103]
Zerovalent iron NPs	30–100 nm	1,200 microgram/kg	*Alpinia calcarata*	Endosulfan	Enhanced removal of Endosulfan	[104]
Silver NPs	2–100 nm	2%	*Plantago major*	Fipronil	Improved the remediation of the fipronil product, fipronil amide.	[107]
Iron NPs	2.46–11.49 nm	–	*Plantago major*	Chlorfenapyr	Degradation of chlorfenapyr was enhanced.	[105]

[112]. Simultaneous application of nano-zerovalent iron and plant growth-promoting bacteria to soil contaminated with antimony had good growth and phytoremediation efficiency by *Trifolium repens*. The accumulation of Sb in the plant increased with nano-zerovalent iron, but a high concentration [1,000 mg/kg] of nano-zerovalent iron negatively impacted plant growth and phytoremediation [113]. Phytoremediation of Cd, Pb, and Zn by sweet sorghum in the presence of nano-zero valent iron and arbuscular mycorrhizal fungi decreased the concentrations of heavy metals in the soil. Nano-zerovalent iron increased the phytoextraction efficiency of the plant, while arbuscular mycorrhizal fungi were responsible for the immobilization of heavy metals [114]. Zand and Tabrizi [115] found that low to moderate concentrations [100–500 mg/kg] of nano-zerovalent iron increased Pb accumulation in the plant *Kochia scoparia*. The arsenic, lead, and mercury remediation potential of the plant *Cosmos bipinnatus* was enhanced in the presence of nano-zerovalent iron when the concentration of the NPs used was low [20 mg/kg] [116]. Raghib et al. [117] reported that simultaneous application of the arbuscular mycorrhizal fungi, *Glomus macrocarpum*, and ZnO NPs to wheat plants increased the plant height, fresh weight, dry weight, chlorophyll content, proline content, increased the activity of superoxide dismutase and catalase enzymes and decreased the accumulation of lead in root and shoot of wheat plants. This means the lead-contaminated soil can be used to successfully cultivate wheat in the presence of the above two amendments. The application of chitosan NPs increased the bioconcentration factor [BCF] and translocation factor [TF] of *Datura stramonium* L. towards Cd, which may be due to the negligible size, low crystalline nature, and large surface area of the chitosan NPs [118]. *Eichhornia crassipes* and *Salvinia molesta* were investigated for their removal efficiency towards Pb, Cd, and Cu. Application of TiO_2 NPs resulted in the efficient removal of Pb, Cd, and Cu and increased BCF, TF, growth, and chlorophyll production in the plants [119].

The effect of silver NPs and plant growth-promoting bacteria on the remediation of municipal wastewater by maize was investigated by Khan and Bano [120]. Individual application of plant growth-promoting bacteria and NPs resulted in the accumulation of Pb and Ni in the rhizosphere soil and shoot. Fe and Co accumulation, leaf area, root length, fresh weight, and activities of antioxidant enzymes were also increased in the plant by silver NPs. However, combined treatment using plant growth-promoting bacteria and silver NPs reduced the accumulation of Pb and Ni in the rhizosphere soil and plant shoot [120]. Many authors have reported the role of silicon in improving plant growth, stress resistance, and light use efficiency [121]. Silicon was also found to alleviate the phytotoxic effects of aluminum in *Fagopyrum esculentum* M. [122]. Vatehova et al. [123] reported that silicon could enhance the photosynthetic efficiency of *Brassica napus* in Cd-contaminated soil. Silicon could also improve the translocation of cadmium to shoots of *Brassica napus*. In a study by Mokarram-Kashtiban et al. [124], *Salix alba* L. was investigated for its heavy metal remediation efficiency in nano-zerovalent iron and arbuscular mycorrhizal fungus, *Rhizophagus irregularis* and *Pseudomonas fluorescens*, a plant growth promoting rhizobacteria. The presence of rhizosphere microbes enhanced plant growth and heavy metal accumulation. The application of a low concentration of nano-zerovalent iron was also found to increase the plant's root length and leaf area and improve the BCF value of cadmium. The list of NMs used in the nano-phytoremediation of heavy metals is provided in Table 2.2.

The above-mentioned application of NPs in phytoremediation promotes the removal of pollutants from the environment in lesser time periods. However, there is definitely an adverse impact on the use of NPs in the context of environmental restoration. The intentional application of NPs in soil can change the soil pH, nutrient availability, soil quality, and plant growth [125]. The normal functioning of the microbial cells present in soil is also disrupted by the presence of NPs [126].

2.6 CONCLUSION AND FUTURE PERSPECTIVES

Nano-phytoremediation has come out as a successful mechanism for the remediation of polluted areas. The blending of NT with phytoremediation has paved the way for a more rapid and efficient mechanism for environmental clean-up. Even though there are a number of studies on nano-phytoremediation, more plants and NM combinations are required to be investigated to know more about the underlying mechanisms and the reasons for the failure of nano-phytoremediation. More studies have to be carried out to identify the way NMs respond to various plant species, soil characteristics, pollutants, and climatic conditions for their application generally. Studies should be conducted in field conditions as well to assess the efficiency of this technology. Integration of nano-phytoremediation with genetic engineering and inoculating rhizospheric microorganisms also needs to be learned effectively to better this reclamation technology. Though NPs have been found to facilitate phytoremediation, the adverse effects of NPs on the environment

TABLE 2.2 NPs Used in Nano-Phytoremediation of Heavy Metals

NPs Used	Size of NP Used	Applied Concentration of NP	Plant Used	Contaminant	Response in Plant	References
TiO_2 NPs	–	–	*Eichhornia crassipes* and *Salvinia molesta*	Pb, Cd, Cu	Increased plant growth, BCF, TF, etc.	[119]
Salicylic acid NPs	200 nm	250 micromolar	*Isatis cappadocica*	As	Increased plant biomass and enhanced accumulation in the shoot and root.	[112]
Nanozerovalent iron	20–100 nm	150 mg/kg	*Salix alba* L.	Cd	Enhanced plant growth, improved BCF	[124]
Nanozerovalent iron	40–100 nm	500 mg/kg	*Trifolium repens*	Sb	Enhanced plant growth and increased accumulation of Sb in the plant.	[113]
Titanium dioxide NPs	15–40 nm	500 mg/kg	*Trifolium repens*	Cd	Enhanced accumulation of Cd	[127]
ZnO NPs	30 nm	100 mg/kg	Wheat plant	Pb	Increased plant biomass, chlorophyll content, proline content, etc.	[117]
Nanozerovalent iron	40–100 nm	500 mg/kg	*Kochia scoparia.*	Pb	Increased Pb accumulation	[115]
Nanozerovalent iron	30–50 nm	50–1,000 mg/kg	Sweet sorghum	Cd, Pb, and Zn	Augmented the phytoextraction efficiency for Cd, Pb, and Zn	[114]
Chitosan NPs	40–50 nm	75–225 mg/kg	*Datura stramonium* L.	Cd	Increased BCF and TF	[118]
Nanozerovalent iron	–	20 ppm	*Cosmos bipinnatus*	As, Pb, and Hg	Enhanced arsenic, lead, and mercury accumulation.	[116]
Titanium dioxide NPs	–	80 mg/L	*Coriandrum sativum*	Cd	Amelioration of Cd stress, enhanced yield.	[128]

cannot be neglected. NPs may become an additional pollutant besides the inorganic or organic pollutant that requires to be removed. NPs can alter the pH of the soil, thereby changing the soil properties and affecting soil health, growth, and development of plants. Therefore, before designing any nano-phytoremediation experiment, the fate of NPs has to be studied keenly.

KEYWORDS

- **environmental cleanup**
- **heavy metal pollution**
- **nano-phytoremediation**
- **pH**
- **plant-NPs interaction**
- **pollutants**
- **remediation**

REFERENCES

1. Landrigan, P. J., Fuller, R., Acosta, N. J. R., Adeyi, O., Arnold, R., Basu, N., et al., (2017). The lancet commission on pollution and health. *Lancet, 391,* 462512. Available from: https://doi.org/10.1016/S0140- 6736[17]32345-0.
2. Ukaogo, P. O., Ewuzie, U., & Onwuka, C. V., (2020). Environmental pollution: Causes, effects, and the remedies. *Microorganisms for Sustainable Environment and Health*, 419–429. doi: 10.1016/b978-0-12-819001-2.00021-8.
3. Arora, V., & Khosla, B., (2021). Conventional and *Contemporary Techniques for Removal of Heavy Metals from Soil.* doi: 10.5772/intechopen.98569.
4. Sharma, S., Tiwari, S., Hasan, A., Saxena, V., & Pandey, L. M., (2018). Recent advances in conventional and contemporary methods for remediation of heavy metal-contaminated soils. *Biotech., 8*, 216. doi: 10.1007/s13205-018-1237-8.
5. Berti, W. R., & Cunningham, S. D., (2000). Phytostabilization of metals. In: Raskin, I., & Ensley, B. D., (eds.), *Phytoremediation of Toxic Metals: Using Plants to Clean-up Cleanup the Environment* (pp. 71–88). New York, NY: John Wiley & Sons, Inc.
6. Jacob, J. M., Karthik, C., Saratale, R. G., Kumar, S. S., Prabakar, D., Kadirvelu, K., et al., (2018). Biological approaches to tackle heavy metal pollution: A survey of literature. *J. Environ. Manage., 217*, 56–70. doi: 10.1016/j.jenvman.2018.03.077.
7. Suman, J., Uhlik, O., Viktorova, J., & Macek, T., (2018). Phytoextraction of heavy metals: A promising tool for cleanup of polluted environment? *Frontiers in Plant Science, 9*, 1476.

8. Katiyar, P., Pandey, N., & Sahu, K. K., (2020). Biological approaches of fluoride remediation: Potential for environmental clean-up. *Environmental Science and Pollution Research, 27*, 13044–13055.
9. Oberdorster, G., Maynard, A., Donaldson, K., Castranova, V., Fitzpatrick, J., Ausman, K., Carter, J., et al., (2005). Principles for characterizing the potential human health effects from exposure to NMs: Elements of a screening strategy. *Part Fibre Toxicol., 2*, 1–35.
10. Singh, D. V., Bhat, R. A., Dervash, M. A., Qadri, H., Mehmood, M. A., Dar, G. H., Hameed, M., & Rashid, N., (2020). Wonders of NT for remediation of polluted aquatic environs. In: Qadri, H., Bhat, R. A., Dar, G. H., & Mehmood, M. A., (eds.), *Freshwater Pollution Dynamics and Remediation* (pp. 319–339). Springer, Singapore.
11. Karn, B., Kuiken, T., & Otto, M., (2009). NT and in situ remediation: A review of the benefits and potential risks. *Environ. Health Perspect., 117*, 1823–1831. https://doi.org/10.1289/ ehp.0900793.
12. Shanthilal, J., & Bhattacharya, S., (2014). NPs and NT in food. In: Bhattacharya, S., (ed.), *Conventional and Advanced Food Processing Technologies*. https://doi.org/10.1002/9781118406281.ch23.
13. Kiparissides, C., & Kammona, O., (2015). NT advances in diagnostics, drug delivery, and regenerative medicine. In: *The Nano-Micro Interface: Bridging the Micro and Nano Worlds:* (2nd edn., pp. 311–340). https://doi.org/10.1002/9783527679195.ch16.
14. Zhang, W. X., (2003). Nanoscale iron particles for environmental remediation: An overview. *J Nanopart Res., 5*, 323–332.
15. Salipira, K., Mamda, B. B., Krause, R. W., Malefetse, T. J., & Durbach, S. H., (2007). Carbon nanotubes and cyclodextrin polymers for removing organic pollutants from water. *Environ Chem Lett., 5*, 13–17. https://doi.org/10.1007/s10311-006-0057-y.
16. Szalai, G., Kellos, T., Galiba, G., & Kocsy, G., (2009). Glutathione is an antioxidant and regulatory molecule in plants under abiotic stress conditions. *J. Plant Growth Regul., 28*, 66–280.
17. Mierziak, J., Kostyn, K., & Kulma, A., (2014). Flavonoids as important molecules of plant interactions with the environment. Molecules, *19*, 16240–16265.
18. Camejo, D., Guzmán-Cedeño, Á., & Moreno, A., (2016). Reactive oxygen species, essential molecules, during plant-pathogen interactions. *Plant Physiol. Biochem., 103*, 10–23.
19. Franzoni, G., Trivellini, A., Bulgari, R., Cocetta, G., & Ferrante, A., (2019). Bioactive molecules as regulatory signals in plant responses to abiotic stresses. In: Khan, M. I. R., Reddy, P. S., Ferrante, A., & Khan, N., (eds.), *Plant Signaling Molecules* (1st edn., Vol. 1, pp. 169–182). Woodhead Publishing: Duxford, UK.
20. Srivastav, A., Yadav, K. K., Yadav, S., Gupta, N., Singh, J. K., Katiyar, R., & Kumar, V., (2018). Nano-phytoremediation of pollutants from contaminated soil environment: Current scenario and future prospects. *Phytoremediation,* 383–401. doi: 10.1007/978-3-319-99651-6_16.
21. Nwadinigwe, A. O., & Ugwu, E. C., (2018). Overview of nano-phytoremediation applications. In: Ansari, A., Gill, S., Gill, R., Lanza, G. R., & Newman, L., (eds.), *Phytoremediation*. Springer, Cham. https://doi.org/10.1007/978-3-319-99651-6_15.
22. Song, B., Xu, P., Chen, M., Tang, W., Zeng, G., Gong, J., & Ye, S., (2019). Using NMs to facilitate the phytoremediation of contaminated soil. *Critical Reviews in Environmental*

SCIENCE and Technology, *49*, 791–824. https://doi.org/10.1080/10643389.2018.1558891.

23. Khan, A. G., (2020). Promises and potential of *in situ* nano-phytoremediation strategy to mycorrhizo-remediate heavy metal contaminated soils using non-food bioenergy crops *(Vetiver zizinoides & Cannabis sativa)*. *International Journal of Phytoremediation*. doi: 10.1080/15226514.2020.1774504.
24. Khajeh, M., Laurent, S., & Dastafkan, K., (2013). Nanoadsorbents: Classification, preparation, and applications (with emphasis on aqueous media). *Chemical Reviews*, *113*(10), 7728–7768.
25. Abbasi, S. A., Abbasi, T., & Anuradha, J., (2012). A Process for the Synthesis of Metal NPs from Aquatic Weeds (p. 166184). *Off. J. Patent Office*. Application No.458/CHE/2012A.
26. Abbasi, S. A., Abbasi, T., Anuradha, J., Neghi, N., Prathiba, S., & Ganaie, S. U., (2011). *Gainful Utilization of Four Otherwise Worthless and Problematic Weeds for Silver NP Synthesis*. Off. J. Patent Office. Application No.2175/CHE/2011A, 11869.
27. Anuradha, J., Abbasi, T., & Abbasi, S. A., (2015). An eco-friendly method of synthesizing gold NPs using an otherwise worthless weed, Ppistia (*Pistia stratiotes* L.). *Journal of Advanced Research*, *6*, 711–720. doi: 10.1016/j. jare.2014.03.006 PMID:27563461.
28. Tai, C. Y., Tai, C., Chang, M., & Liu, H., (2007). *Synthesis of Magnesium Hydroxide and Oxide NPs Using a Spinning Disk Reactor*, 5536–5541.
29. Mohammadi, S., Harvey, A., & Boodhoo, K. V. K., (2014). Synthesis of TiO_2 NPs in a spinning disc reactor. *Chem. Eng. J.*, *258*, 171–184.
30. Kammler, B. H. K., Mädler, L., & Pratsinis, S. E., (2001). *Flame Synthesis of NPs, 24*, 583–596.
31. Amato, R. D., Falconieri, M., Gagliardi, S., Popovici, E., Serra, E., Terranova, G., & Borsella, E., (2013). Synthesis of ceramic NPs by laser pyrolysis: From research to applications. *J. Anal. Appl. Pyrolysis, 104*, 461–469.
32. Amendola, V., & Meneghetti, M., (2009). Laser ablation synthesis in solution and size manipulation of noble metal NPs. *Physical Chemistry Chemical Physics*, 3805–3821.
33. Salavati-niasari, M., Davar, F., & Mir, N., (2008). Synthesis and characterization of metallic copper NPs via thermal decomposition. *Polyhedron, 27*, 3514–8.
34. Rahman, I. A., & Padavettan, V., (2012). Synthesis of silica nanoparticles by sol-gel: Size-dependent properties, surface modification, and applications in silica-polymer nanocomposites—A review. *Journal of Nanomaterials*.
35. Mann, S., Burkett, S. L., Davis, S. A., Fowler, C. E., Mendelson, N. H., Sims, S. D., Walsh, D., & Whilton, N. T., (1997). *Sol – Gel Synthesis of Organized Matter, 4756*, 2300–2310.
36. Agarwal, H., Kumar, S. V., & Rajeshkumar, S., (2017). A review on green synthesis of zinc oxide nanoparticles–an eco-friendly approach. *Resource-Efficient Technologies*, *3*, 406–413.
37. Li, Y., Zhang, P., Du, Q., Peng, X., Liu, T., Wang, Z., Xia, Y., Zhang, W., Wang, K., Zhu, H., et al., (2011). Adsorption of fluoride from aqueous solution by graphene. *J. Colloid Interface Sci.*, *363*, 348–354. doi: 10.1016/j.jcis.2011.07.032.
38. Singh, P., Kim, Y. J., Zhang, D., & Yang, D. C., (2016). Biological synthesis of NPs from plants and microorganisms. *Trends in Biotechnology, 34*, 588–599. doi: 10.1016/j. tibtech.2016.02.006.

39. Duan, H., et al., (2015). Green chemistry for NP synthesis. *Chem. Soc. Rev., 44*, 5778–5792.
40. Roopan, S. M., Bharathi, A., Kumar, R., Khanna, V. G., & Prabhakarn, A., (2012). Acaricidal, insecticidal, and larvicidal efficacy of aqueous extract of *Annona squamosa* L peel as a biomaterial for the reduction of palladium salts into NPs. *Colloids Surf. B Biointerfaces, 92*, 209–212. doi: 10.1016/j.colsurfb.2011.11.044.
41. Roopan, S., Thakur, R., Rahuman, A., Kamaraj, D. C., Annadurai, B., & Tammineni, S., (2013). Low-cost and eco-friendly bio-synthesis of silver NPs using *Cocos nucifera* coir extract and its larvicidal activity. *Ind. Crops Prod., 43*, 31–635. doi: 10.1016/j.indcrop.2012.08.013.
42. Sundrarajan, M., & Gowri, S., (2011). Green synthesis of titanium dioxide NPs by *Nyctanthes arbor-tristis* leaves extract. *Chalcogenide Lett., 8*, 447–451.
43. Thirunavukkarasu, S., Rahuman, A., Chidambaram, J., Rajakumar, G., Marimuthu, S., Kirthi, V., et al., (2013). Green synthesis of titanium dioxide NPs using *Psidium guajava* extract and its antibacterial and antioxidant properties. *Asian Pac. J. Trop. Med., 7*, 968–976. doi: 10.1016/S1995-7645[14] 60171-1.
44. Zahir, A., Chauhan, I., Bagavan, A., Kamaraj, D. C., Elango, G., Shankar, J., et al., (2015). Green synthesis of silver and titanium dioxide NPs using *Euphorbia prostrata* extract shows shift from apoptosis to G_0 /G_1 arrest followed by necrotic cell death in *Leishmania donovani.* Antimicrob. *Agents Chemother., 59*, 4782–4799. doi: 10.1128/AAC.00098-15.
45. Kanayairam, V., Rahuman, A., Rajakumar, G., Thirunavukkarasu, S., Marimuthu, S., Chidambaram, J., et al., (2011). Evaluation of *Catharanthus roseus* leaf extract-mediated biosynthesis of titanium dioxide NPs against *Hippobosca maculata* and *Bovicola ovis. Parasitol. Res., 111*, 2329–2337. doi: 10.1007/s00436-011-2676-x.
46. Singh, R., (2011). Biological approach of zinc oxide NPs formation and its characterization. *Adv. Mater. Lett., 2,* 313–317. doi: 10.5185/amlett.indias.204.
47. Sangeetha, G., Rajeshwari, S., & Rajendran, V., (2011). Green synthesis of zinc oxide NPs by *Aloe barbadensis* miller leaf extract: Structure and optical properties. *Mater. Res. Bull., 46*, 2560–2566. doi: 10.1016/j.materresbull.2011.07.046.
48. Duran, N., & Seabra, A., (2012). Metallic oxide NPs: State of the art in biogenic syntheses and their mechanisms. *Appl. Microbiol. Biotechnol., 95*, 275–288. doi: 10.1007/s00253-012-4118-9.
49. Njagi, E. C., Huang, H., Stafford, L., Genuino, H., Galindo, H. M., Collins, J. B., & Suib, S. L., (2011). Biosynthesis of iron and silver NPs at room temperature using aqueous sorghum bran extracts. *Langmuir, 27*, 264–271.
50. Shah, R., Eldridge, D., Palombo, E., & Harding, I., (2014). Optimization and stability assessment of solid lipid NPs using particle size and zeta potential. *Journal of Physical Science, 25*(1).
51. Kumari, B., & Singh, D. P., (2016). A review of multifaceted application of NPs in the field of bioremediation of petroleum hydrocarbons. *Ecol. Eng., 97,* 98–105.
52. Yadav, K. K., Singh, J. K., Gupta, N., & Kumar, V., (2017). A review of nanobioremediation technologies for environmental cleanup: A novel biological approach. *J. Mater. Environ. Sci., 8*, 740–757.
53. Guerra, F. D., Attia, M. F., Whitehead, D. C., & Alexis, F. (2018). NT for environmental remediation: Materials and applications. *Molecules (Basel, Switzerland), 23*, 1760. https://doi.org/10.3390/molecules23071760.

54. Hooshyar, Z., Rezanejade, B. G., & Ghayeb, Y., (2013). Sonication enhanced the removal of nickel and cobalt ions from polluted water using an iron-based sorbent. *J. Chem., 2013*, 786954. doi: 10.1155/2013/786954.
55. Ebrahim, S. E., Sulaymon, A. H., & Saad, A. H., (2016). Competitive removal of Cu^{2+}, Cd^{2+}, Zn^{2+}, and Ni^{2+} ions onto iron oxide NPs from wastewater. *Desalin. Water Treat.*, *57*, 20915–20929. doi: 10.1080/19443994.2015.11.
56. Guo, M., Weng, X., Wang, T., & Chen, Z., (2017). Biosynthesized iron-based NPs are used as a heterogeneous catalyst for the removal of 2,4-dichlorophenol. *Sep. Purif. Technol.*, *175*, 222–228. doi: 10.1016/j.seppur.2016.11.042.
57. Rasalingam, S., Peng, R., & Koodali, R. T., (2014). Removal of hazardous pollutants from wastewaters: Applications of TiO_2-SiO_2 mixed oxide materials. *J. Nanomater.*, *2014*, 617405. doi: 10.1155/2014/617405.
58. Lien, H. L., & Zhang, W., (2005). Hydrodechlorination of chlorinated ethanes by nanoscale Pd/Fe bimetallic particles. *J. Environ. Eng.*, *131*, 4–10. doi: 10.1061/[ASCE]0733-9372[2005]131:1.
59. Chen, C., Wang, X., Chang, Y., & Liu, H., (2008). Dechlorination of disinfection by-product monochloroacetic acid in drinking water by nanoscale palladized iron bimetallic particle. *J. Environ. Sci.*, *20*, 945–951. doi: 10.1016/S1001-0742[08]62191-9.
60. Wang, X., Chen, C., Chang, Y., & Liu, H., (2009). Dechlorination of chlorinated methanes by Pd/Fe bimetallic NPs. *J. Hazard. Mater.*, *161*, 815–823. doi: 10.1016/j.jhazmat.2008.
61. Zhang, Z., Shen, Q., Cissoko, N., Wo, J., & Xu, X., (2010). Catalytic dechlorination of 2,4-dichlorophenol by Pd/Fe bimetallic NPs in the presence of humic acid. *J. Hazard. Mater.*, *182*, 252–258. doi: 10.1016/j.jhazmat.2010.06.022.
62. Khin, M. M., Nair, A. S., Babu, V. J., Murugan, R., & Ramakrishna, S., (2012). A review on NMs for environmental remediation. *Energy Environ. Sci.*, *5,* 8075–8109. doi: 10.1039/c2ee21818f.
63. Nomura, A., & Jones, C. W., (2013). Amine-functionalized porous silicas as adsorbents for aldehyde abatement. *ACS Appl. Mater. Interfaces*, *5*, 5569–5577. doi: 10.1021/am400810s.
64. Bollini, P., Didas, S. A., & Jones, C. W., (2011). Amine-oxide hybrid materials for acid gas separations. *J. Mater. Chem.*, *21*, 15100–15120. doi: 10.1039/c1jm12522b.
65. Choi, S., Drese, J. H., Eisenberger, P. M., & Jones, C. W., (2011). Application of amine-tethered solid sorbents for direct CO_2 capture from the ambient air. *Environ. Sci. Technol.*, *45*, 2420–2427. doi: 10.1021/es102797w.
66. Drese, J. H., Talley, A. D., & Jones, C. W., (2011). Aminosilica materials as adsorbents for the selective removal of aldehydes and ketones from simulated bio-oil. *ChemSusChem.*, *4,* 379–385. doi: 10.1002/cssc.201000347.
67. Ren, X., Chen, C., Nagatsu, M., & Wang, X., (2011). Carbon nanotubes as adsorbents in environmental pollution management: A review. *Chem. Eng. J.*, *170*, 395–410. doi: 10.1016/j.cej.2010.08.045.
68. Kharisov, B. I., Rasika, D. H. V., & Kharissova, O. V., (2014). NT-based remediation of petroleum impurities from water. *J. Pet. Sci. Eng.*, *122*, 705–718. doi: 10.1016/j.petrol.2014.09.013.
69. Zhang, N., Yang, M. Q., Tang, Z. R., & Xu, Y. J., (2013). CdS-graphene nanocomposites as visible light photocatalyst for redox reactions in water: A green route for selective transformation and environmental remediation. *J. Catal.*, *303*, 60–69.

70. Tungittiplakorn, W., Lion, L. W., Cohen, C., & Kim, J. Y., (2004). Engineered polymeric NPs for soil remediation. *Environ. Sci. Technol., 38*, 1605–1610. doi: 10.1021/es0348997.
71. Diallo, M. S., Christie, S., Swaminathan, P., Johnson, J. H., & Goddard, W. A., (2005). Dendrimer enhanced ultrafiltration. 1. Recovery of Cu[II] from aqueous solutions using PAMAM dendrimers with ethylene diamine core and terminal NH_2 groups. *Environ. Sci. Technol., 39*, 1366–1377. doi: 10.1021/es048961r.
72. Servin, A., Elmer, W., Mukherjee, A., et al., (2015). A review of the use of engineered NMs to suppress plant disease and enhance crop yield. *Journal of NP Research, 17*, 1–21.
73. Kim, D. H., Gopal, J., & Sivanesan, I., (2017). NMs in plant tissue culture: The disclosed and undisclosed. *RSC Advances, 7*, 36492–36505.
74. Nair, R., Varghese, S. H., Nair, B. G., Maekawa, T., Yoshida, Y., & Kumar, D. S., (2010). Nanoparticulate material delivery to plants. *Plant Science, 179*, 154–163.
75. Raliya, R., Franke, C., Chavalmane, S., Nair, R., Reed, N., & Biswas, P., (2016). Quantitative understanding of NP uptake in watermelon plants. *Frontiers in Plant Science, 7*, 1228.
76. Kaphle, A., Navya, P. N., Umapathi, A., & Daima, H. K. (2018). NMs for agriculture, food, and environment: Applications, toxicity, and regulation. *Environmental Chemistry Letters, 16*, 43–58.
77. Lin, D., & Xing, B., (2008). Root uptake and phytotoxicity of ZnO NPs. *Environmental Science & Technology, 42*, 5580–5585.
78. Du, W., Sun, Y., Ji, R., Zhu, J., Wu, J., & Guo, H., (2011). TiO_2 and ZnO NPs negatively affect wheat growth and soil enzyme activities in agricultural soil. *Journal of Environmental Monitoring, 13*, 822–828.
79. Perez-de-Luque, A., (2017). Interaction of NMs with plants: What do we need for real applications in agriculture? *Frontiers in Environmental Science, 5*, 12.
80. Tripathi, D. K., Singh, S., Singh, V. P., Prasad, S. M., Dubey, N. K., & Chauhan, D. K., (2017). Silicon NPs more effectively alleviated UV-B stress than silicon in wheat (*Triticum aestivum*) seedlings. *Plant Physiology and Biochemistry, 110*, 70–81.
81. Wang, Z., Xie, X., Zhao, J., et al., (2012). Xylem-and phloem-based transport of CuO NPs in maize (*Zea mays* L.). *Environmental Science & Technology, 46*, 4434–4441.
82. Ruttkay-Nedecky, B., Krystofova, O., Nejdl, L., & Adam, V., (2017). NPs based on essential metals and their phytotoxicity. *Journal of Nanbiotechnology, 15*, 33.
83. Kole, C., Kole, P., Randunu, K. M., Choudhary, P., Podila, R., & Ke, P. C., (2013). Nanobiotechnology can boost crop production and quality: First evidence from increased plant biomass, fruit yield, and phytomedicine content in bitter melon (*Momordica charantia*). *BMC Biotechnology, 13*, 37.
84. Gopinath, K., Gowri, S., Karthika, V., & Arumugam, A., (2014). Green synthesis of gold NPs from fruit extract of *Terminalia arjuna* for the enhanced seed germination activity of *Gloriosa superba*. *Journal of Nanostructure in Chemistry, 4*, 115.
85. Hafeez, A., Razzaq, A., Mahmood, T., & Jhanzab, H. M., (2015). The potential of copper NPs to increase growth and yield of wheat. *Journal of Nanoscience with Advanced Technology, 1*, 6–11.
86. Gao, F., Hong, F., Liu, C., et al., (2008). Mechanism of nanoanatase TiO_2 on promoting photosynthetic carbon reaction of spinach. *Biological Trace Element Research, 111*, 239–253.

87. Klaine, S. J., Alvarez, P. J. J., Batley, G. E., et al., (2008). NMs in the environment: Behavior, fate bioavailability, and effects. *Environmental Toxicology and Chemistry, 27*, 1825–1851.
88. Linglan, M., Chao, L., Chunxiang, Q., et al., (2008). Rubisco activase mRNA expression in spinach: Modulation by nanoanatase treatment. *Biological Trace Element Research, 122*, 168–178.
89. Wang, P., Lombi, E., Zhao, F. J., & Kopittke, P. M., (2016). NT: A new opportunity in plant sciences. *Trends in Plant Science*, *21*, 699–712.
90. Aken, V. B., (2015). Gene expression changes in plants and microorganisms exposed to NMs. *Current Opinion in Biotechnology*, *33*, 206–219.
91. Gul, M. Z., Rupula, K., & Beedu, S. R., (2022). Nano-phytoremediation for soil contamination: An emerging approach for revitalizing the tarnished resource. In: *Phytoremediation* (pp. 115–138). Academic Press.
92. Mueller, N. C., & Nowack, B., (2010). NPs for remediation: Solving big problems with little particles. *Elements*, *6*, 395–400. doi: 10.2113/gselements.6.6.395.
93. Song, B., Zeng, G., Gong, J., Zhang, P., Deng, J., Deng, C., & Cheng, M., (2017). Effect of multi-walled carbon nanotubes on phytotoxicity of sediments contaminated by phenanthrene and cadmium. *Chemosphere*, *172,* 449–458. doi: 10.1016/j. chemosphere.2017.01.032.
94. Tripathi, D. K., Singh, V. P., Prasad, S. M., Chauhan, D. K., & Dubey, N. K., (2015). Silicon NPs (SiNp) alleviate chromium (VI) phytotoxicity in *Pisum sativum* [L.] seedlings. *Plant Physiology and Biochemistry*, *96*, 189–198. doi: 10.1016/j.plaphy.2015.07.026.
95. Wild, E., & Jones, K. C., (2009). Novel method for the direct visualization of in vivo NMs and chemical interactions in plants. *Environmental Science & Technology*, *43*, 5290–5294. doi: 10.1021/es900065h.
96. Su, Y., Yan, X., Pu, Y., Xiao, F., Wang, D., & Yang, M., (2013). Risks of single-walled carbon nanotubes acting as contaminants-carriers: Potential release of phenanthrene in Japanese medaka (*Oryzias latipes*). *Environmental Science & Technology*, *47,* 4704–4710. doi: 10.1021/es304479w.
97. Glomstad, B., Altin, D., Sørensen, L., Liu, J., Jenssen, B. M., & Booth, A. M., (2016). Carbon nanotube properties influence the adsorption of phenanthrene and subsequent bioavailability and toxicity to *Pseudokirchneriella subcapitata. Environmental Science & Technology*, *50*, 2660–2668. doi: 10.1021/acs.est.5b05177.
98. Ashraf, M. A., (2017). Persistent organic pollutants (POPs): A global issue, a global challenge. *Environ. Sci. Pollut. Res. Int.*, *24*, 4223–4227. doi: 10.1007/s11356-015-5225-9.
99. Gaur, N., Narasimhulu, K., & PydiSetty, P., (2018). Recent advances in the bio-remediation of persistent organic pollutants and their effect on the environment. *J. Clean. Prod.*, *198*, 1602–1631. doi: 10.1016/j.jclepro.2018.07.076.
100. Rosenfeld, P. E., & Feng, L. G. H., (2011). 15- bioaccumulation of dioxins, PCBs, and PAHs. In: Rosenfeld, P. E., & Feng, L. G. H., (eds.), *Risks of Hazardous Wastes* (pp. 201–213). William Andrew Publishing; Boston, MA, USA.
101. Alharbi, O. M. L., Basheer, A. A., Khattab, R. A., & Ali, I. (2018). Health and environmental effects of persistent organic pollutants. *J. Mol. Liq.*, *263*, 442–453. doi: 10.1016/j.molliq.2018.05.029.

102. Bootharaju, M. S., & Pradeep, T., (2012). Understanding the degradation pathway of the pesticide chlorpyrifos by noble metal NPs. *Langmuir*, *28*, 2671–2679. http://dx.doi.org/10.1021/la2050515.
103. Pillai, H. P. S., & Kottekottil, J., (2016). Nano-phytotechnological remediation of endosulfan using zero valent iron NPs. *Journal of Environmental Protection*, *7*, 734–744. doi: 10.4236/jep.2016.75066.
104. Jesitha, K., & Harikumar, P. S., (2018). Application of nano-phytoremediation technology for soil polluted with pesticide residues and heavy metals. *Phytoremediation*, 415–439. doi: 10.1007/978-3-319-99651-6_18.
105. Romeh, A. A., & Saber, R. A. I., (2020). Green nano-phytoremediation and solubility-improving agents for the remediation of chlorfenapyr-contaminated soil and water. *Journal of Environmental Management*, *260*, 110104.
106. Saraswathi, V. S., Kamarudheen, N., BhaskaraRao, K. V., & Santhakumar, K., (2017). Phytoremediation of dyes using *Lagerstroemia speciosa* mediated silver NPs and its biofilm activity against clinical strains *Pseudomonas aeruginosa*. *Journal of Photochemistry and Photobiology B: Biology*, *168*, 107–116.
107. Romeh, A. A. A., (2018). Green silver NPs for enhancing the phytoremediation of soil and water contaminated by fipronil and degradation products. *Water Air Soil Pollut.*, *229*, 147. https://doi.org/10.1007/s11270-018-3792-3.
108. Ma, X., & Wang, C., (2010). Fullerene NPs affect the fate and uptake of trichloroethylene in phytoremediation systems. *Environmental Engineering Science*, *27*, 989–992.
109. Gao, Y. Y., & Zhou, Q. X., (2013). *Application of Nanoscale Zero Valent Iron Combined with Impatiens balsamina to Remediation of E-Waste Contaminated Soils. Advanced Materials Research, 790,* 73–76. doi: 10.4028/www.scientific.net/am.
110. Gupta, M. K., Kumari, K., Srivastava, A., & Shikha, G. (2014). Bioremediation of heavy metal-polluted environment using resistant bacteria. *Journal of Environmental Research and Development*, *8*, 883–889.
111. Coral, M. N. U., Korkmaz, H., & Arikan, B., (2015). Plasmid-mediated heavy metal resistance in *Enterobacter* spp. isolated from Sofulu landfill in Adana, Turkey. *Annales de Microbiologie, 55*, 175–179.
112. Souri, Z., Karimi, N., Sarmadi, M., & Rostami, E., (2017). Salicylic acid NPs [SANPs] improve the growth and phytoremediation efficiency of *Isatis cappadocica* Desv. under stress. *IET Nanobiotechnology*, *11*, 650–655.
113. Daryabeigi, Z. A., Tabrizi, A. M., & Heir, A. V., (2020). The influence of association of plant growth-promoting rhizobacteria and zero-valent iron NPs on the removal of antimony from the soil by *Trifolium repens*. *Environ Sci Pollut Res.*, *27*, 42815–42829. https://doi.org/10.1007/s11356-020-10252-x.
114. Cheng, P., Zhang, S., Wang, Q., Feng, X., Zhang, S., Sun, Y., & Wang, F., (2021). Contribution of nano-zero-valent iron and arbuscular mycorrhizal fungi to phytoremediation of heavy metal-contaminated soil. *NMs*, *11*, 1264.
115. Zand, A. D., & Tabrizi, A. M., (2020). Effect of zero-valent iron NPs on the phytoextraction ability of *Kochia scoparia* and its response in Pb contaminated soil. *Environ. Eng. Res*. https://doi.org/10.4491/eer.2020.227.
116. Majumdar, A., Upadhyay, M. K., Ojha, M., Afsal, F., Giri, B., Srivastava, S., & Bose, S., (2022). Enhanced phytoremediation of metal [loid] s via spiked ZVI NPs: An urban cleanup strategy with ornamental plants. *Chemosphere*, *288*, 132588.

117. Raghib, F., Naikoo, M. I., Khan, F. A., Alyemeni, M. N., & Ahmad, P., (2020). Interaction of ZnO NP and AM fungi mitigates Pb toxicity in wheat by upregulating antioxidants and restricting the uptake of Pb. *Journal of Biotechnology*, *323*, 254–263.
118. Shirkhani, Z., Rad, A. C., & Mohsenzadeh, F., (2021). Improving Cd-phytoremediation ability of *Datura stramonium* L. by chitosan and chitosan NPs. *Biologia*, 1–11.
119. Harikumar, P. S., & Megha, T., (2017). Treatment of cadmium, lead, and copper using phytoremediation enhanced by titanium dioxide NPs. *Eco Chronicle*, *12*(2), 25–34.
120. Khan, N., & Bano, A., (2016). Modulation of phytoremediation and plant growth by the treatment with PGPR, Ag NP, and untreated municipal wastewater. *International Journal of Phytoremediation*, *18*, 1258–1269.
121. Li, P., Song, A., Li, Z. J., Fan, F. L., & Liang, Y. C., (2012). Silicon ameliorates manganese toxicity by regulating manganese transport and antioxidant reactions in rice (*Oryza sativa* L.). *Plant and Soil*, *354*, 407–419.
122. Dar, F. A., Tahir, I., Hakeem, K. R., *et al.*, (2021). Silicon application enhances the photosynthetic pigments and phenolic/flavonoid content by modulating the phenylpropanoid pathway in common buckwheat under aluminum stress. *Silicon*, https://doi.org/10.1007/s12633-021-01501-w.
123. Vatehova, Z., Kollarova, K., Zelko, I., Kucerova, D. R., Bujdos, M., & Liskova, D., (2012). Interaction of silicon and cadmium in *Brassica juncea* and *Brassica napus*. *Biologia*, *67*, 498–504.
124. Mokarram-Kashtiban, S., Hosseini, S. M., Tabari, K. M., et al., (2019). The impact of NPs zero-valent iron [nZVI] and rhizosphere microorganisms on the phytoremediation ability of white willow and its response. *Environ Sci Pollut Res.*, *26*, 10776–10789. https://doi.org/10.1007/s11356-019-04411-y.
125. Fernández, F. G., & Hoeft, R. G., (2009). Managing soil pH and crop nutrients. *Ill. Agron. Handb.*, *24*, 91–112.
126. Ameen, F., Alsamhary, K., Alabdullatif, J. A., & Al-Nadhari, S. (2021). A review on metal-based NPs and their toxicity to beneficial soil bacteria and fungi. *Ecotoxicol. Environ. Saf.*, *213*, 112027.
127. Zand, A. D., Tabrizi, A. M., & Heir, V. A., (2020). Application of titanium dioxide NPs to promote phytoremediation of Cd-polluted soil: Contribution of PGPR inoculation. *Bioremediation Journal, 2020,* 1–19. doi: 10.1080/10889868.2020.179992.
128. Sardar, R., Ahmed, S., & Yasin, N. A., (2022). Titanium dioxide NPs mitigate cadmium toxicity in *Coriandrum sativum* L. by modulating the antioxidant system, stress markers and reducing cadmium uptake. *Environ Pollut.*, *292*(Pt A), 118373. doi: 10.1016/j.envpol.2021.118373.

CHAPTER 3

Microorganisms, Plants, and Nanotechnology for Environmental Remediation: A Sustainable Prospect

VÁZQUEZ-NÚÑEZ EDGAR,[1] PÉREZ-HERNÁNDEZ HERMES,[2] VALLE-GARCÍA JESSICA DENISSE,[3] PÉREZ-MORENO ANDREA,[4] SARABIA-CASTILLO CESAR ROBERTO,[4] VERA-REYES ILEANA,[5] and FERNÁNDEZ-LUQUEÑO FABIÁN[4]

[1]*Departamento de Ingenierías Química, Electrónica y Biomédica, División de Ciencias e Ingenierías, Campus León, Universidad de Guanajuato, León, Guanajuato, México*

[2]*El Colegio de la Frontera Sur, Agroecología, Unidad Campeche, Campeche, Mexico*

[3]*Transdisciplinary Doctoral Program in Scientific and Technological Development for the Society, Cinvestav-Zacatenco, Mexico City, Mexico*

[4]*Sustainability of Natural Resources and Energy Program, Cinvestav-Saltillo, Saltillo, Coahuila, Mexico*

[5]*Department of Plastics in Agriculture, Centro de Investigación en Química Aplicada, Saltillo, Coahuila, Mexico*

ABSTRACT

Currently, thousands of metric tons of engineered nanomaterials (ENMs) are delivered to the ecosystems after they have accomplished their useful life or even before. These ecosystems harbor millions of microorganisms, plants, and non-living components interacting as a system, which nanosized materials could jeopardize with unique properties and characteristics. However, ENMs

Nano-Bioremediation for Water and Soil Treatment: An Eco-Friendly Approach.
Vishnu D. Rajput, Arpna Kumari, and Tatiana M. Minkina (Eds.)

could also be an option to dissipate, degrade, or adsorb pollutants, i.e., ENMs could be a risk and a solution to different human and environmental issues. Nanotechnology could shape sustainable development by pursuing sustainable development goals, but it could also hamper them. Therefore, this chapter discusses new strategies regarding cutting-edge nanobiotechnologies for the dissipation of pollutants from soil, water, or air developed at the laboratory and field scale. Phytoremediation, bioaugmentation, and biostimulation technologies work together with plants, bacteria, fungi, and other organisms. Alongside bionanomaterials, nanobiohybrids, smart materials, and nanomachines are the technological bet for developing sustainable technologies for environmental remediation. Consequently, eco-friendly, cheap, and easily manufactured nanotechnologies have attracted attention during the last years. However, the release of new developments with remediation efficiencies never seen before, reduced costs, and no human or environmental risks is still pending.

3.1 INTRODUCTION

It is well known that there are more than 1×10^6 microbial cells in 1 cm^3 of dry soil and between 3,000 and 7,000 different species. Besides, this small soil volume could contain dozens of seeds from tiny plants and some milligrams of several kinds of natural nanomaterials (NMs) [1]. Therefore, the soil and its microorganisms, plants, and NMs work together with such a system, maintaining balance in ecosystems. This system can dissipate pollutants under a natural procedure named natural attenuation [2]

However, natural attenuation could not degrade the pollutants if the system's microorganisms or substrate die, their biochemical or physiological characteristics change, or their genes in charge of triggering the synthesis of metabolites are not expressed. Besides, other physical, chemical, and biological properties of the substrate (soil, water, or air) could reduce contaminants' natural dissipation [3]. Therefore, different strategies have been developed to remediate contaminated sites. Bioremediation is a set of techniques that use organisms, organic substrates, or organic compounds to dissipate pollutants, i.e., these techniques harness microorganisms or plants' properties to reduce the concentration of xenobiotic, recalcitrant, or carcinogenic compounds [4].

Dealing with soil, water, or air contaminants demands different knowledge areas with contrasting theoretical approaches and perspectives. It is well known that remediation techniques are unique because they necessitate numerous local stakeholders in the decision-making processes and technicians and scientists' participation because the remediation practices have social, economic, and environmental consequences and legal, technological, or institutional outcomes [5].

With technical issues of uncertainties regarding different contaminant characteristics and various properties of the polluted substrate, it is hard to decide when dealing with polluted sites and choose appropriate remediation methods [6].

Vázquez-Núñez et al. [7] stated that the synergy between ENMS, microorganisms, and plants used to remediate polluted sites had been proved. According to the same authors, there is still reduced knowledge about the concomitant effect of ENMs and bio-based technologies such as phytoremediation during nanobioremediation processes. It is well known that nanobioremediation technologies have emerged during the last years to face environmental pollution. Still, more cost-effective methods of ENMs production should arise to use nanotechnologies in remediation processes for decontaminating soil, air, or water.

Polymer nanocomposites (PNCs) are nanofillers dispersed in a polymer matrix; these PNCs represent a feasible alternative to conventional composites for several applications, such as environmental remediation. ENPs such as semiconductive materials, zero-valence metals, or bimetallic are important in the remediation of air, water, and soil due to their high interfacial reactivity with pollutants. The main working principle of ENPs during remediation processes is catalytic degradation or adsorption. Besides, to facilitate the separation and reuse and prevent the release of these ENPs into the ecosystem, a most used strategy is to impregnate or coat the ENPs onto a host such as polymers [8]. These PNCs could work together with organisms to increase the dissipation rate of environmental pollutants, but also these could work alongside other technologies to degrade contaminants. Besides, biocompatible functionalization of nanoclays or other ENPs for improving environmental remediation has also been proposed [9].

It has to be highlighted that modified nanoclays, PNCs, or other nanosized materials promise environmental remediation. Still, their biocompatibility and eco-compliance are not yet guaranteed or remain unknown when used in natural systems. Besides, according to [9, 10], there is insufficient data to conclude whether these never seen nanosized technologies are environmentally biocompatible. Otherwise, various types of functional nanoreactors such as polymers, dendrimers, bacteria, liposomes, organelles, or cells have been designed for diverse applications due to their potential to improve the rate of chemical reactions by space constraints through efficient and cascaded chemical conversions [11]; for this reason, numerous researchers have designed specific nanoreactors expecting to change the chemical reaction behavior with environmental or ecological applications.

This chapter aims to discuss new strategies regarding cutting-edge nanobiotechnologies for the dissipation of pollutants from soil, water, or air developed at the laboratory and field scale.

3.2 ENVIRONMENTAL SUSTAINABILITY AND ITS REGULATORY FRAMEWORK

In 1987, the World Commission on Environment and Development (WCED) published 'Our Common Future,' also known as 'The Brundtland Report.' This report synthesizes global environmental challenges and the concept of sustainable development (SD), which is defined as the SD that allows the fulfillment of human needs without fulfilling the needs of future generations at risk. The concept of SD results from a growing awareness of diverse problems and concerns related to poverty, inequality, and depletion of natural resources and ecosystems. Besides, SD has been amply used in political discourse and policymaking because this concept binds together socioeconomic, institutional, and environmental themes [12]. The SD framework includes human well-being, ecological equilibrium, and interactions. Therefore, SD is an integrated approach to economic, social, and environmental performance to solve poverty while avoiding an ecological crisis. The SD concept has been linked to the corporate world, contributing to enacting a significant number of environmental laws and conduct codes, which have been developed in response to the concern for environmental deterioration. Besides, organizations and institutions have developed environmental management systems and environmental audits as a corporate response [12]. These authors proposed using 18 different environmental regulation indicators through a proposal with relative complexity, reflecting the complexity of environmental regulation itself.

Industrial businesses have generated significant environmental pollution levels; therefore, green or non-polluting innovations have received more attention for achieving environmental sustainability. Qiu et al. [13] stated that environmental regulation promotes green process innovation and green product innovation, improving the company's financial performance. Governments and international organizations are currently focusing on environmental issues and are promoting increasingly strict and diverse environmental regulations. Consequently, the awareness about the environment and public health is increasing. As a result, customers demand more environmentally safe products and environmental liability among companies. Under these dual pressures from the government and market, firms cannot ignore environmental issues in their production and marketing processes.

It is well known that whether production and consumption patterns are not sustainable, undesirable consequences arise. These outcomes may include but are not limited to environmental contamination, climate change, loss of biological diversity, soil degradation, increased inequality, and jeopardizing social well-being and sustainable development [14]. Therefore, the regulatory frameworks of each country might help mitigate pollution and contribute to environmental sustainability. According to Martínez-Hernández et al. [12], business-oriented environmental regulations are expected to significantly mitigate human activity's adverse effects on the natural environment.

The Rio Declaration on Environment and Development in 1992 determined that environmental impact assessment (EIA) should be the tool of choice for achieving sustainable development. According to Ross et al. [15], EIA is defined as a process dedicated to supporting the management of actual and potential environmental impacts associated with developing specific projects; however, a significant drawback is the cost and time of this process. Taking South Africa as a case study, one-third of all the officials responsible for EIA review and decision-making were surveyed by Ross et al. [15] to identify their perceptions of EIA's benefits and their ideas to bridge the gap between potential benefits and realized benefits. Their research demonstrates that the perceptions of the officials responsible for EIA review and decision-making broadly tally with the statement that the protection of biodiversity is considered the main potential and realized benefit for society, which highlights the fact that EIA and life-cycle assessments (LCA) are recognized to contribute mostly to the ecosystem services constituent of the SD.

According to Mathis [16], environmental protection is often seen as a cost factor that impairs international competitiveness, but more stringent environmental regulations can incentivize innovations. Therefore, companies that take advantage of an "early-mover advantage" can gain a long-term competitive advantage. Companies, decision-makers, and stewards are looking to improve their revenues or political recognition by forming the SD.

According to Hou & O'Connor [17], sustainable remediation (SR) is a holistic approach where the environmental, social, and economic benefits of remediation are maximized for all stakeholders inside and outside the site boundary, current and future generations. Therefore, SR must meet five criteria [17]: (i) each viable remediation alternative must be evaluated by an evidence-based sustainability assessment focusing on the possible environmental, social, and economic impacts; (ii) the sustainability benefits of the chosen remedial alternative exceed the local and broader detrimental impacts on an EIA and LCA basis; (iii) relevant and up-to-date best management practices are applied to minimize secondary emissions, waste, energy and resource use, and ecological impacts; (iv) the social impacts to workers and local communities are considered and addressed by stakeholder engagement; and (v) the remediation must minimize life-cycle project costs and maximizes revenues on the broader economy, without prejudice to social and environmental aspects.

The United States Environmental Protection Agency (USEPA) defines green remediation (GR). According to USEPA [18], the GR uses technologies capable of efficiently harnessing natural resources and energy; these technologies must positively impact the environment, reduce the pollution at its source, and minimize waste to the greatest extent possible. Besides, GR has to consider all environmental effects of remedy implementation for polluted sites and include alternatives for maximizing remediation procedures to benefit the environment. According to Hou and O'Connor [17], SR and GR are too similar; these technologies should be synonymous. However, Nathanail et al. [19] stated that the terms SR and GR are not synonymous and recommend that assessors clarify which framework they are adopting.

According to Nathanail et al. [19], the 'ISO 18504:2017 Soil Quality — Sustainable Remediation' started in 2012. This stated that remediation begins with evaluating an option of those strategies that could reduce the risk. Therefore, a feasible remediation strategy should consider integrating remediation technologies that assure high removal efficiencies and reduce subsequent risks. The ISO standard will help assessors identify the most sustainable, valid alternative remediation strategies among the shortlisted. Consequently, this ISO might increase the acceptance of the definition of SR worldwide [19].

As said before, during the last years, stakeholders, decision-makers, scientists, and researchers have worked to lay the foundations and guarantee the best conditions for the increase and spread of good understanding regarding environmental sustainability and its regulatory framework. However, most work is ahead and uphill, considering globalization, the thousands of pollutants spread worldwide, and the unregulated procedures for the final disposal.

3.3 ORGANISMS' COMMUNITIES AND NANOTECHNOLOGIES WORKING TOGETHER TO DISSIPATE POLLUTANTS

Increasing concerns worldwide have arisen due to the constant and continuous soil and water pollution by organic and inorganic pollutants. Anthropogenic activities such as industrial production, population growth, and food demand are essential issues because they belong to the principal sources of contamination in the environment.

Nowadays, to solve this serious problem, tools such as nanotechnology allow for exploring strategies to treat soils and water using sustainable methods more precisely. Nanomaterials' particular characteristics, such as surface area, chemical reactivity, lower cost and energy, and mechanical properties, make them suitable candidates for different applications, including environmental remediation. However, conventional routes for nanomaterials synthesis often require contaminating chemical substances. Simply, nanomaterials have properties, or at specific concentrations, they are toxic to various organisms; therefore, their use creates secondary contamination. This issue has led researchers to economical methods using organisms such as plants, bacteria, fungi, yeasts, and others to synthesize nanomaterials and later use them in environmental remediation. Microorganisms like bacteria, fungi, and actinomycetes can secrete chemical compounds that reduce or oxidize metal ions into magnetic nanoparticles (NPs) or zero-valent elements [20].

These nanomaterials manufactured by living organisms can absorb pollutants or catalyze degradation to convert the pollutants into a non-toxic form. However, according to Gautam et al. [20], overcoming the deficiencies presented by these nanomaterials related to stability, size control, and possible aggregation is necessary, so this field of study has become a challenge. Additionally, it is crucial to study the toxicity of nanomaterials to evaluate and ensure their real applications. Primarily because of the popularity that cell extracts of plants, bacteria, fungi, algae, and other organisms have achieved due to their low cost for manufacturing NPs and nanomaterials (Figure 3.1).

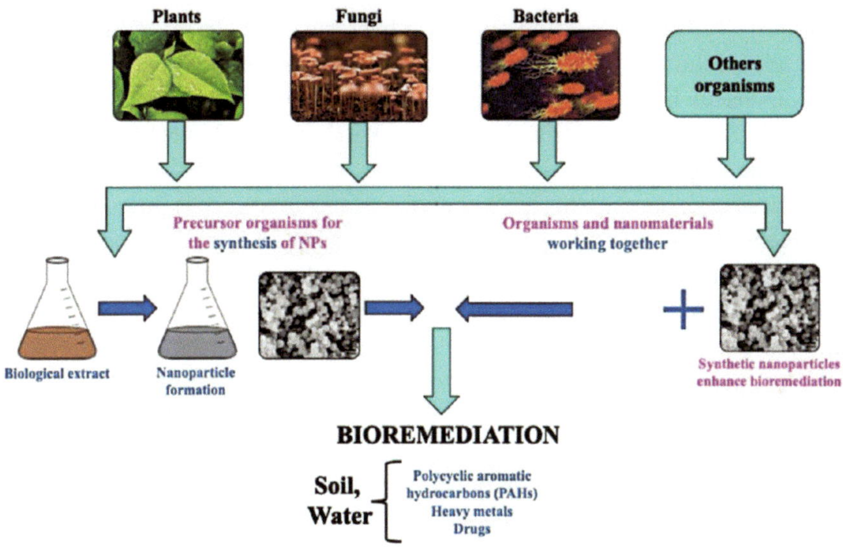

FIGURE 3.1 Schematic representation of organisms and NPs working together for environmental remediation. First, the use of NPs synthesized from a biological extract. On the other hand, the bioremediation of an organism is enhanced by using synthetic NPs. Both of them dissipate pollutants from contaminated sites.

3.3.1 DEGRADATION OF CONTAMINANTS BY BACTERIA

Due to the concern about the different contaminated sites mentioned earlier, several physicochemical remediation strategies have been proposed, developed, and tested; however, all techniques are expensive and not applicable in large, contaminated sites.

Nanotechnology has been of great interest in the last two decades due to its impact on catalysis, water treatment, and soil remediation [21]. The synthesis of NPs by chemical methods for application in contaminated sites has been increasing, but its process is complex and requires expensive equipment. Therefore, scientists and researchers have committed to developing efficient, ecological, less expensive, and sustainable methods for the synthesis of NPs.

Recently, bioremediation technology that uses nanomaterials as an alternative method for environmental remediation has emerged to address global concerns about contamination. Bacteria are the most widely used organisms for the synthesis of NPs due to their production capacity [22]. Scientific research has been reported on the green synthesis of various NPs such as Ag, Au, Cu, Zn, Ti, Pd, Mg, Pt, Fe_3O_4, etc. [23]. Additionally, to synthesis, NPs can work together with bacteria and help microbial activities degrade pollutants. On the other hand, bacteria can play a role as model organisms to find possible harmful effects on ecosystems [24]. Table 3.1 summarizes some bacteria that synthesize NPs that have been applied to remediate contaminated sites.

These microorganisms have shown extraordinary capabilities and performance in removing different pollutants, such as heavy metals, hydrocarbons, and emerging organic compounds, such as pharmaceuticals. The capacity of a microbial population to degrade contaminants in different environmental matrices (soil, air, and water) depends on several factors, such as pH, temperature, water content, type and concentration of pollutants, and diversity and richness of microbial communities [25].

Similarly, bacterial metabolic versatility has been exploited to be employed in concomitance with compatible technologies for remediating polluted sites.

This section will analyze some results related to the bacterial metabolic machinery, focusing on heavy metal transformation and degradation of polycyclic aromatic hydrocarbons (PAHs) and emerging pollutants, i.e., pharmaceutical compounds.

3.3.1.1 BACTERIAL BIOTRANSFORMATION OF HEAVY METALS

It has been proved that bacteria can carry out chemical transformations of heavy metals. Among the reported processes are oxidation, reduction, methylation, and demethylation [26]; some of the byproducts produced after the transformation of heavy metals are produced as a stress response of bacteria; these byproducts can include metallic NPs [27].

Some criteria to identify microorganisms for biosynthesizing NPs are reported by Pat-Espadas & Cervantes [28]. They published a methodology for considering the main elements for the synthesis of NPs by bacteria. The first step is to select the microorganism of interest; later, the target metal element is selected. it is necessary to use a precursor for metal and the concentration. Finally, an incubation period is reached in which the metal, bacteria, and growth medium coexist. It is necessary to know essential physicochemical characteristics for synthesis: pH, medium concentration, and temperature, among others.

Bacteria have been used in the bioremediation of metals due to their capability to survive high concentrations and their metabolic strategies to tolerate the accumulation of metal ions. Also, many toxic substances have been reported; arsenic is an element with great potential to jeopardize humanity through groundwater poisoning. Additionally, chronic diseases caused by As, such as lung and skin cancer, are of particular concern. To date, different conventional methods have been developed to remove As(V), such as adsorption, precipitation, coagulation, and others. With the scope of nanotechnology, magnetic iron oxide NPs (FeNPs) have been tested for the adsorption of As—as a pollutant due to its high affinity for this element. The process is also efficient and sustainable.

The data described led Kim et al. [29] to develop a fast and efficient removal method of As(V) by adsorption using the *Deinococcus radiodurans* strain. The FeNPs were synthesized by the *D. radiodurans* strain and immobilized in the cell to apply it to eliminate As(V). This method presented advantages over conventional physicochemical methods since it sought to avoid secondary contamination resulting from immobilization of the strain with NPs using magnets to recover Fe.

Finally, it has been reported that carbon nanotubes (CNTs) can enter cells through damage to the cell membrane, which subsequently facilitates the entry of heavy metals, induces toxicity, and can harm microorganisms that play an essential role in bioremediation [30, 31].

3.1.1.2 BACTERIAL BIODEGRADATION OF PHARMACEUTICAL COMPOUNDS

Debnath et al. [32] performed an experiment where *Pseudomonas aeruginosa* bacterium was used to synthesize zirconium oxide NPs through a green technology for the bioremediation of water contaminated with tetracycline. Other emerging pollutants, i.e., pharmaceuticals and personal care products, have been treated in recent decades due to many applications and discharges in aquatic bodies [33]. Tetracycline is frequently used as an antibiotic for diseases such as pests, cholera, acne, respiratory infections, etc., which makes it a potential contaminant. According to Chen et al. [34], the products used to degrade this drug cause harmful impacts on human health, and in recent years, new techniques have been used to eliminate it from wastewater; adsorption is the most effective. This is why different types of NPs have been used to remove drugs in water in recent years.

The mechanism of the extracellular synthesis of metallic NPs using *Pseudomonas aeruginosa* is not completely understood. Still, it is considered that some reducing enzymes secreted by the microorganisms help in the bio-reduction of metallic ions to metallic NPs. In the case of the synthesis of zirconium oxide NPs using *P. aeruginosa*, the results of the equilibrium isotherm analysis and the Langmuir model allowed us to conclude that bacterial synthesis is an efficient alternative to conventional methods of this type of NPs and possibly of many others [32].

3.1.1.3 BACTERIAL BIODEGRADATION OF POLYCYCLIC AROMATIC HYDROCARBONS (PAHS)

On the other hand, there is research about improving bacterial strains by applying NPs to remove contaminants. El-Seshtawy et al. [51] used four bacteria strains: *Halomonas xianhensis, Halomonas zincidurans, Pseudomonas stutzeri*, and *Halomonas salifodinae* isolated from a soil polluted with PAHs.

PAHs have a harmful effect due to their hydrophobicity, allowing their persistence in the environment and toxicity to humans. These pollutants are adsorbed in the soil adhering to the organic fraction, which complicates their bioavailability for bacterial strains and restricts bioremediation [52, 53]; for this reason, the use of biosurfactants that improve the degradation of oil and other hydrocarbons has been implemented [54]. In their work, El-Sheshtawy

TABLE 3.1 List of Microorganisms Used to Synthesize NPs and Their Application in Environmental Remediation

Microorganism	Synthesized Nanoparticle	Application	References
Bacillus amyloliquefaciens	TiO_2	Photocatalytic degradation of reagent red 31.	[35]
Geobacter sulfurreducens	α-FeOOH $Fe_3O_4FeCO_3$	Cr(VI) remediation	[36]
Desulfovibrio desulfuricans	Pd^0	Treatment of industrial wastewater.	[37]
Rhodococcus erythropolis	Fe_3O_4	Dibenzothiophene degradation.	[38]
Deinococcus radiodurans	Fe	As(V) removal.	[29]
Pseudomonas aeruginosa	ZrO_2	Tetracycline removal.	[32]
Penicillium pimiteouiense	Iron oxide	Chromium removal.	[39]
Escherichia sp.	Cu NPs	Azo dye degradation.	[40]
Aspergillus flavus	Silver NPs	Malachite green dye nano-sorption.	[41]
Leuconostoc lactis	Silver NPs	Methyl orange and Congo red dye degradation.	[42]
Raoultella sp.	Tellurium NPs	Tellurite reduction	[43]
Escherichia sp.			
Bacillus subtilis	Co_3O_4 NPs	Transformation of lithium.	[44]
Pseudomonas aeruginosa JP-11	CdS NPs	Adsorption of Cd(II)	[45]
Gordonia amicalis HS-11	Au and Ag NPs	Catalytic degradation of hexadecane.	[46]
Desulfovibrio vulgaris	Pt and Pd NPs	Removal of 17β-estradiol, sulfamethoxazole, and ciprofloxacin.	[47]
Pseudomonas aeruginosa	Zirconia NPs	Chemisorption of tetracycline in liquid media.	[32]
Pseudomonas putida	Mn oxide NPs	Oxidation of estrone, 17-α ethinylestradiol, diclofenac, and ibuprofen.	[48]
Aspergillus niger	ZnS nanohybrids	Removal of BTX hydrocarbons.	[49]
Deinococcus radiodurans	Au NPs	Bioremediation of radioactive iodine.	[50]

et al. [51] used the four aforementioned bacterial strains, biosurfactant, and NiO NPs for PAHs' bioremediation. After seven days, they found a removal efficiency of 90% compared to 80% without using NPs. In another similar study, *Bacillus cereus*, *Acidovorax wohlfahrtii*, and *Bacillus thuringiensis* were isolated from soil contaminated with PAHs. With hematite NPs, a better percentage of pyrene removal from the soil 96%, 87%, and 82% was obtained for each corresponding strain, compared to 88%, 59%, and 37% without using NPs [55].

In another investigation, Mu et al. [56] propose a strategy to eliminate oil contamination using autochthonous bacterial communities improved with graphene oxide quantum dots (GOQDs) that improve the degradation of PAHs using *Bacillus cereus*, which has been reported to metabolize xenobiotic aromatic compounds.

Regarding the biological mechanisms involved in the bacterial response to NPs, Tang et al. [57] hypothesized that a periphytic film exposed to cerium oxide NPs would maintain the pollutant removal performance due to the self-regulation of its population and diversity. However, there is an irregular distribution of CeO_2 NPs in the peripheral biofilm and subsequent microbial community changes after the results. Peripheral biofilm could maintain its total COD, phosphorus, and nitrogen removal performance by adapting the microbial community composition, although exposure to Ce NPs negatively affected sensitive bacterial growth. This study was a complete process examining how CeO_2 NPs affect peripheral biofilm and provides essential information to understand better the role NPs can play with bacterial systems.

Due to nanotechnology's wide use and applications, it is common to ignore all the effects and impacts that nanomaterials can present, so more studies are needed to provide a broader knowledge about the use of these materials. Some studies report antagonistic effects by NPs directly related to microorganisms that degrade a toxic compound, and as a result, the removal efficiency is lower. More studies are required to increase the understanding of the effects of NMs on the biotic and abiotic elements of the environment.

3.3.2 DEGRADATION OF CONTAMINANTS BY FUNGI

Previously, the different possible sources of intentional or accidental release of pollutants into the environment were described. According to Harms et al. [58], the biological degradation or detoxification of a chemical pollutant requires its exposure to microorganisms to be catalyzed. Still, sometimes, these chemicals precipitate and are no longer bioavailable. Thus, it is vital to acquire the appropriate machinery for the remediation of different sites. Fungi, alone or collaborating with other organisms, could be essential components to design methods to remediate polluted environmental matrices.

Fungi van exists as parasites, saprobes, or mutualists, for instance, in plants or algae. This kingdom includes species that can live, such as molds, lichens, and yeast, with diverse life cycles. It has been described that 1,00,000 species of the 1.5 million are estimated. These organisms have been described as eukaryotic heterotrophs that possess a tubular body called mycelium and reproduce by sporulation [59]. Besides, fungi can be considered macroorganisms packed into microscopic units; thus, they present a particular lifestyle that adapts to different environments. Fungi, by themselves, can transform a wide range of contaminants. Most pollutant degraders belong to the Ascomycota and Basidiomycota phylum, followed by the mucoromycota subphyla [58].

Today, with advances and approaches in nanotechnology, the complexity of the degradation of contaminants has been reduced, and it is also possible to implement environmentally sustainable techniques.

Through the years, fungi have increased due to their cost and the ability to synthesize NPs with high control of size and stability. Some authors have called "Myconanotechnology," a new field of synthesis by fungi. It was mentioned that fungi are diverse and have excellent tolerance to accumulating metals, just like bacteria. They can create mechanisms to biotransform metals into mycogenic NPs [60–62].

Otherwise, according to Yadav et al. [63], some fungi species are known to secrete polypeptides, proteins, and other types of metabolites capable of synthesizing, stabilizing, and even immobilizing metallic NPs. Ji et al. [64] used the *Pleurotus ostreatus* fungus following a bottom-up enzymatic synthesis strategy. Subsequently, they applied functionalized TiO_2 NP to perform the purification/immobilization of the crude enzyme extracts in a single step.

The argument for choosing these two compounds is that they are found in water bodies and have an ecotoxicological impact on aquatic organisms. They also have excellent resistance to biodegradation. It has been reported that *P. ostreatus* mainly secretes laccase; therefore, these authors used the crude extract of this enzyme to immobilize the TiO_2 NPs, which can be applied in water treatment.

P. ostreatus is suitable for degrading some aromatic compounds due to their ubiquity in diverse ecosystems and high production of extracellular enzymes, which perform low substrate specificity [65, 66]. Like TiO_2 NPs, iron oxide NPs are prevalent in the environment. They can function for environmental remediation, for which it is necessary to comprehend the interaction and effects between NPs, fungi, and pollutants of interest. Li & Zhang [67] proposed a mechanism by which NPs of γ-Fe_2O_3 facilitate the biodegradation of bisphenol A (BPA) using *P. ostreatus,* and this occurred due to oxidation by Fe called via Fenton.

Other fungi are capable of synthesizing different types of NPs themselves. It has been reported that *Fusarium oxysporum* can carry out an extracellular synthesis of Ag NPs used in the treatment of effluents from textile plants. Also, *Itajahia sp.* has been used to synthesize iron NPs used as stabilizers to remove lindane from water [68, 69]. In this field of "Myconanotechnology," gold NPs have been the most studied for their ease in removing aromatic contaminants. Two research groups used *Aspergillus* sp. to develop biocompatible Au NPs that could be used to remove a wide range of aromatic contaminants from water, mainly 4-nitrophenol [70, 71].

On the other side, the research group of Bhargava et al. [72] carried out a study using *Cladosporium oxysporum* because of its potential to accomplish the extracellular synthesis of Au NPs high catalytic efficiency towards the reduction of rhodamine B dye mediated by $NaBH_4$.

Finally, Gupta et al. [24] highlighted the importance and interest in investigating the toxicity of NPs synthesized from fungi. Today, there is a wide range of studies only on NPs synthesized by chemical or physical methods. This group of researchers carried out the mycosynthesis of Ag NPs using *Fusarium culmorum*, *Fusarium moniliforme*, *Fusarium tricinctum,* and *Phoma glomerata*, which are resistant to different drugs. They could evaluate the activity of the synthesized NPs against *Pseudomonas putida* KT2440, resulting in the death of the bacteria at a concentration of 0.4–0.8 $\mu g\ mL^{-1}$.

Added to this, the importance of future research is highlighted to be able to make scientific contributions that can determine and evaluate the different effects that NPs synthesized by fungi present, and their impact is not counterproductive; in this case, it does not generate damage to microorganisms that contribute to the degradation of pollutants. However, it is good to remember and consider that the biosynthesis of NPs using fungi shows welfare and is preferred because the culture is cost-effective and requires simple growth conditions. Furthermore, it collaborates with sustainable environmental remediation.

3.3.3 DEGRADATION OF CONTAMINANTS BY OTHER ORGANISMS

To offer new technologies for the removal of contaminants in the environment and that these are of less impact, the combined use of nanotechnology and other conventional techniques of contaminant elimination, such as phytoremediation, has been proposed, giving rise to phytonanoremediation, which is proposed as a green technology, in which plants are used to eliminate contaminants or to reduce their toxicity, taking advantage of the intrinsic characteristics of nanomaterials (NMs), such as their large surface area and high energy, so they can absorb or catalyze reactions quickly, helping in the joint action with plants, in the elimination, degradation or reduction of organic and inorganic pollutants released into the environment [73].

The need for the elimination, reduction, or degradation of these pollutants is because they are a threat to the environment and all its components as they tend to be persistent, accumulate, and then enter the food chain, affecting food safety [74]. Therefore, a way has been sought to use plants with potential for phytonanoremediation, such as energy crops, which are used for energy production without interfering with or affecting food security, making it a sustainable alternative in phytonanoremediation processes [75].

However, even though energy crops are a promising option for the recovery of contaminated sites, they are little used due to the low capacity of accumulation or degradation of pollutants that they possess. Table 3.2 shows some recent examples where NMs-assisted phytoremediation was used. It is evident that the common use of plants and NMs is a promising strategy since they can increase the rates of accumulation and translocation of contaminants (organic and inorganic) in plant tissues and decrease toxicity, making the plants used in phytoremediation increase their antioxidant activity and thus promote the growth and development of these plants, making the remediation of contaminated sites more useful.

3.4 ORGANISM-ENVIRONMENT INTERACTIONS ACROSS POLLUTED ECOSYSTEMS

The damage to the environment, such as pollution, has been on the rise mainly because of the increase in the human population and excessive industrial development. This has caused pollution to become one of the most far-reaching health risks, affecting not only humanity but all biotic and abiotic components of ecosystems, causing, to a large extent, a destabilization of the ecological balance.

TABLE 3.2 Recent Examples of Phytonanoremediation for the Removal of Contaminants

Plant Species	NM Employed	Contaminant	Result	References
Lolium perene L.	nZVI	Pb	It was observed that the presence of nZVI at concentrations of 100 to 500 mg kg^{-1} promoted the growth of the plant, as well as favoring the absorption of lead by the plants, obtaining accumulation capacities of up to 1175.4 µg of Pb, and being able to stabilize the residual Pb in the sediments.	[76]
Glycine max (L.) Merr.	TiO_2	^{133}Cs	The addition of TiO_2 NPs significantly improved the uptake and accumulation of ^{133}Cs in *Glycine max* (L.) Merr plants obtain 731.7 µg g^{-1} in the shoots and 597.8 µg g^{-1} in the roots.	[77]
Arabidopsis thaliana (L.) Heynh	N-doped carbon dots	Cd	Cd's bioaccumulation in the plants was improved by 58.3% concerning the control and increased the germination rate, more significant development of biomass and roots, and decreased Cd's toxicity towards the plants.	[78]
Isatis cappadocica Desv.	Salicylic acid NPs (SANPs)	As	As by plants, the absorption capacity increased by adding SANPs, reaching an accumulation of 705 and 1,188 mg kg^{-1} in roots and shoots, respectively; besides, the height of shoots, the length of roots, and the amount of biomass were favored.	[79]
Alpinia calcarata (Haw.) Roscoe, *Ocimum sanctum* L. y *Cymbopogon citratus* (DC.) Stapf	nZVI	Endosulfan	The joint action of nZVI NPs and plants of *A. calcarata* (Haw.) Roscoe, *C. citratus* L. and *O. sanctum* (DC.) Stapf caused the elimination of about 100%, 86.16%, and 76.28% of Endosulfan in the soil at 28 days of the experiment, finding, in all cases, accumulation of this contaminant in roots, shoots, and leaves.	[80]
Tradescantia spathacea Sw. y Alternanthera dentata (Moench) Stuchlík ex RE P.	nZVI	Pb y Cd	Effective remediation of heavy metals was detected in soils contaminated with Pb and Cd, with the joint action of nZVI and the plants used, in which an accumulation of 84.4% of Pb and 64.8% of Cd was detected in *T. spathacea* Sw., and an accumulation of 73.7% of Pb and 71.3% of Cd in *A. dentata* (Moench) Stuchlík, being the roots the part of the plants that accumulated more heavy metals.	[80]

TABLE 3.2 *(Continued)*

Plant Species	NM Employed	Contaminant	Result	References
Prosopis juliflora (Sw.) DC.	Fe_3O_4	Fluorine	The addition of up to 200 mg kg^{-1} of Fe_3O_4 NPs increased the rate of fluoride accumulation by 63.07% distributed in the stem and root tissues. Similarly, this concentration of NPs was favorable for increasing the study plant's root and shoot length.	[81]
Boehmeria nivea (L.) Gaudich	MWCNT	Cd	The addition of up to 500 mg kg^{-1} of MWCNT increased the rate of accumulation and translocation of Cd in *B. nivea* (L.) Gaudich plants grown in polluted river sediments reduced the toxic effects of this metal on the plant, which was reflected in increased growth, reduced oxidative stress, and increased antioxidant content.	[82]
Sorgum. Bicolor (L.) Moench	TiO_2	Sb	TiO_2 NPs at concentrations from 50 to 250 mg kg^{-1} increased the absorption and translocation of Sb mainly in the plant shoots, reaching an accumulation of 4034.3 μg per pot; besides, the total biomass production was favored.	[83]

Therefore, it can be said that the excessive release of pollutants due to human actions has caused animals, plants, microorganisms, human beings, and ecosystems, in general, to be exposed to these pollutants, but how is the behavior of these organisms when faced with a contaminated environment? According to Saaristo et al. [84], to know this behavior, it is necessary to research with a multidisciplinary approach to evaluate the impacts resulting from the interaction between an individual-environment or community-environment in the presence of pollutants and be carried out in the short and long term and identify if there are adaptation processes in response to the modification of their environment or if it is affected in such a way that populations tend to decrease.

In this sense, Li et al. [85] studied the interaction of soil microbial communities in the presence of heavy metals. Their results state that microorganisms can adapt and survive in environments contaminated with heavy metals in the long term. Besides, they observed interactions between microorganisms such as archaea and bacteria, where archaea were more adaptable since they showed a greater capacity to reduce metal ions by decreasing their toxicity, which favored bacteria and other microorganisms to resist and adapt to this environment. On the other hand, Kavitha et al. [86] evaluated the interaction and response of the earthworm *Lampito mauritii* and its intestinal microbiota to an environment contaminated with the monocrotophos pesticide. The results showed that exposure of these organisms to this contaminant led to a reduction in the proliferation of the intestinal microbiota of *L. mauritii*, in addition to causing changes in the intestine, which can lead to a decrease in earthworm populations and consequently a decrease in the activity related to the formation of humic substances that help to improve soil fertility and quality.

In contrast, Sievers et al. [87] claim that assessing environmental impacts using metrics related to changes in abundance or species richness at a contaminated site reveals little information about how species respond to pollution, so they suggest assessing the behavioral responses of individuals under the new habitat conditions, The study of animal behavior has made it possible to understand how the populations of some species diminish due to alterations in their behavior, which to a great extent affect their physical condition, diminishing their abilities to perceive signs of danger, diminishing their sensory acuity and reducing their defense mechanisms, which increases the risk of being preyed upon.

In this sense, Mitchkash et al. [88] evaluated the exposure of *Ambystoma maculatum* in an environment contaminated with carbaryl or atrazine, where they found that resistance was not affected by any of the contaminants. However, individuals exposed to carbaryl caused increased fatigue, resulting in these individuals being less able to escape attack by predators; likewise, Moore et al. [89] evaluated the behavior *of Lithobates sylvatica* tadpoles exposed to Roundup™, finding a decrease in the basal movement rate of the tadpoles and a decrease in sensory perception related to the anti-predator response, making them more susceptible to predation, so that pollution, in general, represents a significant threat to the conservation and prevalence of the species and their interactions with the environment.

3.5 BIONANOMATERIALS FOR ENVIRONMENTAL DECONTAMINATION

The bionanomaterials are comparable in size to ENMs, ranging from 1 to 100 nm. Nonetheless, these biological materials can mainly be obtained from plants, fungi, bacteria, algae, and actinomycetes [90]. Remarkably, molecular compounds are composed of proteins, lipids, enzymes, viruses, cells, RNA, and DNA fractions [91].

For a little over 20 years, numerous documents have expressed interest in using biological materials for soil, water, and air remediation. The most widely mentioned and investigated bionanomaterials for decontamination of organic pollutants are dendrimers and CNTs; these include the multi-walled carbon nanotubes (MWCNTs), and single-walled carbon nanotubes (SWCNTs), CNTs combined with Ti, Fe, and Zn NPs, chitosan (CS), graphene oxide (GO), CS combined with GO, biopolymers (nanocellulose and chitin, principally), and metallic NPs and metal oxides synthesized from plants, organic residues, and organisms [92]. According to a search carried out in the "Web Science" database, many documents are related to bionanomaterials and environmental remediation. The OG, dendrimers, CNTs, chitosan, and biopolymers (nanocellulose) are the most reported environmental decontamination with 703, 532, 429, and 435 documents, respectively. However, the green synthesis of NMs has progressed in the last 20 years. According to the search, there are at least 4,408 documents that relate the words "green synthesis" with "nanomaterials" and "nanoparticles" to obtain new biological materials for the area of environmental remediation, medicine, and pharmaceutical, among others. The search was performed in the topic section between 2000 and 2020 until July 18, 2020.

Biopolymers are biocompatible and biodegradable materials that do not harm biological systems [90]. These biopolymers included cellulose nanofibrils (CNF) and cellulose nanocrystals (CNC). They have been used as matrices to uptake molecular guests, catalyst support, and filters in separation, among others [93]. In this sense, the CNC, CNF, and chitin nanocrystals (ChCN) have been shown to be effective in removing Ag^+ ions from contaminated water. Therefore, following the order of effectiveness, the CNC, ChCN, and CNF eliminated 64% (34.4 mg g^{-1}), 37% (20 mg g^{-1}), and 27% (15.5 mg g^{-1}) of Ag^+ ions, respectively [94]. On the other hand, Zhan et al. [95] found that graphene-chitosan sponges (GO-CSs) functionalized by di-n-octylamine (DNOA) showed up to 98% removal of radioactive waste present in water samples, $^{99}TcO_4^-$/perrhenate anions (ReO_4^-).

Although in this section, we focus on presenting bionanomaterials and their effects on environmental decontamination, different studies show that combining organic nanomaterials and inorganic NMs can help remove water contaminants. For instance, the GO/Fe_3O_4 nanocomposite showed better photocatalytic activity for removing methyl orange. Besides, they inhibited the growth of the bacteria *Staphylococcus aureus* to above 95% [96]. Recently, the porous CNF modified with aluminum hydroxides [CNF-Al $(OH)_3$] has been tested as fluoride absorbers in aqueous media [97]. The results demonstrated an absorption capacity of 20 to 35 mg g^{-1} of fluoride. Therefore, the authors suggest CNF-Al $(OH)_3$ as a promising fluoridated water absorber [97]. On the other hand, polymeric composites such as nanocellulose reinforced with inorganic NPs have also effectively purified contaminated water [98]. Inorganic NPs include metallic, metallic, nano-clay, and zeolite oxides, with low cost and reusable advantages [99]. For instance, Hosseini et al. [98] demonstrated that the CNF extracted from date palm tree waste (*Phoenix dactylifera* L.), the incorporation of graphene oxide (GO) as a natural compound, and Fe_3O_4 NPs (CNF/GO/F_3O_4), obtained adsorption rates of 126.58, 36.70, and 73.52 mg g^{-1} of Pb(II), Hg(II), and Cr(VI), respectively. Therefore, the authors suggested that bionanomaterial compounds could be considered efficient and reusable for four cycles in heavy metals' adsorption.

The green synthesis of NPs has increased in recent years [100]. To mention just a few studies, Majumdar et al. [23] synthesized iron oxide NPs (IONP) from banana peel extract. The results demonstrated a high adsorption capacity of As(V) from an aqueous medium by IONP. A similar study demonstrated that the synthesis of a green biochar/iron oxide compound using banana peel extract and $FeSO_4$ was able to absorb 862 mg g^{-1} of methylene blue (MB) from water samples. The study demonstrated that biochar/iron's electronic attraction is the cause of the adsorption of MB molecules with a capacity of up to 5 cycles of use [101]. Also, the green synthesis of Ag and Cu NPs from *Azadirachta indica* and *Coriandrum sativum* demonstrated a greater than 95% removal of naphthalene from contaminated wastewater [102].

Regarding the decontamination of the soil by bionanomaterials, the carboxylated graphene oxide-chitosan (GO-COOH/CS) spheres showed effectiveness for the immobilization of Cu^{2+} (78 mg g^{-1}) from the soil. In this same study, the authors showed that when wheat plants were used, the metal's bioaccumulation in the roots was reduced due to Cu fixation in the GO-COOH/CS spheres without causing toxic effects in wheat plants. Therefore, combining these bionanomaterials for copper decontamination of soils can be considered promising [103]. Also, the studies by Yi et al. [104] showed that the combination of chitosan and activated carbon (CS-CA) and *Brassica juncea* plants at a dose of 75 g of the mixture of bionanomaterials per pot, adsorption of cadmium (Cd) decreased by 22%, also reducing the bioavailability of the metal by 27% compared to the control.

In-situ remediation and laboratory experiments have shown the use of manufactured nanomaterials such as nanoscale zero-valent iron (nZVI), magnetite (Fe_3O_4), and maghemite (γ-Fe_2O_3) are successful in removing various pollutants from soil and groundwater [105]. However, studies show that the residues of these NPs can cause toxic effects on terrestrial and aquatic organisms and plants [106]. In this sense, iron nanomaterials' green synthesis for soil remediation can be an environmentally friendly technology. For instance, Shanker et al. [107] demonstrated the formation of NPs of iron hexacyanoferrates (FeHCF) from *Sapindus mukorossi* for the photocatalytic degradation of PAHs. The results demonstrated that in the presence of 50 mg L^{-1} of PAHs with a dose of catalyst 25 mg with solar irradiation at neutral pH, the anthracene, phenanthrene, chrysene, fluorene, and benzo(a) pyrene were transformed to non-toxic products above 80% in both water and soil. The authors suggest that FeHCF are promising and excellent adsorbents for organic compounds. Also, other research has shown efforts to prevent damage to the soil biota.

The synergy of materials is another nanotechnological strategy for removing heavy metals in soils. Mandal et al. [108] developed nano-zero-valent iron (nZVI) impregnated magnetic green tea biochar (nZVI @ GTBC). The results demonstrated that the combination of nZVI @ GTBC caused an efficiency of soil immobilization of lead (Pb) of 73.3% compared to pristine nZVI of 55.41% in 30 days. The pH factor plays an important role in immobilization. The best efficiency is achieved with a natural soil pH of 6.8, while with a pH increase of 8.0, the efficiency is reduced due to an increased surface charge and repulsion effect of the nZVI @ GTBC functional groups.

Particulate matter (PM), with a particle size of 2.5 μm (PM 2.5) associated with a mixture of harmful airborne pollutants, is the leading cause of mortality due to its diffusion in the bronchi and lungs. However, particles the size of 500 nm represent a feasible inhalation for humans. These particles are products of combustion, ash dust, industrial complexes, and automotive activity, among others [109, 110].

Regarding applying bionanomaterials for air remediation, our working group has discussed the importance of various NPs of plant and animal origin to decontaminate closed environments (hospitals and offices, among others) [111]. However, few studies have evaluated NPs in closed environments, much less in free areas. The challenge is enormous to find the effectiveness of bionanomaterials. In this sense, fibrous filters are being manufactured and evaluated for air decontamination until the date. For instance, Zhu et al. [112] proposed the manufacture of bio-based air filtration membranes. They evaluated the chitosan/polyvinyl alcohol (CS-PVA) filtration membrane with superhydrophobic silica NPs (SiO_2) and silver NPs ($AgNO_3$) introduced into the nanofibers (CS-PVA @ SiO_2-Ag NP_S). The compound nanofibers were chosen with 4% by weight SiO_2 and 0.3% by weight $AgNO_3$. According to the results, the CS-PVA @ SiO_2-Ag NPs nanofibers are 97% and 99% efficient for removing sodium chloride and aerosol particles (300–500 nm), respectively. Besides, composite nanofibers show antibacterial activity against *Escherichia coli* and *Bacillus subtilis.* Along these same lines, researchers have compared nanofiber filters with commercial filters. For example, Lv et al. [113] found that loading ZnO NP on nanofiber membranes based on polyvinyl alcohol (PVA) and konjac glucomannan (KGM) obtained an antibacterial efficiency with photocatalytic activity. These membranes are close to 100% particle filtration (particle size of 300 nm) compared to commercial filter buildings with non-biodegradable materials. In another study of green technology application, Liu et al. [114] demonstrated that soy protein isolate (SPI) that contains functional groups and bacterial cellulose (BC) are capable of removing up to almost 100% of $PM_{2.5}$ and PM_{10} particulate contaminants. Therefore, the results suggest that bionanomaterials of this type can be used in ecological air filters. For other examples of green nanotechnology applied to air remediation, see the information published by Garcia-Mayagoitia et al. [111].

Although green synthesis has proven to be a feasible solution to avoid environmental contamination, there are some drawbacks. For example, controlling size and shape is related to pH and reaction temperature [100]. On the other hand, many authors, including us, consider a high amount of biomass required to produce the necessary amount of bionanomaterials. This factor is considered one of the main drawbacks of remedying a high water and soil volume and air. Therefore, a joint scientific and technical effort is necessary to propose and evaluate the protocols and generate knowledge to advance in the solution of all the inconveniences related to the production of green NMs so that environmental and human damages are avoided in the future and applied on a larger scale.

3.5.1 NANOBIOHYBRIDS FOR RECOVERING POLLUTED SITES

Along with the industrial and economic growth, many new compounds have been released and represent a potential risk to the environment, affecting soil, water, and air. These pollutants can be effectively treated and removed using physical, chemical, or biological methods.

Combining the methods mentioned above has been successfully used to increase the removal effectiveness and enhance the recovery of the contaminated sites. In recent years, the incorporation of nanomaterials (NMs) into the remediation fields has taken relevance in the global scenario due to their remarkable physical and chemical characteristics, i.e., size, large surface area, optical, thermal, and magnetic properties [115], conferring them great applicability in different sectors such as medicine, electronics, construction, packaging, agriculture, and environment.

Nanomaterials have emerged as a versatile and practical option to remediate polluted soils with various pollutants, i.e., organic and inorganic compounds, emerging organic contaminants, and radionuclides materials [7].

These materials have been developed from single-particle synthesis to multi-component assemblies or hierarchical structures. Thus, a new category of nanomaterials emerges, the so-called nanohybrids (NHs), which combine two or more pre-synthesized nanomaterials to extract multifunctionality [116]. The most relevant characteristics of NHs are given due to the modulation of their properties during the synthesis [117].

According to Saleh et al. [118], NHs can be classified based on their parent materials, i.e., organic/inorganic or metallic/carbonaceous; therefore, these materials are grouped into four classes: carbon-carbon NHs (CCNH), carbon-metal NHs (CMNH), metal-metal NHs (MMNH) and organo-metal-carbon NHs (OMCNH).

Nanohybrid materials offer particular properties that can be used in a wide range of applications, for instance, biomedical, catalysis, construction industry, energy, electronics, and environmental remediation.

3.5.1.1 CARBON-CARBON NHS

Carbonaceous nanomaterials (CNs) have high electric and electronic applications due to their conductive properties. In the environmental field, the CNs present essential advantages such as chemical inertness, stability, and tunable electrical and structural properties, which have allowed their use in pollutant removal from water and soil via photocatalytic reactions [119].

Guo et al. [120] reported the first study where CNTs and graphene oxide (GO) hybrid hydrosol were used to remove Gd(III) in an aqueous model, a rare-earth that has increased its use due as a consequence of its emergent new applications. This element has also been associated with non-beneficial effects on human health when it enters water supplies and food chains via water, especially in areas where mining and metal refineries are present.

The study evaluated the absorption capacity of this carbonaceous NH compared to the GO only. The results showed an increment of the absorption (86.42%) of Gd(III) by the NH (534.76 mg L^{-1} compared to the GO. The remarkable synergistic adsorption improvement effect for Gd(III) was demonstrated due to the inhibited restacking of steric resistance of GO nanosheets by CNTs during the absorption process. The study revealed that Gd(III)'s adsorption on CNTs/GO hybrid hydrosols fit the pseudo-second order and Langmuir isotherm adsorption models. These NHs sealed in dialysis bags could have extensive and promissory results for treatments of other heavy metal ions-containing wastewater and no second pollution.

3.5.1.2 CARBON-METAL NHS

The most common combinations to manufacture nanohybrids are the blend of carbon nanomaterials (e.g., CNTs, graphene family NMs, carbon dots, and graphitic carbon nitride) and metal/metal oxide NPs (e.g., Ag, Au, CdS, Cu_2O, MoS_2, TiO_2, and ZnO). These nanomaterials exhibit interesting and promising properties to face problematic environmental pollution and energy [121].

CMNHs have demonstrated effective, cheap, and fast removal of inorganic/organic pollutants in water and microorganisms' inactivation. The involved mechanisms in degradation processes are adsorption and photocatalytic/redox degradation [122]. It has been reported generally >90% degradation [123]; among the evaluated contaminants are polychlorinated biphenyls (PCBs), emerging contaminants (EC) (pharmaceuticals, endocrine disrupting substances, and perfluorochemical compounds) [124], inorganic contaminants such as heavy metals (Cr, Hg, Pb, and As) [125] and radionuclides (U, Am, Eu) [126].

Tan et al. [127] observed ~75% degradation of the organic contaminate glyphosate after two h in the presence of hydrogen peroxide (H_2O_2); the study used CNTs in composite with Al to accelerate the electron transfer and accelerate the removal process; the experiment was carried out in aqueous solution. Later, Lin et al. [128] reported the synergistic action of CNTs, Ag_3PO_4, and polyaniline (PANI) to facilitate the rapid electron migration in an aqueous matrix to photodegrade ~ 100% of phenol and p-nitrophenol in 20 minutes.

Regarding EC's treatment, the removal of chlortetracycline, tetracycline, and oxytetracycline was tested by Liu et al. [129]. The authors used CNTs decorated uniformly with NPs of Fe_3O_4. The combination of dielectric and magnetic losses of the material contributed to its stronger microwave absorption and the ability to produce hot spots, which favored the oxidation of these antibiotics via conjugated action of hydroxy (•OH) with superoxide radicals (•O_2–), this author evaluated the removal efficiency of oxytetracycline as well in an aqueous system using CNTs–Fe_3O_4–ZnO nanocomposites, observing that ~98.6% of the antibiotic was oxidized at an initial concentration of 100 mg/L by two-step process (adsorption and oxidation degradation). The authors attribute the interactions π-π and hydrogen bond interaction for the adsorptive removal of oxytetracycline and the Fenton reaction as responsible for the oxidation process.

An exciting study reported using magnetic Fe/Zn layered double oxide decorated CNTs (M-Fe/Zn-LDO@CNTs) composites to eliminate U(VI) under different experimental conditions such as temperature, pH, ionic strength, time of contact, and coexisting ions [130]. The results proved that the U(VI) adsorption performance of M-Fe/Zn-LDO@CNTs was significantly higher than bare CNTs. The process's thermodynamics reveals that U(VI) adsorption on the nanocomposite was endothermic and spontaneous.

This study concluded that M-Fe/Zn-LDO@CNTs could remove ~95.9% of 241 Am(III) from solution at pH=8.0, which proved that the nanocomposite has demonstrated to be an excellent material for the efficient removal of long-lived lanthanides and actinides from aqueous environmental solutions in radionuclides' pollution remediation.

Finally, as was mentioned before, the presence of microorganisms (MOs) in the water represents a public health concern due to their impact on human wellness; for this reason, some experiments have been carried out to eliminate the MOs; Engel et al. [131] has proposed a new carbon-based nanomaterial formed of SWCNTs and iron oxides to inactivate microorganisms in water. This nanocomposite exhibited high antimicrobial activity against *Escherichia coli*, and it was also possible to reuse the nanomaterial by washing it with calcium chloride and distilled water. Either reactive oxygen species (ROS) [132] or direct physical puncture might be the antimicrobial mechanism.

3.5.1.3 METAL-METAL NHS

These nanomaterials result from metal and metal oxides conjugated, forming multi-metallic ensembles. According to their functionalities, these metals are grouped as follows: plasmonic (Au, Ag, Pt) [133], quantum dots (CdSe, CdTe, PbS) [134], semiconducting oxides (TiO_2) [135] and, magnetic (Fe_3O_4, Fe_2O_3) [136].

Their unique properties, such as magnetic, electrical, mechanical, catalytic, and sensing ability, have provided them with many applications in diverse fields, such as chemical catalysis, optoelectronics, solar cells, and chemical sensing, biomedical imaging, and environmental pollution monitoring and mitigation.

Regarding decontamination processes, a seminal work by Wang et al. [137] reported the high photocatalytic activity of a porous titania network containing gold NPs. The UV/vis reflection spectra of the Au/TiO_2 material showed strong absorbance near 580 nm, indicating successful uptake of Au species. The tests demonstrated that the photocatalytic activity of the nanocomposite depended on the Au particle size and gold quantity. This study concluded that the highest photocatalytic activity was observed for the Au/TiO_2 containing 2.0 wt.% gold measured as the photodecomposition of methylene blue.

The photodegradation of organic pollutants using MO NHs is described as follows. When incident light is radiated with photon energy higher than the bandgap of Mos, electrons in the valence band (VB) are injected into the conduction band (CB), leaving the same number of holes in the VB. The electrons captured by O_2 dissolved in solution form $\bullet O_2$—superoxide radicals, subsequently transforming into •OH radicals. The photo-generated holes react with H_2O adsorbed on the surface of MOs to produce •OH. The $\bullet O_2$—and •OH are highly reactive and can completely decompose most organic waste materials into less toxic inorganic small molecules [138].

Several authors have evaluated TiO_2/ZnO nanocomposites' removal efficiency to remove contaminants by testing their photocatalytic activity. TiO_2/ZnO NHs prepared by Xiao et al. [139] showed enhanced photocatalytic activity due to the interfacial integration of TiO_2 nanotubes and ZnO nanorods towards the organic pollutants under UV radiation. In another study where TiO_2/ZnO was evaluated, Araújo et al. [140] stated that 90% of the rhodamine B (RhoB) dye was degraded after 1 hour and 10 minutes.

Yang et al. [141] reported the enhanced photocatalytic activity of Zn-TiO_2 composites assisted with ozonation to degrade organic pollutants. The evaluation was conducted in aqueous solutions using salicylic acid, methyl orange, and phenol as organic compounds. A synergistic effect of ozonation and photocatalytic activity of the metal-metal NH composite was observed, providing some guidance in developing more efficient methods to decontaminate water.

Zhang et al. [142] reported a novel Ti-Mn binary oxide sorbent efficiently synthesized that showed high efficiency in oxidizing As(III) to As(V) and removing the oxidative formed As(V) from water. It has been suggested that the As(III) uptake can be achieved primarily by combining oxidation and sorption processes. This study is relevant as it provides evidence for using this bi-functional adsorbent in drinking water treatment and removal of pollutants from the environment.

3.5.1.4 ORGANIC MOLECULE COATED NHS

Due to the organic molecules' presence, the coated NHs have been widely used for environmental purposes since it is assumed that they do not require a systematic and independent environmental assessment for accurate risk evaluation. These nanomaterials have been functionalized with organic polymers such as polyethylene glycol (PEG) [143] and poly(vinyl pyrrolidone) (PVP) [144] to facilitate their solubility in biological systems.

For instance, Akin et al. [145] synthesized magnetic Fe_3O_4-chitosan nanocomposites (m-Fe_3O_4-CNs) to evaluate their dye removal ability. The bromothymol blue (BB) was tested as an organic compound in an aqueous solution. Adsorption of BB on m-Fe_3O_4-CNs was studied in a batch reactor under various experimental conditions such as adsorbent dose, pH value, contact time, initial concentration of BB, and temperature. The results showed that the nanocomposite's adsorption capacity increased with BB concentration, adsorbent dosage, and temperature reduction. It was also shown that the adsorption process is spontaneous and exothermic. This study opened a broad spectrum of applications for treating water polluted with organic compounds.

The use of organic molecule coated NHs (OMCNHs) to eliminate EOC was evaluated by Liu et al. [146] when determining the removal of tetracycline (TC) from aqueous solution by Fe_3O_4 incorporated polyacrylonitrile (PAN) nanofiber mat (Fe-NFM). The author conducted a series of adsorption experiments to evaluate the efficiency of TC removal by FE-NFM. In this experiment, the highest adsorption capacity was observed at an initial solution pH of 4, while relatively high adsorption performance was obtained from an initial pH of 4 to 10. It was confirmed that the adsorption resulted from both electrostatic interaction and complexation between TC and Fe-NFM. This new nanomaterial shows considerable potential in drinking and wastewater treatment for micro-pollutant removal.

Hammouda et al. [147] published a novel magnetic heterogeneous catalyst by incorporation of iron (II) and magnetic functionalized NPs Fe_3O_4 in alginate beds to oxidize the three methyl-indole (3-MI) via Fenton reaction. The authors found that a critical aspect of being chosen is related to the Fenton catalysis to avoid excessive iron release. Under controlled laboratory conditions, the magnetic catalyst exhibited good performance. According to the results, for a 3-MI concentration of 20 mg L^{-1}, it is recommended to work with a pH of around three and a hydrogen peroxide concentration of 9.8 mmol L^{-1}. Under these conditions, low iron leaching (0.7 mg L^{-1}) was observed, and complete removal of 3-MI and 80% of the initial total organic carbon (TOC) after 120 min of reaction. This new catalyst would be of potential application in the environmental remediation segment due to its simplicity, low cost, and good recoverability and stability.

- **Final Remark:** Besides the variety of synthesis methods for fabricating NHs, these nanomaterials show a wide range of applications, including those addressed to removing contaminants from the environment. Their versatility has been demonstrated, emphasizing the importance of specific size, shape, and surface charge to remove toxic compounds effectively. Apart from the organic molecules coated NHs, most of the NHs evaluated have not been proved safe under natural environmental conditions, representing the most crucial issue to elucidate before further large-scale applications.

Shortly, the next generations of NHs might present more complicated hierarchical structures from more than two conjugated nanomaterials and diverse chemical origins; thus, a possibility of infinite combinations and functionalization would be accessible to be used in diverse environmental applications.

3.6 SMART MATERIALS AND NANOMACHINES TO DISSIPATE HAZARDOUS CHEMICAL COMPOUNDS

Smart materials integrate multiple synergistic and advanced functions in a single material; their properties allow controlled molecular changes in response to external stimuli, allowing varied applications and better performance [148]. The engineering of surfaces is a crucial tool for designing and manufacturing intelligent materials that can adsorb and disintegrate environmental pollutants, providing them with hydrophobic, hydrophilic, photocatalytic properties, etc. It is highly useful for the design of micro/nanomachines. The latter refers to motors, robots, or swimmers with directed autonomous movement, which they achieve by translating free energy into their environment [149]. Catalytic reactions promoted by Pt, MnO_2, or Pd are generally used, which decompose hydrogen peroxide into oxygen bubbles to provide movement (Figure 3.2); recently, in the search for new methods for propulsion, the supply of external energy has been included using electrical and magnetic sources, as well as self-diffusion and immunoelectrophoresis mechanisms [150, 151]. For the most part, these micro/nano-machines' design corresponds to tubular and spherical geometric shapes developed in coating nanotechnology [152]. In the particular case of spherical structures, these are covered with catalyst metal that promotes autophoresis, contrary to tubular ones, whose internal surface houses chemical reactions to supply energy for movement. The external wall is modified for specific chemical functions, such as disintegrating pollutants (Figure 3.2) [151]. These micro/nanomachines can improve the reaction's yields and reduce treatment times, overcoming the reactions of limited diffusion and improving the interactions between its active surface and the target contaminants [150].

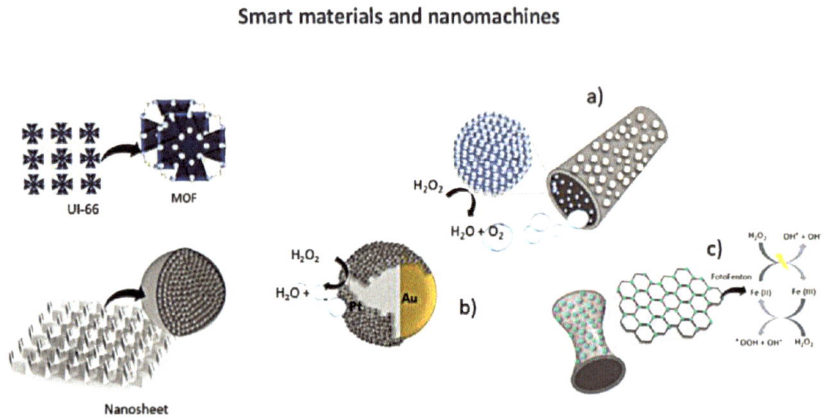

FIGURE 3.2 Primary structures of smart nanomaterials and nanomachines: (a) Spherical structure nanomachine; (b) tubular micromotor; and (c) Fenton's reaction micromotor.

Source: adapted by authors own work from (no copyright issue)https://www.sciencedirect.com/science/article/abs/pii/B9780323994460000088

Oil spills due to various accidents worldwide and the discharge of wastewater with high oil content are serious problems that increase the pollutants load in water bodies. In this type of contamination, water, and oil can be in different interfacial conditions that allow the sustainable application of porous super-wetting materials to extract the contaminant [153] effectively. Wettability is a surface property that is governed by both the chemical composition and morphology of the material to filter or adsorb oils from mixtures or emulsions; these materials can exhibit superhydrophobic and superoleophilic wettability according to their surface energy, which must be in a range between the value of water and the oil (72.8 – <30 mN m^{-1}), in turn, the construction of the rough structure responds to a hierarchical order from micro to nano [154]. However, the challenge centers on the design of bi-directional separation smart films. Cheng et al. [153] designed a copper mesh film whose pore size was adjusted with $Cu(OH)_2$ nanorods with diameters of 100 to 250 nm, created on the mesh surface. The water-on-film test reported a water contact angle (WCA) of 154°, indicating little superhydrophobic adhesion of the film.

In contrast, in the test with oil (1,2-dichloroethane), it spread quickly on the film at a contact angle of 0°, highlighting its oleophilic property. The O_2 plasma treatment readjusted the properties, endowing the film with superhydrophilic/superoleophobic properties; furthermore, the film's restoration was feasible after the heat treatment at 120°C, returning its initial state of humidification, allowing separation processes—bidirectional and good film recyclability [153].

Regarding the application of hydrophobic properties, Guix et al. [155] designed a self-propelled Au/Ni/PEDOT/Pt superhydrophobic catalytic micro-submarine with a gold surface modified by alkaniols in self-assembled monolayers (SAM), functional for the adsorption of oils, managing to collect around 40 drops of oil in a period of 80 s. However, this machine lost speed as the load of collected oil increased, reporting a speed loss from 26 to 12 μm/s when increasing the number of drops from 7 to 30 and then to 11 μm/s for 43 drops [155]. Later, Mou et al. [156] developed $MnFe_2O_4$@OA NPs micromotors with a simple monolayer structure for direct use in the remediation of hydrocarbons in water, without any additional modification on the surface due to its hydrophobic nature endowed by the long chains of oleic acid (OA) [156]. Zhan et al. [95] recently synthesized nanocrystals of MOF UIO-66-F4 to later develop the hydrophobic hybrid material UIO-66-F4 @ rGO with a superhydrophobic/superoleophilic character. The hybrids UIO-66-F4 @ rGO/MS and UIO-66-F4@rGO/FP exhibited high contact angles with water: 169 ± 0.6° and 155 ± 1.2°, being UIO-66-F4@rGO/MS, the one with the highest absorption capacity (26 ~ 61 g/g) for various organic solvents/oils [95].

The palladium and iron bimetallic catalytic NPs (Pd/Fe) have been synthesized in the pores of PVDF membranes for microfiltration tasks in the remediation of groundwater contaminated with toxic chlorinated aliphatic compounds, such as chloroform (CF), carbon tetrachloride (CTC), trichloroethylene (TCE) and tetrachlorethylene (PCE). The results report 99% degradation of CTC and 76% dechlorination in a reaction time of 4 h, detecting dichloromethane as a by-product [157]. On the other hand, both doped and straightforward metal oxide NPs allow the generation of nanocomposite photocatalysts, which are highly useful for the degradation of a great variety of pollutants, as shown in Table 3.3. In particular, iron oxide NPs have high selectivity and specificity, even as a chemical sensor, hexagonal α-Fe_2O_3 NPs have reported high selectivity against 4-nitrophenol, with values of ~ 367.6 $\mu A\ (mol\ L^{-1})^{-1}\ cm^{-2}$ with a practical detection limit of ~ $1.56 \times 10^{-3}\ mol\ L^{-1}$ in a response time of approximately 10.0 s [158]. It should be noted that a simple method to establish control over the movement of tubular micro/nanomotors is to add a ferromagnetic layer. Villa et al. [159] combined the oxidative potential of metal oxides, Fe_2O_3 and MnO_3, in a FeSiMnOx tubular microjet configuration self-propulsion speed of 485 ± 32 $\mu m\ s^{-1}$. The γ-Fe_2O_3 NPs increase the nanomotors' movement speed and allow them to be magnetically guided towards areas of high contamination, facilitating in situ

TABLE 3.3 Smart Materials for Photocatalytic Remediation of Hazardous Chemical Compounds

Material	Synthesis	Pollutant	Conditions	Results	References
Ce/ZnO@Ce/ TiO_2–N/S/RGO	Core-shell nanostructures in nanosheets.	Phenol, bisphenol, and methylene blue (MB).	Treatment via a sol-gel route, whit UV-Vis 400 W, pH at 2.5.	Photocatalytic degradation of phenol 95.09%, bisphenol 74.99%, and MB 76.37% after 120 min.	[162]
$Bi_2O_2CO_3$/g-C_3N_4	Z-scheme structure	Sulfamethazine (SMT)	Visible light illumination	90.31% for degradation of SMT in 90 min.	[163]
TiO_2–GR, TiO_2–CNTs, and TiO_2–C_{60} nanocomposites	Synthesized of sol-gel, TiO_2-carbon: graphene (GR), nanotube (CNTs), and fullerene (C_{60}).	Benzyl alcohol	Photocatalytic selective oxidation whit irradiation to 300 W.	Maximum yield 50 to 60% with 0.1% GR, 0.5% CNTs and 1.0% C_{60}.	[164]
TiO_2–GR	Synthesized hydrothermal reaction.	Benzene gas-phase and liquid-phase.	Reactor surrounded by four 4 W.	The higher addition of graphene in TiO_2–GR leads to a decreased photocatalytic activity.	[165]
VO_2(B) NPs	Synthesized hydrothermal reaction.	Cr(VI)	Adsorption process, desired change in pH (2–12).	Maximum adsorption NPs capacity was found to be 85 mg g^{-1}.	[166]

processes and the recovery of the material with the contained contaminant [159]. However, due to their magnetic properties, they are susceptible to pH, compromising their removal efficiency. In the case of the CB-R250 dye, the positive surface charge of the Fe_3O_4 NPs, tested in the Fe_3O_4-Si-IIL hybrid configuration (19 ± 6 mV), changed to slightly negative values after adsorption of the dye under acidic conditions.

Similarly, phenol's presence produces negative charges under primary conditions [160]. Metalorganic structures combined with NPs have been used with a synergistic effect. Xu et al. [161] assembled nano-iron in the MOF+ Ui-66-COOH material, which is hydrolytically stable, increasing sorption and accelerating the elimination kinetics of the radioactive contaminant uranium hexavalent that MOF^+ commonly presents. The Fe_0@UiO-66-COOH material reported an elimination capacity of 504 mg U g^{-1} of material, with elimination kinetics of 10 min, considered fast at an acid pH [161].

Regarding the micro/nanomachines with photocatalytic action, Wang et al. [167] configured a nanomotor made of TiO_2 nanotubes decorated internally and externally with Pt NPs, obtaining superhydrophilic surfaces and self-propulsion without using surfactants. In its evaluation, its decontamination capacity was determined with rhodamine B (RB), which was dissipated entirely using 0.6 mL of 0.2% H_2O_2 in a 10 mL solution, 1×10^{-4} mol L^{-1} under lighting. UV (500 W) [167]. In this type of nanomachines, the movement is induced by UV illumination, achieving efficiencies in both peroxide and water. Even though its design has allowed the elimination of microplastics, the synthesis of micromotors with photocatalytic particles of TiO_2 and endowed with magnetic properties with nickel, coated with Au to reduce its environmental toxicity (Au @ Ni @ TiO) was reported with 67% efficiency for the removal of microplastics in suspension [168]. On the other hand, artificial colorants and toxic metal cations, including Co, Zn, Pb, U, Sr, etc. They can be eliminated using iron oxides' properties in generating photo Fenton reactions (Table 3.4).

Concerning gel phase nanomaterials, these can put the solid phase in contact with a liquid in an environmental environment; that is, the nanofibers come into interaction with the gel pollutants, resulting in an absorption material with extraction selective due to its increased surface area and rapid internal diffusion kinetics [175]. The use of stimulus-sensitive nanogels in surface modifying systems is an innovative strategy for designing selective sensors by preparing composite materials from nanoparticle mixtures with inorganics and biopolymers [176]. Yan et al. [177] designed a highly selective detector for trace elements in wastewater based on commercial multi-pore

TABLE 3.4 Micro/Nanomachines for Photocatalytic Remediation of Hazardous Chemical Compounds

Material	Pollutant	Velocity	Efficiently	References
Microbots GOx tubular	Pb(III) ions	500 μms^{-1}	Cleaning water from 1,000 ppb down to below 50 ppb in 60 min.	[169]
Micromotor photocatalytic ZIF-8@ZnONPs on the surface and Fe_3O_4@AgNPs at the bottom.	Rhodamine B	~75 to ~1,063 μms^{-1} (5 to 30% H_2O_2)	With H_2O_2/UV removal of 98.6% within 60 min.	[170]
Micromotors photocatalytic CdS-polyaniline-Pt and ZnS-polyaniline-Pt.	Bisphenol A	2,500 μm s^{-1} (5% H_2O_2)	Degradation of ~80% and 100% after a 10 min treatment of 25 and 70 μg/L.	[171]
Micromotors photocatalytic ZnO/ZnO_2/Pt spherical Janus.	Picric acid and methyl orange.	350 μm s^{-1} (<5% H_2O_2)	Efficiency under visible light irradiation.	[172]
TiO_2@mSiO_2 bilayer tubular motor	Rodamina 6G (R6G)	970 ± 354 μm s^{-1}	Degradation of ~98% at R6G	[173]
Fe_3O_4-rGO/Pt microjets heterogeneous Fenton-like reaction.	Methylene blue (MB)	1,300 μm s^{-1}	Degradation of 62% at MB, which remains even after the fifth recycling.	[174]

membranes, on which smart nanogels with polydopamine adhesion were immobilized. In a detection process, these nanogels swell, changing the membrane flux so that the quantitative detection of trace elements' concentration is achieved by measuring the transmembrane flux. Ultrasensitive detection can be achieved in the case of Pb_2^+ at concentrations of 1–10 mol L^{-1} [177]. In a capsule membrane system, Pb_2^+ ions present in aqueous environments become encapsulated by forming 18-crown-6/Pb_2^+ complexes to be detected by nanogels, causing a phase transition, hydrophobic to hydrophiles [178]. Radwan et al. [179] obtained a nanometric carboxymethyl cellulose (CMC) gel functionalized with acrylamide and acrylic acid, whose adsorption potential was evaluated in the removal of Cd(III) and RR195 dye in an aqueous solution. The results reveal that 0.5 g L^{-1} of the functionalized CMC gel can almost eliminate Cd(II) at a concentration of 20 mg L^{-1} under acidic pH conditions in a reaction time equal to 5 min. While 2 g L^{-1} of the gel is able to remove up to 90% of the RR195 dye solution at a concentration of 10 mg L^{-1} in acidic pH as a result of the electronegative chemical functionalities of the functional groups [179].

In particular, micro/nanomachines show great potential for the remediation of contaminants in water bodies, such as removing oils the degradation of organic pollutants and heavy metals, which is an excellent advance in oxidative technologies. These machines' interaction and collection efficiency are closely related to intelligent materials that allow increasingly versatile and independent designs.

3.7 THE SUSTAINABLE PROSPECT OF BIONANOTECHNOLOGY

The development of genetic engineering technologies to improve the remediation of contaminated soils by combining plants and microorganisms has become the most popular research topic. Phyto-microbial remediation is based on the interaction between plants and microorganisms that break down organic pollutants in soils and sediments [180]. Considered to be a "green technology," biobased remediation technologies have the potential to remove various pollutants, such as heavy metals, organic pollutants, hydrocarbons, radioisotopes, and drugs from the soil while minimizing the generation of secondary waste [181]. Further, bioremediation is an operationally simple, feasible, and economically viable technology [182].

However, there are factors such as the physical and chemical properties of the soil that influence the joint plant-microorganism remediation system as well as the activity of the soil's microorganisms, microbial species, plant species, the growth period of the plant, and the plant's sensitivity to soil contaminants [183, 184]. Among the new technologies that obtain better results and overcome the limitations of bioremediation, there are techniques such as: (i) immobilization of beneficial microorganisms [185, 186], (ii) genetic modification of microorganisms [187], and (iii) genetic modification of plants for remediation, i.e., direct modification of a plant for use without applying a microbial inoculant [188].

The application of these biotechnological alternatives requires the development of synthetic biological models of bioremediation, for which it is essential to understand the existing metabolic pathways that include data on the structure, sequence, and function of proteins and the chemical structure and interaction of organic compounds; genomics; environmental microbiology; among others. These studies will promote an understanding of interactions and networks at the molecular, cellular, community, and ecosystem levels. Combining the results is possible through the various "omics" tools (i.e., genomics, transcriptomics, proteomics, and metabolomics); this approach provides crucial insights into the survival, metabolism, and interactions of microorganisms.

Although genetic engineering technologies are promising, there is limited literature on the efficiency of removal of organic pollutants in polluted soils through the combined remediation of genetically modified (GM) plants and genetically engineered microorganisms (GEMs) [189] due to the possible risks of genetic contamination of the native microbial consortium to be treated. Regulations for the use of genetically and metabolically modified bacteria are currently being generated to help assess the possible risks concerning environmental safety [190] by organizations such as the Organization for Economic Cooperation and Development (OECD) because of the competitiveness between native and genetically modified species can lead to a decrease in non-target microflora [191, 192].

Furthermore, the use of biosensors will allow the detection of specific signal molecules, which can provide information for monitoring pollutants and nutrients (nitrogen, phosphorus, oxygen) and provide information on the status of the plant [193, 194]. Compared to traditional methods, the higher specificity and biosensing systems are due to bioreceptors (natural elements) combined with suitable transducers that generate signals after interacting with the target molecule of interest [195]. Biosensor sensitivity and performance can be improved by incorporating nanoscale materials into the biosensor due to the surface area to volume ratio [196].

Advances in the synthesis of a wide range of NMs, such as NPs, nanotubes, nanocomposites, and nanoencapsulation, together with the many methodologies for their production [197], have the potential for the inclusion of NMs in environmental applications, such as environmentally friendly products [198], pollutant removal [199] and sensors [200].

Among the advantages of NMs are (i) their small dimensions, which allow the particles to penetrate the systems effectively, and (ii) a large surface area that could even be modified to increase reactivity or absorption capacity. Nanotechnology for water and soil restoration represents a sophisticated and innovative solution that guarantees the rapid and efficient removal of pollutants. NMs as adsorbent systems offer an extensive reactive surface area with a small mass and are inexpensive to manufacture. Furthermore, Au, Ag, and SiO_2 NPs and CNTs have been used to detect low levels of contaminants [201, 202]. Various modifications to the surface of NMs have also been reported to increase their sensitivity and specificity [203, 204]. NMs can degrade contaminants through processes such as photocatalysis, dechlorination, and catalytic reduction or as adsorbent matrices [7]. Their size, selectivity, and sensitivity are advantages offered by materials and biosensors based on nanotechnology over conventional techniques.

However, NMs cannot be quickly recovered once dispersed due to their size. Hence, assessing environmental and human risks linked to the use of NMs is still a matter of debate, which is why nano-remediation is still considered an emerging technology. Depending on the environment in which the presence of NMs is considered, the fate of NMs can change by interacting with the characteristics of the environment and their unique properties (size, shape, distribution, surface area, charge, catalytic activity, etc.) [205]. It has been reported that NMs can affect the biota present or be adsorbed physicochemically on the surface of soil particles, and their aggregation and colloidal stability will depend on the texture, pH, humic acids, and soil chemistry [206]. Furthermore, the toxicity of NMs is a critical concern for human health [207].

To implement nano-remediation, an exhaustive evaluation of its ecological safety must be carried out that addresses aspects of (i) prediction of the behavior of NMs in the media to be corrected while considering the nature of the contaminants and the characteristics of the contaminated area; and (ii) identification of the ecotoxicity of NMs in different species of the ecosystem [208]. Similarly, there should be the development of new NMs combined with other technologies, such as bioremediation and real-time monitoring and control techniques of the applied NMs [209].

3.8 CONCLUSION

Soils perform several crucial functions, making them environmentally, economically, and socially significant. This chapter provides insights into cutting-edge environmental technologies and procedures to dissipate pollutants from soil, air, and water. However, social concerns and ecological and economic drawbacks regarding these technologies should be acknowledged. Simultaneously, eco-friendly and cheaper solutions should be adequately addressed to shape sustainable development, considering the regulators' perspectives.

There is evidence of some drawbacks caused by several nanotechnologies during the remediation processes and the possibility of obtaining low-toxic or eco-friendly nanobioremediation technologies. Therefore, the concepts of sustainable or green remediation should be advocated in developing strategies, materials, and organisms for more effective remediation practices. It has to be stated that the use of new generation environmentally friendly nanobioremediation technologies should not result in the potential loss of diversity at a polluted site.

It is reasonable to expect reliable strategies for industrial mass production of these ENMs, nanobiotechnologies, and commercial applications regarding recalcitrant and ubiquitous dissipation pollutants through technologies without an ecological or technological drawback. This goal may face many difficulties or even be impossible to achieve, but in this pursuit process, we became more aware of the potential of these techniques. However, challenges and opportunities continue to evolve. This means that there are many opportunities to develop cutting-edge technologies for nanobioremediation through unprecedented technologies. The examples listed here represent some of the available options.

Therefore, given many pollutants and non-remediated sites, it is tempting to think that we are waiting for a new dimension in the field of remediation to be discovered. However, this will not be an easy task, and there are some challenges to overcome. In this sense, it is necessary to make more vigorous efforts must be made to develop new nanomaterials with the required biocompatibility, oxidation or degradation potential, and structural robustness. This will help clean up the contaminated sites without any additional disadvantages.

ACKNOWLEDGMENTS

This research was founded by 'Ciencia Básica SEP-CONACyT' project 287225, 'Fondo FONCYT-COECYT-Convocatoria 2021-C15, Bio-y nano-remediación de suelo y agua, contaminados con metales pesados, del municipio de San Juan de Sabinas, Coahuila de Zaragoza, México, (COAH-2021-C15-C095),' and the Sustainability of Natural Resources and Energy Programs from Cinvestav-Saltillo.

KEYWORDS

- **decontamination**
- **environmental and human health**
- **hazardous chemical compounds**
- **nanoscience**
- **phytonanoremediation**
- **remediation**
- **soil degradation**
- **soil microorganisms**
- **soil pollution**
- **sustainability**

REFERENCES

1. Neal, A. L., (2008). What can be inferred from bacterium–nanoparticle interactions about the potential consequences of environmental exposure to nanoparticles? *Ecotoxicology, 17*, 362–71.
2. Fernández–Luqueño, F., López–Valdez, F., Pérez–Morales, C., García–Mayagoitia, S., Sarabia–Castillo, C. R., & Pérez–Ríos, S. R., (2017). Enhancing decontamination of PAHs–polluted soils: Role of organic and mineral amendments. In: *Enhancing Cleanup of Environmental Pollutants* (pp. 339–368). doi.org/10.1007/978–3–319–55423–5_11.
3. Balseiro–Romero, M., Monterroso, C., & Casares, J. J., (2018). Environmental fate of petroleum hydrocarbons in soil: Review of multiphase transport, mass transfer, and natural attenuation processes. *Pedosphere, 28*, 833–847.
4. Sarabia–Castillo, C. R., Pérez–Moreno, A., Galindo–Ortiz, A., Fraga–Pecina, N., Pérez–Hernández, H., Medina–Pérez, G., & Fernández–Luqueño, F., (2020).

Phytonanotechnology and environmental remediation. In: *Phytonanotechnology* (pp. 159–185). doi.org/10.1016/B978–0–12–822348–2.00009–7.
5. Tilla, I., & Blumberga, D., (2018). Qualitative indicator analysis of a sustainable remediation. *Energy Proced.*, *147*, 588–93.
6. Naseri–Rad, M., Berndtsson, R., Persson, K. M., & Nakagawa, K., (2020). INSIDE: An efficient guide for sustainable remediation practice in addressing contaminated soil and groundwater. *Sci. Total Environ.*, *740*, 139879. doi.org/10.1016/j.scitotenv.2020.139879.
7. Vázquez–Núñez, E., Molina–Guerrero, C. E., Peña–Castro, J. M., Fernández–Luqueño, F., & De La Rosa–Álvarez, M., (2020). Use of nanotechnology for the bioremediation of contaminants: A review. *Processes*, *8*, 826.
8. Wei, H. G., Wang, H., Li, A., Cui, D. P., Zhao, Z. N., Chu, L. Q., Wei, X., et al., (2020)., Multifunctions of polymer nanocomposites: Environmental remediation, electromagnetic interference shielding, and sensing applications. *ChemoNanoMat, 6*, 174–184. doi.org/10.1002/cnma.201900588.
9. Biswas, B., Warr, L. N., Hilder, E. F., Goswami, N., Rahman, M. M., Churchman, J. G., Vasilev, K., et al., (2019). Biocompatible functionalization of nanoclays for improved environmental remediation. *Chem. Soc. Rev.*, *48*, 3740–3770. doi.org/10.1039/c8cs01019f.
10. Biswas, B., Labille, J., & Prelot, B., (2020). Clays and modified clays in remediating environmental pollutants. *Environ. Sci. Pollut. R.*, *27*, 38381–38383. doi.org/10.1007/s11356–020–09828–4.
11. Zhou, H., Tan, J., & Zhang, X., (2019). Nanoreactors for chemical synthesis and biomedical applications. *Chem–Asian J.*, *14*, 3240–3250. doi.org/10.1002/asia.201900967.
12. Martínez, H. J. J., Sánchez–Medina, P. S., & Díaz–Pichardo, R., (2020). Business–oriented environmental regulation: Measurement and implications for environmental policy and business strategy from a sustainable development perspective. *Bus. Strateg. Environ.*, *30*, 507–521. doi.org/10.1002/bse.2634.
13. Qiu, L., Hu, D., & Wang, Y., (2020). How do firms achieve sustainability through green innovation under external pressures of environmental regulation and market turbulence? *Bus. Strateg. Environ.*, *29*, 2695–2714. doi.org/10.1002/bse.2530.
14. Omojolaibi, J., & Nathaniel, S. P., (2020). Assessing the potency of environmental regulation in maintaining environmental sustainability in MENA countries: An advanced panel data estimation. *J. Publ. Aff.*, E2526. doi.org/10.1002/pa.2526.
15. Roos, C., Cilliers, D. P., Retief, F. P., Alberts, R. C., & Bond, A. J., (2020). Regulators' perceptions of environmental impact assessment (EIA) benefits in a sustainable development context. *Environ. Impact. Asses., 81,* 106360. doi.org/10.1016/j.eiar.2019.106360.
16. Mathis, K. (2017). Sustainable Development, Economic Growth and Environmental Regulation. In: Mathis, K., Huber, B. (eds). Environmental Law and Economics. *Economic Analysis of Law in European Legal Scholarship, vol. 4.* Springer, Cham. https://doi.org/10.1007/978-3-319-50932-7_1.
17. Hou, D., & O'Connor, D., (2020). Green and sustainable remediation: Past, present, and future developments. In: *Sustainable Remediation of Contaminated Soil and Groundwater*, *19–42.* doi.org/10.1016/b978–0–12–817982–6.00002–1.
18. USEPA, (2008). *Green Remediation: Incorporating Sustainable Environmental Practices into Remediation of Contaminated Sites.*

19. Nathanail, C. P., Bakker, L. M. M., Bardos, P., Furukawa, Y., Nardella, A., Smith, G., Smith, J. W. N., & Goetsche, G., (2017). Towards an international standard: The ISO/DIS 18504 standard on sustainable remediation. *Remediation J., 28*, 9–15. doi.org/10.1002/rem.21538.
20. Gautam, P. K., Singh, A., Misra, K., Sahoo, A. K., & Samanta, S. K., (2019). Synthesis and applications of biogenic nanomaterials in drinking and wastewater treatment. *J. Environ. Manage., 231*, 734–748. doi.org/10.1016/j.jenvman.2018.10.104.
21. Akbarzadeh, A., Zare, D., Farhangi, A., Mehrabi, M. R., Norouzian, D., Tangestaninejad, S., Majid, M., & Nasim, B., (2009). Synthesis and characterization of gold nanoparticles by tryptophane. *Am. J. Appl. Sci., 6*, 691. doi.org/10.3844/ajas.2009.691.695.
22. Singh, H., Du, J., Singh, P., & Yi, T. H., (2018). Extracellular synthesis of silver nanoparticles by *Pseudomonas* sp. THG–LS1.4 and their antimicrobial application. *J. Pharm. Anal., 8*, 258–264. doi.org/10.1016/j.jpha.2018.04.004.
23. Majumdar, M., Biswas, S. C., Choudhury, R., Upadhyay, P., Adhikary, A., Roy, D. N., & Misra, T. K., (2019). Synthesis of gold nanoparticles using citrus macroptera fruit extract: Anti–biofilm and anticancer activity. *ChemistrySelect, 4*, 5714–5723. doi.org/10.1002/slct.201804021.
24. Gupta, I. R., Anderson, A. J., & Rai, M., (2015). Toxicity of fungal–generated silver nanoparticles to soil–inhabiting *Pseudomonas putida* KT2440, a rhizospheric bacterium responsible for plant protection and bioremediation. *J. Hazard Mater., 286*, 48–54. doi.org/10.1016/j.jhazmat.2014.11.044.
25. Kebede, G., Tafese, T., Abda, E. M., Kamaraj, M., & Assefa, F., (2021). Factors influencing the bacterial bioremediation of hydrocarbon contaminants in the soil: Mechanisms and impacts. *J. Chem., 2021*.
26. Sun, W., Cheng, K., Sun, K. Y., & Ma, X., (2021). Microbially mediated remediation of contaminated sediments by heavy metals: A critical review. *Curr. Pollut. Rep., 7*, 201–212.
27. Kang, S., Herzberg, M., Rodrigues, D. F., & Elimelech, M., (2008). Antibacterial effects of carbon nanotubes: Size does matter! *Langmuir, 24*, 6409–6413.
28. Pat–Espadas, A. M., & Cervantes, F. J., (2018). Microbial recovery of metallic nanoparticles from industrial wastes and their environmental applications. *J. Chem. Technol. Biot., 93*, 3091–3112. doi.org/10.1002/jctb.5681.
29. Kim, H. K., Jeong, S. W., Yang, J. E., & Choi, Y. J., (2019). Highly efficient and stable removal of arsenic by live cell fabricated magnetic nanoparticles. *Int. J. Mol. Sci., 20*, 3566. doi.org/10.3390/ijms20143566.
30. Wang, F., Yao, J., Liu, H., Liu, R., Chen, H., Yi, Z., Yu, Q., & Xing, B., (2015). Cu and Cr enhanced the effect of various carbon nanotubes on microbial communities in an aquatic environment. *J. Hazard Mater., 292*, 137–145. doi.org/10.1016/j.jhazmat.2015.03.032.
31. Zhang, H., Wu, F., Chen, W., Zhang, X., Alvarez, P. J. J., Ortega–Calvo, J. J., Yang, Y., et al., (2018). Carbon nanomaterials differentially impact mineralization kinetics of phenanthrene and indigenous microbial communities in a natural soil. *NanoImpact, 11*, 146–155. doi.org/10.1016/j.impact.2018.08.001.
32. Debnath, B., Majumdar, M., Bhowmik, M., Bhowmik, K. L., Debnath, A., & Roy, D. N., (2020). The effective adsorption of tetracycline onto zirconia nanoparticles synthesized by novel microbial green technology. *J. Environ. Manage., 261*, 110235. doi.org/10.1016/j.jenvman.2020.110235.

33. Zhang, D., Gersberg, R. M., Ng, W. J., & Tan, S. K., (2014). Removal of pharmaceuticals and personal care products in aquatic plant–based systems: A review. *Environ. Pollut., 184*, 620–639. doi.org/10.1016/j.envpol.2013.09.009.
34. Chen, Y., Wang, F., Duan, L., Yang, H., & Gao, J., (2016). Tetracycline adsorption onto rice husk ash, an agricultural waste: Its kinetic and thermodynamic studies. *J. Mol. Liq., 222*, 487–494. doi.org/10.1016/j.molliq.2016.07.090.
35. Khan, R., & Fulekar, M. H., (2016). Biosynthesis of titanium dioxide nanoparticles using *Bacillus amyloliquefaciens* culture and enhancement of its photocatalytic activity for the degradation of a sulfonated textile dye reactive red 31. *J. Colloid Interf. Sci., 475,* 184–191. doi.org/10.1016/j.jcis.2016.05.001.
36. Byrne, J. M., Telling, N. D., Coker, V. S., Pattrick, R. A. D., Van Der, L. G., Arenholz, E., Tuna, F., & Lloyd, J. R., (2011). Control of nanoparticle size, reactivity, and magnetic properties during the bioproduction of magnetite by geobacter sulfurreducens. *Nanotechnology, 22*, 455709. doi.org/10.1088/0957–4484/22/45/455709.
37. Yong, P., Rowson, N. A., Farr, J. P. G., Harris, I. R., & Macaskie, L. E., (2002). Bioreduction and biocrystallization of palladium by *Desulfovibrio desulfuricans* NCIMB 8307. *Biotecnol. Bioeng., 80*, 369–379. doi.org/10.1002/bit.10369.
38. Ansari, F., Grigoriev, P., Libor, S., Tothill, I. E., & Ramsden, J. J., (2009). DBT degradation enhancement by decorating *Rhodococcus* erythropolis IGST8 with magnetic Fe_3O_4 nanoparticles. *Biotecnol. Bioeng., 102*, 1505–1512. doi.org/10.1002/bit.22161.
39. Mahanty, S., Bakshi, M., Ghosh, S., Gaine, T., Chatterjee, S., Bhattacharyya, S., Das, S., et al., (2019). Mycosynthesis of iron oxide nanoparticles using manglicolous fungi isolated from Indian sundarbans and its application for the treatment of chromium containing solution: Synthesis, adsorption isotherm, kinetics, and thermodynamics study. *Environ. Nanotechnol. Monit. Manag., 12*, 100276.
40. Noman, M., Shahid, M., Ahmed, T., Niazi, M. B. K., Hussain, S., Song, F., & Manzoor, I., (2020). Use of biogenic copper nanoparticles synthesized from a native *Escherichia* sp. as photocatalysts for azo dye degradation and treatment of textile effluents. *Environ. Pollut., 257*, 113514.
41. Pandian, A. M. K., Karthikeyan, C., & Rajasimman, M., (2016). Isotherm and kinetic studies on nano–sorption of malachite green onto *Aspergillus flavus* mediated synthesis of silver nano particles. *Environ. Nanotechnol. Monit. Manag., 6*, 139–151.
42. Saravanan, A., Kumar, P. S., Devi, G. K., & Arumugam, T., (2016). Synthesis and characterization of metallic nanoparticles impregnated onto activated carbon using leaf extract of *Mukia maderasapatna*: Evaluation of antimicrobial activities. *Microb. Pathogenesis, 97*, 198–203.
43. Nguyen, V., Choi, W., Ha, Y., Gu, Y., Lee, C., Park, J., Jang, G., et al., (2019). Microbial tellurite reduction and production of elemental tellurium nanoparticles by novel bacteria isolated from wastewater. *J. Ind. Eng. Chem., 78*, 246–256. doi.org/10.1016/j.jiec.2019.06.006.
44. Shim, H. W., Jin, Y. H., Seo, S. D., Lee, S. H., & Kim, D. W., (2011). Highly reversible lithium storage in *Bacillus subtilis*–directed porous Co_3O_4 nanostructures. *ACS Nano, 5*, 443–449. doi.org/10.1021/nn1021605.
45. Raj, R., Dalei, K., Chakraborty, J., & Das, S., (2016). Extracellular polymeric substances of a marine bacterium mediated synthesis of CdS nanoparticles for removal of cadmium from aqueous solution. *J. Colloid Interf. Sci., 462*, 166–175.

46. Sowani, H., Mohite, P., Munot, H., Shouche, Y., Bapat, T., Kumar, Kulkarni, M., & Zinjarde, S., (2016). Green synthesis of gold and silver nanoparticles by an actinomycete *Gordonia amicalis* HS–11: Mechanistic aspects and biological application. *Process Biochem.*, *51*, 374–83.
47. Martins, M., Mourato, C., Sanches, S., Noronha, J. P., Crespo, M. B., & Pereira, I. A., (2017). Biogenic platinum and palladium nanoparticles as new catalysts for the removal of pharmaceutical compounds. *Water Res., 108*, 160–8. doi.org/10.1016/j.procbio.2015.12.013.
48. Furgal, K. M., Meyer, R. L., & Bester, K., (2015). Removing selected steroid hormones, biocides and pharmaceuticals from water by means of biogenic manganese oxide nanoparticles in situ at ppb levels. *Chemosphere., 136*, 321–6. doi.org/10.1016/j.chemosphere.2014.11.059.
49. Priyanka, U., & Lens, P. N., (2022). Enhanced removal of hydrocarbons BTX by light–driven *Aspergillus Niger* ZnS nanobiohybrids. *Enzyme Microb. Tech.*, *157*, 110020. doi.org/10.1016/j.enzmictec.2022.110020.
50. Kang, C. K., Jeong, S. H., Kim, H. K., Kim, S. B., & Choi, Y. J., (2018). *Bioremediation of Radioactive Iodine Using Biogenic Gold Nanoparticles in Deinococcus Radiodurans R1* (p. 263). 한국생물공학회 학술대회.
51. El–Sheshtawy, H. S., Khalil, N. M., Ahmed, W., & Nabila, A. A., (2017). Enhancement the bioremediation of crude oil by nanoparticle and biosurfactants. *Egypt. J. Chem., 60*, 835–844. doi.org/10.21608/EJCHEM.2017.1321.1082.
52. Hassanshahian, M., (2014). Isolation and characterization of biosurfactant producing bacteria from Persian Gulf (Bushehr provenance). *Mar. Pollut. Bull., 86*, 361–366. doi.org/10.1016/j.marpolbul.2014.06.043.
53. Wu, G., Zhu, X., Ji, H., & Chen, D., (2015). Molecular modeling of interactions between heavy crude oil and the soil organic matter coated quartz surface. *Chemosphere, 119*, 242–249. doi.org/10.1016/j.chemosphere.2014.06.030.
54. Batista, S. B., Mounteer, A. H., Amorim, F. R., & Tótola, M. R., (2006). Isolation and characterization of biosurfactant/bioemulsifier–producing bacteria from petroleum contaminated sites. *Bioresource Technol., 20*, 274–288. doi.org/10.1016/j.biortech.2005.04.020.
55. Jorfi, S., Samaei, M. R., Darvishi, C. S. R., Talaie, K. A., Ahmadi, M., Barzegar, G., Reshadatian, N., & Mehrabi, N., (2017). Enhancement of the bioremediation of pyrene–contaminated soils using a hematite nanoparticle–based modified Fenton oxidation in a sequenced approach. *Soil Sediment Contam., 26,* 141–156. doi.org/10.1080/15320383.2017.1255875.
56. Mu, L., Zhou, Q., Zhao, Y., Liu, X., & Hu, X., (2019). Graphene oxide quantum dots stimulate indigenous bacteria to remove oil contamination. *J. Hazard Mater., 366,* 694–702. doi.org/10.1016/j.jhazmat.2018.12.044.
57. Tang, J., Zhu, N., Zhu, Y., Zamir, S. M., & Wu, Y., (2018). Sustainable pollutant removal by periphytic biofilm via microbial composition shifts induced by uneven distribution of CeO_2 nanoparticles. *Bioresource Technol., 247,* 1173–1176. doi.org/10.1016/j.biortech.2017.07.064.
58. Harms, H., Schlosser, D., & Wick, L. Y., (2011). Untapped potential: Exploiting fungi in bioremediation of hazardous chemicals. *Rev. Argent. Microbiol., 34,* 157–162. doi.org/10.1038/nrmicro2519.

59. Allen, M. F., (2007). Mycorrhizal fungi: Highways for water and nutrients in arid soils. *Vadose Zone J.*, *6*, 291. doi.org/10.2136/vzj2006.0068.
60. Das, S. K., Liang, J., Schmidt, M., Laffir, F., & Marsili, E., (2012). Biomineralization mechanism of gold by zygomycete fungi *Rhizopous oryzae*. *ACS Nano*, *6,* 6165–6173. doi.org/10.1021/nn301502s.
61. Dhillon, G. S., Brar, S. K., Kaur, S., & Verma, M., (2012). Green approach for nanoparticle biosynthesis by fungi: Current trends and applications. *Crc. Cr. Rev. Biotechn.*, *32*, 49–73. doi.org/10.3109/07388551.2010.550568.
62. Gade, A., Ingle, A., Whiteley, C., & Rai, M., (2010). Mycogenic metal nanoparticles: Progress and applications. *Biotechnol. Lett.*, *32*, 593–600. doi.org/10.1007/s10529–009–0197–9.
63. Yadav, A., Kon, K., Kratosova, G., Duran, N., Ingle, A. P., & Rai, M., (2015). Fungi as an efficient mycosystem for the synthesis of metal nanoparticles: Progress and key aspects of research. *Biotechnol. Lett.*, *37*, 2099–2120. doi.org/10.1007/s10529–015–1901–6.
64. Ji, C., Nguyen, L. N., Hou, J., Hai, F. I., & Chen, V., (2017). Direct immobilization of laccase on titania nanoparticles from crude enzyme extracts of *P. ostreatus* culture for micro–pollutant degradation. *Sep. Purif. Technol.*, *178*, 215–223. doi.org/10.1016/j.seppur.2017.01.043.
65. Kim, S. H., Park, J. H., Hong, Y., & Lee, C. Y., (2014). Removal of BTX using granular octyl–functionalized mesoporous silica nanoparticle. *Int. Biodeter. Biodegr.*, *95,* 219–224. doi.org/10.1016/j.ibiod.2014.04.016.
66. Quintero, J. C., Lú–Chau, T. A., Moreira, M. T., Feijoo, G., & Lema, J. M., (2007). Bioremediation of HCH present in soil by the white–rot fungus *Bjerkandera adusta* in a slurry batch bioreactor. *Int. Biodeter. Biodegr.*, *60*, 319–326. doi.org/10.1016/j.ibiod.2007.05.005.
67. Li, M., & Zhang, C., (2016). γ–Fe_2O_3 nanoparticle–facilitated bisphenol A degradation by white rot fungus. *Sci. Bull.*, *61*, 468–472. doi.org/10.1007/s11434–016–1021–2.
68. Durán, N., Marcato, P. D., De Souza, G. I. H., Alves, O. L., & Esposito, E., (2007). Antibacterial effect of silver nanoparticles produced by fungal process on textile fabrics and their effluent treatment. *J. Biomed. Nanotechnol.*, *3,* 203–208. doi.org/10.1166/jbn.2007.022.
69. Paknikar, K. M., Nagpal, V., Pethkar, A. V., & Rajwade, J. M., (2005). Degradation of lindane from aqueous solutions using iron sulfide nanoparticles stabilized by biopolymers. *Sci. Technol. Adv. Mat.*, *6*, 370–374. doi.org/10.1016/j.stam.2005.02.016.
70. Qu, Y., Pei, X., Shen, W., Zhang, X., Wang, J., Zhang, Z., Li, S., et al., (2017). Biosynthesis of gold nanoparticles by *Aspergillum* sp. WL–Au for degradation of aromatic pollutants. *Physica E, 88*, 133–141. doi.org/10.1016/j.physe.2017.01.010.
71. Shen, W., Qu, Y., Pei, X., Li, S., You, S., Wang, J., Zhan, Z., & Zhou, J., (2017). Catalytic reduction of 4–nitrophenol using gold nanoparticles biosynthesized by cell–free extracts of *Aspergillus* sp. WL–Au. *J. Hazard Mater.*, *321,* 299–306. doi.org/10.1016/j.jhazmat.2016.07.051.
72. Bhargava, A., Jain, N., Khan, M. A., Pareek, V., Dilip, R. V., & Panwar, J., (2016). Utilizing metal tolerance potential of soil fungus for efficient synthesis of gold nanoparticles with superior catalytic activity for degradation of rhodamine B. *J. Environ. Manage.*, *183*, 22–32. doi.org/10.1016/j.jenvman.2016.08.021.

73. Nwadinigwe, A. O., & Ugwu, E. C., (2019). Overview of nano–phytoremediation applications. In: *Phytoremediation: Management of Environmental Contaminants*, 377–382. doi.org/10.1007/978–3–319–99651–6_15.
74. Sharma, P., Tripathi, S., & Chandra, R., (2020). Phytoremediation potential of heavy metal accumulator plants for waste management in the pulp and paper industry. *Heliyon, 6*, e04559. doi.org/10.1016/j.heliyon.2020.e04559.
75. Pandey, V. C., Bajpai, O., & Singh, N., (2016). Energy crops in sustainable phytoremediation. *Renew. Sust. Energ. Rev., 54*, 58–73. doi.org/10.1016/j.rser.2015.09.078.
76. Huang, D., Qin, X., Peng, Z., Liu, Y., Gong, X., Zeng, G., Huang, C., et al., (2018). Nanoscale zero–valent iron assisted phytoremediation of Pb in sediment: Impacts on metal accumulation and antioxidative system of *Lolium perenne*. *Ecotox. Environ. Safe., 153*, 229–237. doi.org/10.1016/j.ecoenv.2018.01.060.
77. Singh, J., & Lee, B. K., (2018). Effects of nano–TiO_2 particles on bioaccumulation of 133Cs from the contaminated soil by soybean (glycine max). *Process Saf. Environ., 116*, 301–311. doi.org/10.1016/j.psep.2018.02.016.
78. Chen, Q., Liu, B., Man, H., Chen, L., Wang, X., Tu, J., Guo, Z., et al., (2020). Enhanced bioaccumulation efficiency and tolerance for Cd (II) in *Arabidopsis thaliana* by amphoteric nitrogen–doped carbon dots. *Ecotox. Environ. Safe., 190,* 110108. doi.org/10.1016/j.ecoenv.2019.110108.
79. Souri, Z., Karimi, N., Sarmadi, M., & Rostami, E., (2017). Salicylic acid nanoparticles (SANPs) improve growth and phytoremediation efficiency of *Isatis cappadocica* Desv., under As stress. *IET Nanobiotechnology, 11*, 650–655. doi.org/10.1049/iet–nbt.2016.0202.
80. Jesitha, K., & Harikumar, P. S., (2019). Application of nano–phytoremediation technology for soil polluted with pesticide residues and heavy metals. In: *Phytoremediation: Management of Environmental Contaminants*, 415–439. doi.org/10.1007/978–3–319–99651–6_18.
81. Kumari, S., & Khan, S., (2018). Effect of Fe_3O_4 NPs application on fluoride (F) accumulation efficiency of *Prosopis juliflora*. *Ecotox. Environ. Safe., 166,* 419–426. doi.org/10.1016/j.ecoenv.2018.09.103.
82. Gong, X., Huang, D., Liu, Y., Zeng, G., Wang, R., Xu, P., Zhang, C., et al., (2019). Roles of multiwall carbon nanotubes in phytoremediation: Cadmium uptake and oxidative burst in: *Boehmeria nivea* (L.) Gaudich. *Environ. Sci–Nano, 6*, 851–862. doi.org/10.1039/c8en00723c.
83. Zand, A. D., & Heir, A. V., (2020). Phytoremediation: Data on effects of titanium dioxide nanoparticles on phytoremediation of antimony polluted soil. *Data in Brief, 31*, 105959. doi.org/10.1016/j.dib.2020.105959.
84. Saaristo, M., Brodin, T., Balshine, S., Bertram, M. G., Brooks, B. W., Ehlman, S. M., McCallum, E. S., et al., (2018). Direct and indirect effects of chemical contaminants on the behavior, ecology, and evolution of wildlife. *P. Roy. Soc. B–Biol. Sci., 285,* 20181297. doi.org/10.1098/rspb.2018.1297.
85. Li, X., Meng, D., Li, J., Yin, H., Liu, H., Liu, X., Cheng, C., et al., (2017). Response of soil microbial communities and microbial interactions to long–term heavy metal contamination. *Environ. Pollut., 231,* 908–917. doi.org/10.1016/j.envpol.2017.08.057.
86. Kavitha, V., Anandhan, R., Alharbi, N. S., Kadaikunnan, S., Khaled, J. M., Almanaa, T. N., & Govindarajan, M., (2020). Impact of pesticide monocrotophos on microbial

populations and histology of intestine in the Indian earthworm Lampito mauritii (Kinberg). *Microb. Pathogenesis, 139*, 103893. doi.org/10.1016/j.micpath.2019.103893.
87. Sievers, M., Hale, R., Swearer, S. E., & Parris, K. M., (2018). Contaminant mixtures interact to impair predator–avoidance behaviors and survival in a larval amphibian. *Ecotox. Environ. Safe., 161*, 483–488. doi.org/10.1016/j.ecoenv.2018.06.028.
88. Mitchkash, M. G., Mcpeek, T., & Boone, M. D., (2014). The effects of 24–h exposure to carbaryl or atrazine on the locomotor performance and overwinter growth and survival of juvenile spotted salamanders (*Ambystoma maculatum*). *Environ. Toxicol. Chem., 33*, 548–552. doi.org/10.1002/etc.2451.
89. Moore, H., Chivers, D. P., & Ferrari, M. C. O., (2015). Sub–lethal effects of RoundupTM on tadpole anti–predator responses. *Ecotox. Environ. Safe., 211*, 281–285. doi.org/10.1016/j.ecoenv.2014.10.014.
90. Mishra, R. K., Ha, S. K., Verma, K., & Tiwari, S. K., (2018). Recent progress in selected bio–nanomaterials and their engineering applications: An overview. *J. Sci. Adv. Mater. Dev., 3*, 263–288. doi.org/10.1016/j.jsamd.2018.05.003.
91. Honek, J. F., (2013). Bionanotechnology and bionanomaterials: John Honek explains the good things that can come in very small packages. *BMC Biochem., 14*, 1–2. doi.org/10.1186/1471–2091–14–29.
92. Das, S., Chakraborty, J., Chatterjee, S., & Kumar, H., (2018). Prospects of biosynthesized nanomaterials for the remediation of organic and inorganic environmental contaminants. *Environ. Sci–Nano, 5,* 2784–2808. doi.org/10.1039/C8EN00799C.
93. Cheng, H., Lijie, L., Wang, B., Feng, X., Mao, Z., Vancso, G. J., & Siu, X., (2020). Multifaceted applications of cellulosic porous materials in environment, energy, and health. *Prog. Polym. Sci., 106*, 101253. doi.org/10.1016/j.progpolymsci.2020.101253.
94. Liu, P., Sehaqui, H., Tingaut, P., Wichser, A., Oksman, K., & Mathew, A. P., (2014). Cellulose and chitin nanomaterials for capturing silver ions (Ag+) from water via surface adsorption. *Cellulose, 21,* 449–461. doi.org/10.1007/s10570–013–0139–5.
95. Zhan, Y., He, S., Hu, J., Zhao, S., Zeng, G., Zhou, M., Zhang, G., & Senguota, A., (2020). Robust super–hydrophobic/super–oleophilic sandwich–like UIO–66–F4@rGO composites for efficient and multitasking oil/water separation applications. *J. Hazard Mater., 388,* 121752. doi.org/10.1016/j.jhazmat.2019.121752.
96. Arshad, A., Iqbal, J., & Mansoor, Q., (2019). Graphene/Fe_3O_4 nanocomposite: Solar light driven Fenton like reaction for decontamination of water and inhibition of bacterial growth. *Appl. Surf. Sci., 474*, 57–65. doi.org/10.1016/j.apsusc.2018.05.046.
97. Mahfoudhi, N., & Boufi, S., (2020). Porous material from cellulose nanofibrils coated with aluminum hydroxyde as an effective adsorbent for fluoride. *J. Environ. Chem. Eng., 8*, 103779. doi.org/10.1016/j.jece.2020.103779.
98. Hosseini, M., Zaki, D. H., Taghavi, M., & Babaei, A. A., (2020). Preparation of ultra–lightweight and surface–tailored cellulose nanofibril composite cryogels derived from date palm waste as powerful and low–cost heavy metals adsorbent to treat aqueous medium. *Ind. Crop. Prod., 154*, 112696. doi.org/10.1016/j.indcrop.2020.112696.
99. Kausar, A., (2019). Inorganic nanomaterials in polymeric water decontamination membranes. *Int. J. Past. Technol., 23*, 1–11. doi.org/10.1007/s12588–019–09230–x.
100. Wang, Y., O'Connor, D., Shen, Z., Lo, I. M. C., Tsang, D. C. W., Pehkonen, S., & Hou, D., (2019). Green synthesis of nanoparticles for the remediation of contaminated waters and soils: Constituents, synthesizing methods, and influencing factors. *J. Clean. Prod., 226,* 540–549. doi.org/10.1016/j.jclepro.2019.04.128.

101. Zhang, P., O'Connor, D., Wang, Y., Jiang, L., Xia, T., Wang, L., Tsang, D. C. W., et al., (2020). A green biochar/iron oxide composite for methylene blue removal. *J. Hazard Mater., 384*, 121286. doi.org/10.1016/j.jhazmat.2019.121286.
102. Abbas, S., Nasreen, S., Haroon, A., & Ashraf, M. A., (2020). Synhesis of silver and copper nanoparticles from plants and application as adsorbents for naphthalene decontamination. *Saudi J. Biol. Sci., 27*, 1016–1023. doi.org/10.1016/j.sjbs.2020.02.011.
103. Zhao, L., Guan, X., Yu, B., Ding, N., Liu, X., Ma, Q., Yang, S., et al., (2019). Carboxylated graphene oxide–chitosan spheres immobilize Cu^{2+} in soil and reduce its bioaccumulation in wheat plants. *Environ. Int., 133,* 105208. doi.org/10.1016/j.envint.2019.105208.
104. Yi, N., Wu, Y., Fan, L., & Hu, S., (2019). Remediating Cd–contaminated soils using natural and chitosan–introduced zeolite, bentonite, and activated carbon. *Pol. J. Environ. Stud., 28*, 1461–1468. doi.org/10.15244/pjoes/89577.
105. Cai, Z., Zhao, X., Duan, J., Zhao, D., Dang, Z., & Lin, Z., (2020). Remediation of soil and groundwater contaminated with organic chemicals using stabilized nanoparticles: Lessons from the past two decades. *Front. Env. Sci. Eng., 14*, 1–20. doi.org/10.1007/s11783–020–1263–8.
106. Kaloyianni, M., Dimitriadi, A., Ovezik, M., Stamkopoulou, D., Feidantsis, K., Kastrinaki, G., Gallios, G., et al., (2020). Magnetite nanoparticles effects on adverse responses of aquatic and terrestrial animal models. *J. Hazard Mater., 385, 121204.* doi.org/10.1016/j.jhazmat.2019.121204.
107. Shanker, U., Jassal, V., & Rani, M., (2017). Green synthesis of iron hexacyanoferrate nanoparticles: Potential candidate for the degradation of toxic PAHs. *J. Environ. Chem. Eng., 5*, 4108–4120. doi.org/10.1016/j.jece.2017.07.042.
108. Mandal, S., Pu, S., Shangguan, L., Liu, S., Ma, H., Adhikari, S., & Hou, A., (2020). Synergistic construction of green tea biochar supported nZVI for immobilization of lead in soil: A mechanistic investigation. *Environ. Int., 135,* 105374. doi.org/10.1016/j.envint.2019.105374.
109. Di, Q., Wang, Y., Zanobetti, A., Wang, Y., Koutrakis, P., Choirat, C., Dominici, F., & Schwartz, J. D., (2017). Air pollution and mortality in the medicare population. *New Engl. J. Med., 376*, 2513–2522. doi.org/10.1056/nejmoa1702747.
110. Podgórski, A., Ba\lazy, A., & Gradoń, L., (2006). Application of nanofibers to improve the filtration efficiency of the most penetrating aerosol particles in fibrous filters. *Chem. Eng. Sci., 61*, 6804–6815. doi.org/10.1016/j.ces.2006.07.022.
111. García–Mayagoitia, S., Pérez–Hernández, H., Medina–Pérez, G., Campos–Montiel, R. G., & Fernández–Luqueño, F., (2020). Bio–nanomaterials in the air pollution treatment. In: *Nanomaterials for Air Remediation*, 227–248. doi.org/10.1016/B978–0–12–818821–7.00012–9.
112. Zhu, M., Xiong, R., & Huang, C., (2019). Bio–based and photo crosslinked electrospun antibacterial nanofibrous membranes for air filtration. *Carbohy. Polym., 205*, 55–62. doi.org/10.1016/j.carbpol.2018.09.075.
113. Lv, D., Wang, R., Tang, G., Mou, Z., Lei, J., Han, J., De Smedt, S., et al., (2019). Ecofriendly electrospun membranes loaded with visible–light–responding nanoparticles for multifunctional usages: Highly efficient air filtration, dye scavenging, and bactericidal activity. *ACS Appl. Mater. Inter., 11*, 12880–12889. doi.org/10.1021/acsami.9b01508.
114. Liu, X., Souzandeh, H., Zheng, Y., Xie, Y., Zhong, W. H., & Wang, C., (2017). Soy protein isolate/bacterial cellulose composite membranes for high efficiency particulate

air filtration. Composites *Science and Technology*, *135*, 76–82. doi.org/10.1016/j.compscitech.2016.11.022.
115. Khan, I., Saeed, K., & Khan, I., (2019). Nanoparticles: Properties, applications, and toxicities. *Arab. J Chem.*, *12*, 908. doi.org/10.1016/j.arabjc.2017.05.011.
116. Lu, D., Zhou, J., Hou, S., Xiong, Q., Chen, Y., Pu, K., Ren, J., & Duan, H., (2019). Functional macromolecule–enabled colloidal synthesis: From nanoparticle engineering to multifunctionality. *Adv. Mater.*, *31*, 1902733. doi.org/10.1002/adma.201902733.
117. Farhadi, S., Mahmoudi, F., Dusek, M., Eigner, V., & Kucerakova, M., (2017). A new inorganic–organic nanohybrid based on a copper(II) semicarbazone complex and the PMo12O403−polyanion: Synthesis, characterization, crystal structure and photocatalytic activity for degradation of cationic dyes. *Polyhedron*, *122*, 247–256. doi.org/10.1016/j.poly.2016.11.034.
118. Saleh, N. B., Aich, N., Plazas–Tuttle, J., Lead, J. R., & Lowry, G. V., (2015). Research strategy to determine when novel nanohybrids pose unique environmental risks. *Environ. Sci. Nano*, *2*, 11–18. doi.org/10.1039/c4en00104d.
119. Yang, S., Li, Y., Wang, S., Wang, M., Chu, M., & Xia, B., (2018). Advances in the use of carbonaceous materials for the electrochemical determination of persistent organic pollutants. A review. *Microchimica Acta*, *185*, 1–14. doi.org/10.1007/s00604–017–2638–9.
120. Guo, L., Xu, Y., Zhuo, M., Liu, L., Xu, Q., Wang, L., Shi, C., et al., (2018). Highly efficient removal of Gd (III) using hybrid hydrosols of carbon nanotubes/graphene oxide in dialysis bags and synergistic enhancement effect. *Chem. Eng. J.*, *348*, 535–545. doi.org/10.1016/j.cej.2018.04.212.
121. Wang, D., Saleh, N. B., Sun, W., Park, C. M., Shen, C., Aich, N., Peijnenburg, W. J. G. M., et al., (2019). Next–generation multifunctional carbon–metal nanohybrids for energy and environmental applications. *Environ. Sci. Technol.*, *53*, 7265–7287. doi.org/10.1021/acs.est.9b01453.
122. Kumar, A., Kumar, A., Sharma, G., Naushad, M., Stadler, F. J., Ghfar, A. A., Dhiman, P., & Saini, R. V., (2017). Sustainable nano–hybrids of magnetic biochar supported g–C_3N_4/$FeVO_4$ for solar powered degradation of noxious pollutants– synergism of adsorption, photocatalysis & photo–ozonation. *J. Clean. Prod.*, *165*, 431–451. doi.org/10.1016/j.jclepro.2017.07.117.
123. Anjum, M., Oves, M., Kumar, R., & Barakat, M. A., (2017). Fabrication of ZnO–ZnS@polyaniline nanohybrid for enhanced photocatalytic degradation of 2–chlorophenol and microbial contaminants in wastewater. *Int. Biodeter. Biodegr.*, *119*, 66–77. doi.org/10.1016/j.ibiod.2016.10.018.
124. Hasija, V., Raizada, P., Sudhaik, A., Sharma, K., Kumar, A., Singh, P., Jonnalagadda, S. B., & Thakur, V. K., (2019). Recent advances in noble metal free doped graphitic carbon nitride based nanohybrids for photocatalysis of organic contaminants in water: A review. *Applied Materials Today*, *15*, 494–524. doi.org/10.1016/j.apmt.2019.04.003.
125. Maleki, A., Hayati, B., Najafi, F., Gharibi, F., & Joo, S. W., (2016). Heavy metal adsorption from industrial wastewater by PAMAM/TiO_2 nanohybrid: Preparation, characterization, and adsorption studies. *J. Mol. Liq.*, *224*, 95–104. doi.org/10.1016/j.molliq.2016.09.060.
126. Xu, L., & Wang, J., (2017). The application of graphene–based materials for the removal of heavy metals and radionuclides from water and wastewater. *Critical Reviews in*

Environmental Science and Technology, *47*, 1042–1105. doi.org/10.1080/10643389.2017.1342514.
127. Tan, N., Yang, Z., Gong, X., Wang, Z., Fu, T., & Liu, Y., (2019). In situ generation of H2O2 using MWCNT–Al/O_2 system and possible application for glyphosate degradation. *Sci. Total Environ.*, *650*, 2567–2576. doi.org/10.1016/j.scitotenv.2018.09.353.
128. Lin, Y., Wu, S., Yang, C., Chen, M., & Li, X., (2019). Preparation of size–controlled silver phosphate catalysts and their enhanced photocatalysis performance via synergetic effect with MWCNTs and PANI. *Appl. Catal. B–Environ.*, *245*, 71–86. doi.org/10.1016/j.apcatb.2018.12.048.
129. Liu, Y., Tan, N., Wang, B., & Liu, Y., (2019). Stepwise adsorption–oxidation removal of oxytetracycline by Zn^0–CNTs–Fe_3O_4 from aqueous solution. *Chemical Engineering Journal*, *375*, 121963. doi.org/10.1016/j.cej.2019.121963.
130. Chen, H., Zhang, Z., Wang, X., Chen, J., Xu, C., Liu, Y., Yu, Z., & Wang, X., (2018). Fabrication of magnetic Fe/Zn layered double Oxide@Carbon nanotube composites and their application for U(VI) and 241Am (III) removal. *ACS Applied Nano Materials*, *1*, 2386–2396. doi.org/10.1021/acsanm.8b00528.
131. Engel, M., Hadar, Y., Belkin, S., Lu, X., Elimelech, M., & Chefetz, B., (2018). Bacterial inactivation by a carbon nanotube–iron oxide nanocomposite: A mechanistic study using *E. coli* mutants. *Environ. Sci–Nano*, *5*, 372–380. doi.org/10.1039/c7en00865a.
132. Rajavel, K., Gomathi, R., Manian, S., & Rajendra, K. R. T., (2014). *In vitro* bacterial cytotoxicity of CNTs: Reactive oxygen species mediate cell damage edges over direct physical puncturing. *Langmuir*, *30*, 259–601. doi.org/10.1021/la403332b.
133. Povolotckaia, A., Pankin, D., Petrov, Y., Vasileva, A., Kolesnikov, I., Sarau, G., Christiansen, S., et al., (2019). Plasmonic carbon nanohybrids from laser–induced deposition: Controlled synthesis and SERS properties. *Journal of Materials Science*, *54*. 8177–8186. doi.org/10.1007/s10853–019–03478–9.
134. Sharma, G., Kumar, A., Naushad, M., Kumar, A., Al–Muhtaseb, A. H., Dhiman, P., Ghfar, A. A., et al., (2016). Photoremediation of toxic dye from aqueous environment using monometallic and bimetallic quantum dots-based nanocomposites. *J. Clean. Prod.*, *172*, 2919–2930. doi.org/10.1016/j.jclepro.2017.11.122.
135. Nallal, M., Anantha, I. G., & Pill–Lee, K., (2017). New titanium dioxide–based heterojunction nanohybrid for highly selective photoelectrochemical–electrochemical dual–mode sensors. *ACS Appl. Mater. Inter.*, *9*, 371666–37183. doi.org/10.1021/acsami.7b10519.
136. Jiang, B., Lian, L., Xing, Y., Zhang, N., Chen, Y., Lu, P., & Zhang, D., (2018). Advances of magnetic nanoparticles in environmental application: Environmental remediation and (bio)sensors as case studies. *Environ. Sci. Pollut. R.*, *25*, 30863–30879. doi.org/10.1007/s11356–018–3095–7.
137. Wang, X., Mitchell, D. R. G., Prince, K., Atanacio, A. J., & Caruso, R. A., (2008). Gold nanoparticle incorporation into porous titania networks using an agarose gel templating technique for photocatalytic applications. *Chem. Mater.*, *20*, 3917–3926. doi.org/10.1021/cm703509f.
138. Ma, R., Zhang, S., Wen, T., Gu, P., Li, L., Zhao, G., Niu, F., et al., (2019). A critical review on visible–light–response CeO_2–based photocatalysts with enhanced photooxidation of organic pollutants. *Catal. Today*, *335*. 20–30. doi.org/10.1016/j.cattod.2018.11.016.

139. Xiao, H., Zhang, W., Wei, Y., & Chen, L., (2018). Carbon/ZnO nanorods composites templated by TEMPO–oxidized cellulose and photocatalytic activity for dye degradation. *Cellulose, 25,* 1809–1819. doi.org/10.1007/s10570–018–1651–4.
140. Araújo, E. S., Da Costa, B. P., Oliveira, R. A. P., Libardi, J., Faia, P. M., De Oliveira, H. P., (2016). TiO_2/ZnO hierarchical heteronanostructures: Synthesis, characterization, and application as photocatalysts. *J. Environ. Chem. Eng., 4,* 2820–2829. doi.org/10.1016/j.jece.2016.05.021.
141. Yang, T., Peng, J., Zheng, Y., He, X., Hou, Y., Wu, L., & Fu, X., (2018). Enhanced photocatalytic ozonation degradation of organic pollutants by ZnO modified TiO_2 nanocomposites. *Appl. Catal. B–Environ., 221,* 223–234. doi.org/10.1016/j.apcatb.2017.09.025.
142. Zhang, W., Zhang, G., Liu, C., Li, J., Zheng, T., Ma, J., Wang, L., et al., (2018). Enhanced removal of arsenite and arsenate by a multifunctional Fe–Ti–Mn composite oxide: Photooxidation, oxidation and adsorption. *Water Res., 147,* 264–275. doi.org/10.1016/j.watres.2018.10.001.
143. Tang, T. Y., Wang, H. L., Yao, C. T., Yang, K. C., Ho, R. M., & Tsai, D. H., (2018). A facile method to functionalize gold nano–tripods with high suspension stability in an aqueous environment. *Nanoscale, 10,* 7352–7356. doi.org/10.1039/c8nr00212f.
144. Modi, A., & Bellare, J., (2019). Efficiently improved oil/water separation using high flux and superior antifouling polysulfone hollow fiber membranes modified with functionalized carbon nanotubes/graphene oxide nanohybrid. *J. Environ. Chem. Eng., 7,* 102944. doi.org/10.1016/j.jece.2019.102944.
145. Akın, D., Yakar, A., & Gündüz, U., (2014). Synthesis of magnetic Fe_3O_4–chitosan nanoparticles by ionic gelation and their dye removal ability. *Water Environment Research, 87,* 425–436. doi.org/10.2175/106143014x14062131178673.
146. Liu, Q., Zheng, Y., Zhong, L., & Cheng, X., (2015). Removal of tetracycline from aqueous solution by a Fe_3O_4 incorporated PAN electrospun nanofiber mat. *J. Environ. Sci., 28,* 29–36. doi.org/10.1016/j.jes.2014.04.016.
147. Hammouda, S. B., Adhoum, N., & Monser, L., (2015). Synthesis of magnetic alginate beads based on Fe_3O_4 nanoparticles for the removal of 3–methylindole from aqueous solution using Fenton process. *J. Hazard Mater., 294,* 128–136. doi.org/10.1016/j.jhazmat.2015.03.068.
148. Wang, P., Lu, X., Jin, J., Ngai, T., Zeng, H., & Zacharia, N., (2016). Smart materials for advanced environmental applications. *Roy. Soc. Ch.*
149. Zhang, Y., Yuan, K., & Zhang, L., (2019). Micro/nanomachines: From functionalization to sensing and removal. *Adv. Mater. Technol., 4,* 1800636. doi.org/10.1002/admt.201800636.
150. Parmar, J., Vilela, D., Villa, K., Wang, J., & Sánchez, S., (2018). Micro– and nanomotors as active environmental microcleaners and sensors. *J. Am. Chem. Soc., 140,* 9317–9331. doi.org/10.1021/jacs.8b05762.
151. Xu, B., Zhang, B., Wang, L., Huang, G., & Mei, Y., (2018). Tubular micro/nanomachines: From the basics to recent advances. *Adv. Funct. Mater., 140,* 9317–9331. doi.org/10.1002/adfm.201705872.
152. Katuri, J., Ma, X., Stanton, M. M., & Sánchez, S., (2017). Designing micro–and nanoswimmers for specific applications. *Accounts Chem. Res., 50,* 2–11. doi.org/10.1021/acs.accounts.6b00386.

153. Cheng, Z., Li, C., Lai, H., Du, Y., Liu, H., Liu, M., Sun, K., et al., (2016). Recycled superwetting nanostructured copper mesh film: Toward bidirectional separation of emulsified oil/water mixtures. *Adv. Mater. Interfaces, 3*, 1600370. doi.org/10.1002/admi.201600370.
154. Zhang, W., Liu, N., Cao, Y., Lin, X., Liu, Y., & Feng, L., (2017). Superwetting porous materials for wastewater treatment: From immiscible oil/water mixture to emulsion separation. *Adv. Mater. Interfaces, 4*, 16000029. doi.org/10.1002/admi.201700029.
155. Guix, M., Orozco, J., Garcia, M., Gao, W., Sattayasamitsathit, S., Merkoči, A., Escarpa, A., & Wang, J., (2012). Superhydrophobic alkanethiol–coated microsubmarines for effective removal of oil. *ACS Nano, 6*, 4445–4451. doi.org/10.1021/nn301175b.
156. Mou, F., Pan, D., Chen, C., Gao, Y., Xu, L., & Guan, J., (2015). Magnetically modulated pot–like $MnFe_2O_4$ micromotors: Nanoparticle assembly fabrication and their capability for direct oil removal. *Adv. Funct. Mater., 25*, 6173–6181. doi.org/10.1002/adfm.201502835.
157. Wan, H., Islam, M. S., Briot, N. J., Schnobrich, M., Pacholik, L., Ormsbee, L., & Bhattacharyya, D., (2020). Pd/Fe nanoparticle integrated PMAA–PVDF membranes for chloro–organic remediation from synthetic and site groundwater. *J. Membrane Sci., 594*, 117454. doi.org/10.1016/j.memsci.2019.117454.
158. Umar, A., Akhtar, M. S., Dar, G. N., & Baskoutas, S., (2013). Low–temperature synthesis of α–Fe_2O_3 hexagonal nanoparticles for environmental remediation and smart sensor applications. *Talanta, 262*, 588–596. doi.org/10.1016/j.talanta.2013.08.026.
159. Villa, K., Parmar, J., Vilela, D., & Sánchez, S., (2018). Metal–oxide–based microjets for the simultaneous removal of organic pollutants and heavy metals. *ACS Appl. Mater. Inter., 10*, 20478–20486. doi.org/10.1021/acsami.8b04353.
160. Sabeela, N. I., Almutairi, T. M., Al–Lohedan, H. A., Ezzat, A. O., & Atta, A. M., (2019). New smart magnetic ionic liquid nanocomposites based on chemically bonded imidazole silica for water treatment. *ACS Omega, 4*, 21288–21301. doi.org/10.1021/acsomega.9b02820.
161. Xu, L., Zhang, D., Ma, F., Zhang, J., Khayambashi, A., Cai, Y., Chen, L., et al., (2019). Nano–MOF^+ technique for efficient uranyl remediation. *ACS Appl. Mater. Inter., 11*, 21619–21626. doi.org/10.1021/acsami.9b06068.
162. Rostami, M., (2019). Photodecomposition and adsorption of hazardous organic pollutants by Ce–doped ZnO@Ce–doped TiO_2 –N/S–dual doped RGO ternary nano–composites photocatalyst for water remediation. *J. Mol. Struct., 1135*, 43–54. doi.org/10.1016/j.molstruc.2019.02.094.
163. Wang, Z., Wang, H., Zeng, Z., Zeng, G., Xu, P., Xiao, R., Huang, D., et al., (2020). Metal–organic frameworks derived $Bi_2O_2CO_3$/porous carbon nitride: A nanosized Z–scheme systems with enhanced photocatalytic activity. *Appl. Catal. B–Environ., 267*, 118700. doi.org/10.1016/j.apcatb.2020.118700.
164. Yang, M. Q., Zhang, N., & Xu, Y. J., (2013). Synthesis of fullerene–, carbon nanotube–, and graphene–TiO_2 nanocomposite photocatalysts for selective oxidation: A comparative study. *ACS Appl. Mater. Inter., 5*, 1156–1164. doi.org/10.1021/am3029798.
165. Zhang, Y., Tang, Z. R., Fu, X., & Xu, Y. J., (2010). TiO_2–graphene nanocomposites for gas–phase photocatalytic degradation of volatile aromatic pollutant: Is TiO_2–graphene truly different from other TiO_2–carbon composite materials? *ACS Nano, 4*, 7303–7314. doi.org/10.1021/nn1024219.

166. Kumar, P., Saini, M., Kumar, V., Singh, M., Dehiya, B. S., Umar, A., Khan, A., & Alhuwaymel, T. F., (2020). Removal of Cr (VI) from aqueous solution using VO_2(B) nanoparticles. *Chem– Phys. Lett.*, *739*, 136934. doi.org/10.1016/j.cplett.2019.136934.
167. Wang, S., Jiang, Z., Ouyang, S., Dai, Z., & Wang, T., (2017). Internally/externally bubble–propelled photocatalytic tubular nanomotors for efficient water cleaning. *ACS Appl. Mater. Inter.*, *9*, 23974–23982. doi.org/10.1021/acsami.7b06402.
168. Wang, L., Kaeppler, A., Fischer, D., & Simmchen, J., (2019). Photocatalytic TiO_2 micromotors for removal of microplastics and suspended matter. *ACS Appl. Mater. Inter.*, *11*, 32937–32944. doi.org/10.1021/acsami.9b06128.
169. Vilela, D., Parmar, J., Zeng, Y., Zhao, Y., & Sánchez, S., (2016). Graphene–based microbots for toxic heavy metal removal and recovery from water. *Nano Letters*, *16*, 2860–2866. doi.org/10.1021/acs.nanolett.6b00768.
170. Chen, L., Zhang, M. J., Zhang, S. Y., Shi, L., Yang, Y. M., Liu, Z., Ju, X. J., et al., (2020)., Simple and continuous fabrication of self–propelled micromotors with photocatalytic metal–organic frameworks for enhanced synergistic environmental remediation. *ACS Appl. Mater. Inter.*, *12*, 35120–32131. doi.org/10.1021/acsami.0c11283.
171. Jurado–Sánchez, B., Wang, J., & Escarpa, A., (2016). Ultrafast nanocrystals decorated micromotors for on–site dynamic chemical processes. *ACS Appl. Mater. Inter.*, *8*, 19618–19625. doi.org/10.1021/acsami.6b05824.
172. Pourrahimi, A. M., Villa, K., Ying, Y., Sofer, Z., & Pumera, M., (2018). ZnO/ZnO_2 /Pt janus micromotors propulsion mode changes with size and interface structure: Enhanced nitroaromatic explosives degradation under visible light. *ACS Appl. Mater. Inter.*, *10*, 42688–42697. doi.org/10.1021/acsami.8b16217.
173. Liang, C., Zhan, C., Zeng, F., Xu, D., Wang, Y., Zhao, W., Zhang, J., et al., (2018). Bilayer tubular micromotors for simultaneous environmental monitoring and remediation. *ACS Appl. Mater. Inter.*, *10*, 35099–35107. doi.org/10.1021/acsami.8b10921.
174. Shi, H., Chen, X., Liu, K., Ding, X., Liu, W., & Xu, M., (2020). Heterogeneous Fenton ferroferric oxide–reduced graphene oxide–based composite microjets for efficient organic dye degradation. *J. Colloid Interf. Sci.*, *572*, 39–47. doi.org/10.1016/j.jcis.2020.03.073.
175. Okesola, B. O., & Smith, D. K., (2016). Applying low–molecular weight supramolecular gelators in an environmental setting–self–assembled gels as smart materials for pollutant removal. *Chem. Soc. Rev.*, *45*, 4226–4251. doi.org/10.1039/c6cs00124f.
176. Mohamed, A. L., El–Naggar, M. E., Shaheen, T. I., & Hassabo, A. G., (2016). Novel nano polymeric system containing biosynthesized core shell silver/silica nanoparticles for functionalization of cellulosic based material. *Microsyst. Technol.*, *22*, 979–992. doi.org/10.1007/s00542–015–2776–0.
177. Yan, P. J., He, F., Wang, W., Zhang, S. Y., Zhang, L., Li, M., Liu, Z., Ju, X. J., Xie, R., & Chu, L. Y., (2018). Novel membrane detector based on smart nanogels for ultrasensitive detection of trace threat substances. *ACS Appl. Mater. Inter.*, *10*, 36425–36434. doi.org/10.1021/acsami.8b12615.
178. Liu, W. Y., Ju, X. J., Faraj, Y., He, F., Peng, H. Y., Liu, Y. Q., Liu, Z., et al., (2020). Capsule membranes encapsulated with smart nanogels for facile detection of trace lead(II) ions in water. *J. Membrane Sci.*, *613*, 118523. doi.org/10.1016/j.memsci.2020.118523.
179. Radwan, E. K., Kafafy, H., El–Wakeel, S. T., Shaheen, T. I., Gad–Allah, T. A., El–Kalliny, A. S., & El–Naggar, M. E., (2018). Remediation of Cd(II) and reactive red 195

dye in wastewater by nanosized gels of grafted carboxymethyl cellulose. *Cellulose, 25*, 6645–6660. doi.org/10.1007/s10570–018–2003–0.
180. Kotoky, R., Rajkumari, J., & Pandey, P., (2018). The rhizosphere microbiome: Significance in rhizoremediation of polyaromatic hydrocarbon contaminated soil. *J. Environ. Manage., 217*, 858–870. doi.org/10.1016/j.jenvman.2018.04.022.
181. Mishra, A., Mishra, S. P., Arshi, A., Agarwal, A., & Dwivedi, S. K., (2020). Plant–microbe interactions for bioremediation and phytoremediation of environmental pollutants and agro–ecosystem development. In: *Bioremediation of Industrial Waste for Environmental Safety*. doi.org/10.1007/978–981–13–3426–9_17.
182. Shah, V., & Daverey, A., (2020). Phytoremediation: A multidisciplinary approach to clean up heavy metal contaminated soil. *Environ. Technol. Innov.,* doi.org/10.1016/j.eti.2020.100774.
183. Grzegórska, A., Rybarczyk, P., Rogala, A., & Zabrocki, D., (2020). Phytoremediation–from environment cleaning to energy generation–current status and future perspectives. *Energies, 13*, 2905. doi.org/10.3390/en13112905.
184. Zhang, H., Yuan, X., Xiong, T., Wang, H., & Jiang, L., (2020). Bioremediation of co–contaminated soil with heavy metals and pesticides: Influence factors, mechanisms, and evaluation methods. *J. Hazard Mater., 301*, 1–7. doi.org/10.1016/j.cej.2020.125657.
185. Li, S., Yang, M., Wang, H., & Zhao, Y., (2020). Dynamic characteristics of immobilized microorganisms for remediation of nitrogen–contaminated groundwater and high–throughput sequencing analysis of the microbial community. *Environ. Pollut., 267*, 114875. doi.org/10.1016/j.envpol.2020.114875.
186. Sun, X., Meng, J., Huo, S., Zhu, J., & Zheng, S., (2020). Remediation of heavy metal pollution in soil by microbial immobilization with carbon microspheres. *Int. J. Env. Sci. Dev., 11*, 43–47. doi.org/10.18178/ijesd.2020.11.1.1223.
187. Jaiswal, S., & Shukla, P., (2020). Alternative strategies for microbial remediation of pollutants via synthetic biology. *Front. Microbiol., 11*, 808. doi.org/10.3389/fmicb.2020.00808.
188. Tonelli, F. M. P., & Tonelli, F. C. P., (2020). Pollution remediation by way of using genetically modified plants (GMPs). In: *Bioremediation and Biotechnology: Sustainable Approaches to Pollution Degradation* (pp. 305–318). doi.org/10.1007/978–3–030–35691–0_15.
189. Gu, W., Zhao, Y., Li, Q., & Li, Y., (2020). Plant–microorganism combined remediation of polychlorinated naphthalenes contaminated soils based on molecular directed transformation and Taguchi experimental design–assisted dynamics simulation. *J. Hazard Mater., 396*, 122753. doi.org/10.1016/j.jhazmat.2020.122753.
190. Lai, R. W. S., Yeung, K. W. Y., Yung, M. M. N., Djurišić, A. B., Giesy, J. P., & Leung, K. M. Y., (2018). Regulation of engineered nanomaterials: Current challenges, insights, and future directions. *Environ. Sci. Pollut. R., 25*, 3060–3077. doi.org/10.1007/s11356–017–9489–0.
191. Pastor–Jáuregui, R., Paniagua–López, M., Martínez–Garzón, J., Martín–Peinado, F., & Sierra–Aragón, M., (2020). Evolution of the residual pollution in soils after bioremediation treatments. *Appl. Sci., 10*, 1006. doi.org/10.3390/app10031006.
192. Rycroft, T., Hamilton, K., Haas, C. N., & Linkov, I., (2019). A quantitative risk assessment method for synthetic biology products in the environment. *Sci. Total Environ., 696*, 133940. doi.org/10.1016/j.scitotenv.2019.133940.

193. Coppedè, N., Janni, M., Bettelli, M., Maida, C. L., Gentile, F., Villani, M., Ruotolo, R., et al., (2017). An in vivo biosensing, biomimetic electrochemical transistor with applications in plant science and precision farming. *Sci. Rep.*, *7*, 16195. doi.org/10.1038/s41598–017–16217–4.
194. Kumar, V., & Arora, K., (2020). Trends in nano–inspired biosensors for plants. *Mater. Sci. Energ. Technol.*, *3*, 255–273. doi.org/10.1016/j.mset.2019.10.004.
195. Akkilic, N., Geschwindner, S., & Höök, F., (2020). Single–molecule biosensors: Recent advances and applications. *Biosens. Bioelectron.*, *151*, 111944. doi.org/10.1016/j.bios.2019.111944.
196. Konwarh, R., Gollavelli, G., & Palanisamy, S. B., (2020). Designing of novel nanosensors for environmental aspects. In: *Nanofabrication for Smart Nanosensors Applications* (pp. 51–87). doi.org/10.1016/b978–0–12–820702–4.00003–9.
197. Yin, Y., (2019). *Handbook of Synthetic Methodologies and Protocols of Nanomaterials* (Vol. 4, 18). *World Scientific*.
198. Sankarganesh, E., Rajeshwaran, B., & Sravani, B. L., (2020). Nanotechnology for pest management: A new era in plant protection. *Res. Today*, *25*, 296–297.
199. Yarima, A., Ali, R., Abdullahi, A. A., & Idris, Z., (2020). Nanotechnology: Review on emerging techniques in remediating water and soil pollutions. *J. Appl. Sci. Environ. Manage.*, *24*, 933–941. doi.org/10.4314/jasem.v24i5.31.
200. Mohamed, H. M., (2020). Sensors and biosensors for environment contaminants. In: *Nanosensor Technologies for Environmental Monitoring* (pp. 109–134). Springer.
201. Filik, H., & Avan, A. A., (2020). Review on applications of carbon nanomaterials for simultaneous electrochemical sensing of environmental contaminant dihydroxybenzene isomers. *Arabian J. Chem.*, *13*, 6092–6105. doi.org/10.1016/j.arabjc.2020.05.009.
202. Kumar, D. K., Reddy, K. R., Sadhu, V., Shetti, N. P., Reddy, C. V., Chouhan, R. S., & Navee, S., (2020). Metal oxide–based nanosensors for healthcare and environmental applications. In: *Nanomaterials in Diagnostic Tools and Devices*, *1*, 113–129.
203. Aashima, M. S. K., (2020). Impact of functionalized nanomaterials towards the environmental remediation: Challenges and future needs. In: *Handbook of Functionalized Nanomaterials for Industrial Applications* (pp. 505–524). doi.org/10.1016/b978–0–12–816787–8.00018–1.
204. Jawed, A., Saxena, V., & Pandey, L. M., (2020). Engineered nanomaterials and their surface functionalization for the removal of heavy metals: A review. *J. Water Process. Eng.*, *33*, 101009. doi.org/10.1016/j.jwpe.2019.101009.
205. Uddin, M. N., Desai, F., & Asmatulu, E., (2020). Engineered nanomaterials in the environment: Bioaccumulation, biomagnification, and biotransformation. *Environ. Chem. Lett.*, *18*, 1073–1083. doi.org/10.1007/s10311–019–00947–0.
206. Piplai, T., Parsai, T., Kumar, A., & Alappat, B. J., (2020). understanding interactions of nanomaterials with soil: Issues and challenges ahead. *Environ. Nanotechnol.*, *3*, 117–141. doi.org/10.1007/978–3–030–26672–1_4.
207. Iavicoli, I., Leso, V., Beezhold, D. H., & Shvedova, A. A., (2017). Nanotechnology in agriculture: Opportunities, toxicological implications, and occupational risks. *Toxicol. Appl. Pharm.*, *329*, 96–11. doi.org/10.1016/j.taap.2017.05.025.
208. Corsi, I., Winther–Nielsen, M., Sethi, R., Punta, C., Della, T. C., Libralato, G., Lofrano, G., et al., (2018). Ecofriendly nanotechnologies, and nanomaterials for environmental applications: Key issue and consensus recommendations for sustainable

and ecosafe nanoremediation. *Ecotox. Environ. Safe.*, *154*, 237–244. doi.org/10.1016/j.ecoenv.2018.02.037.

209. Qian, Y., Qin, C., Chen, M., & Lin, S., (*2020*). Nanotechnology in soil remediation – applications vs. implications. *Ecotox. Environ. Safe.*, *201*, 110815. doi.org/10.1016/j.ecoenv.2020.110815.

CHAPTER 4

Nanomaterials-Assisted Decontamination of Various Pollutants From Water Resources

PREETI RAINA,[1] GAURI SHARMA,[1] AKANKSHA JASROTIA,[1] AKSHI BHARDWAJ,[1] PUSHAP RAJ,[1] RITU BALA,[2] and RAJINDER KAUR[1]

[1]Department of Botanical and Environmental Sciences, Guru Nanak Dev University, Amritsar, Punjab, India

[2]Department of Chemistry, Guru Nanak Dev University, Amritsar, Punjab, India

ABSTRACT

In aquatic systems, the presence of various materials like organic, inorganic, radioactive, and biological pollutants beyond their permissible limits is dangerous to humans and aquatic life because of their bioaccumulation and biomagnification properties. These pollutants contaminate the water bodies and impact living organisms directly due to their toxic and carcinogenic effects. For this, several nanomaterials-assisted techniques are used for the treatment of polluted aquatic systems, which include processes like adsorption, coagulation, chemical precipitation, electrochemical treatment, photocatalysis, and membrane technologies. Among different nanomaterial-assisted techniques for the removal of pollutants, the adsorptive removal of contaminants metal and its oxides, carbon-based complexes, and zeolites have shown the maximum potential. This is due to their properties like more active surface area, having number of functional groups, high chemical stability, high thermal stability, production of non-toxic side-products, cost-effectiveness, and

Nano-Bioremediation for Water and Soil Treatment: An Eco-Friendly Approach.
Vishnu D. Rajput, Arpna Kumari, and Tatiana M. Minkina (Eds.)

low-cost operational unit and simple design. Some of the studied sorbents used in these processes are hydrogel nanocomposite, water-soluble Fe_3O_4 nanoparticles, polyacrylic acid/graphene oxide/Fe_3O_4 (PAA/GO/Fe_3O_4) nanocomposites, ultrafiltration-adsorption membranes (UFAMs) coated with polydopamine (PDA) nanoparticles, amphiphilic nanoparticles with poly(methyl methacrylate) (PMMA) cores, poly(ethyleneimine) (PEI), polyamidoamine (PAMAM) dendrimers, nano-NH_2-MCM-41 nanoparticle, nano-MCM-41 nanoparticle, silicate-chitosan composite, etc., which showed good results in removing various pollutants from water like heavy metal ions (Cd, Cr, Pb, Zn, etc.), organic pollutants like pesticides, PCBs, and various xenobiotics. They also showed effective removal of various harmful microbes like *Escherichia coli*, *Pseudomonas aeruginosa*, and *Bacillus subtilis.* This chapter includes the various nanomaterial-assisted techniques for the removal of various pollutants from contaminated water bodies.

4.1 INTRODUCTION

Water is an essential natural resource that is important to sustain life, ecosystems, and human civilization [1]. Water is a necessity for all living species on this planet, and it is the world's most valuable resource. Unsafe water kills over 2.2 million people every year, as reported by World Health Organization–2019. Water scarcity can be avoided by managing water resources sustainably [2]. Household sewage, industrial effluent, commercial, and agricultural runoff all contribute to water contamination. The explosive growth of sectors such as textiles, mining, fertilizers, insecticides, tanneries, batteries, and paper industries in developed and developing nations is a serious problem, as it adds effluent containing contaminants such as heavy metals, various organic and inorganic particles like pesticide and fertilizers, toxins, drugs, hormones, and other dangerous compounds directly into the waterbodies [3]. The steady growth of the population demands resources, which leads to their overutilization, which results in the addition of various pollutants in waterbodies and alters their physical and chemical composition, texture, and level of toxicity [4]. Industrial effluent is an issue that affects the entire world's environment. Numerous organic contaminants, such as pesticides, heavy metals, and various non-biodegradable compounds, are found at elevated levels. At the moment, effective pollutant removal from industrial wastewater using various prior disposal techniques and effective contaminant removal has become a serious concern [5]. Due to the complexity of the composition of wastewater, its treatment challenges are primarily a complicated process as it demands the use of technologies that are not cost-effective and generally take a long operational time [4]. Advancement in engineering has led to new avenues in the treatment technologies for wastewater, one of which is the removal of pollutants using nanotechnology. Nanotechnology has proven to improve water quality by using techniques like photocatalysis, adsorption of pollutants, coagulation, and membrane processes [6]. Nanotechnology is a cost-effective and efficient method of increasing water quality. In this sector of technology, the industrial applications for making this technology cost effective low,ering energy consumption, and improving quality and environmental controls are expected to increase [7]. Nanomaterials are a combination of common materials with improved properties and newly developed high-performance materials. Furthermore, nanomaterials can be made from any material, for example, metals, polymers, ceramics, semiconductors, nanoengineered materials, and biomaterials. Metal oxide nanoparticles, carbon nanotubes, membranes, and membrane processes are some of the nanomaterials used in the treatment of water. Metal nanoparticles are used for adsorbing various heavy metals like such as lead (Pb), arsenic (As), cadmium (Cd), mercury (Hg), copper (Cu), chromium (Cr), and nickel (Ni) and have demonstrated a remarkable ability to overcome usage of activated carbon [8]. Carbon nanotubes (CNTs) have shown a superior adsorption capacity due to their exceptionally large surface area and multiple interactions with various organic pollutants [9]. Pb, Cr, Cd, Ni, Cu, Fe, and Zn are efficiently removed with the help of CNTs [10]. Treatment of contaminated water includes getting rid of the things that don't belong in the water. Membranes that stop these things from getting into the water are automated and use very little space and reagents to prevent water contamination [11]. Different types of hydrophilic metal oxide nanoparticles, like Al_2O_3, TiO_2, nano-Ag, CNTs, photocatalytic nanomaterials, and zeolite, are used in the decontamination of water bodies. Adding nanoparticles of metal oxides like zeolite and TiO_2 to polymeric ultrafiltration membranes makes their surface more hydrophilic in nature and also increases the membrane permeability, antifouling properties, and also the amount of water that can pass through it [12].

4.2 NANOMATERIAL-BASED REMEDIATION TECHNIQUES FOR VARIOUS CONTAMINANTS' REMOVAL FROM WATER RESOURCES (FIGURE 4.1)

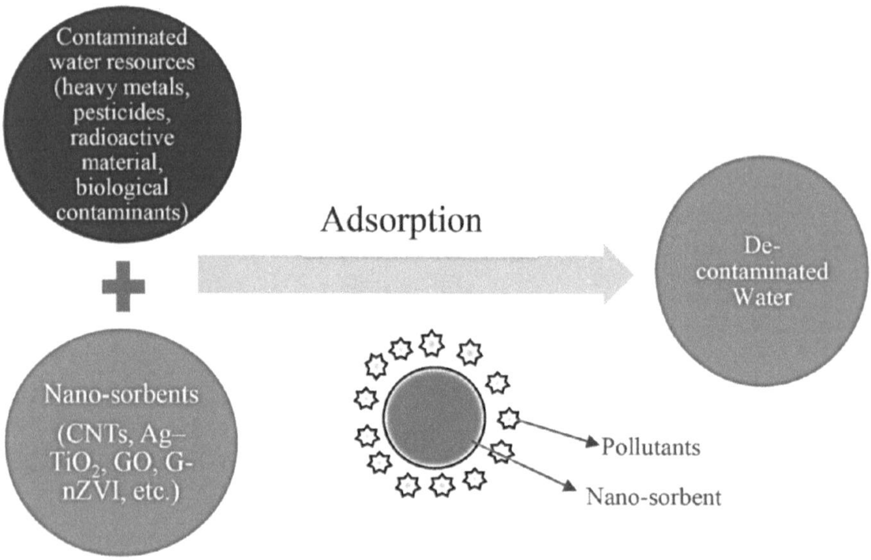

FIGURE 4.1 A general illustration of the interaction between water pollutants and nano-sorbents.

4.2.1 HEAVY METAL REMOVAL USING NANOMATERIALS

Inorganic pollutants are a major class of pollutants that include trace elements, heavy metals, minerals, and other compounds that are persistently found in the environment. They are non-biodegradable and toxic to humans, flora, and fauna. Chemicals released from the fertilizer industry, textiles, refineries, and pharmaceutical industries contribute majorly to adding these pollutants to water resources. Nanotechnology is a trendy emerging technique that makes the use of nanomaterials to remove inorganic contaminants from water [13].

To remove the heavy metal pollutants from water/wastewater, nano-sorbents such as carbon nanotubes (CNTs), zeolites, and dendrimers have been introduced, which have proved to exhibit outstanding adsorption properties compared to other nano-sorbents. In a few investigations removal of heavy metal ions was found to be effectively done using CNTs composites that contain iron oxide and cerium oxide [14]:

1. **Adsorption of Anionic Pollutants:** Anionic contaminants such as fluoride (F^-), arsenic (As), and chromium (Cr) are found abundantly in wastewater. The primary species of fluoride that is found in an aqueous solution is F^-. Surface water is dominated by As(V), whereas groundwater systems are dominated by As(III). Arsenate (As(V)) and arsenite (As(III)), inorganic arsenic species, are more hazardous than organic arsenic species [18, 19]. Arsenic can be adsorbed by CeO_2 nanoparticles on carbon nanotubes. CNTs have highly accessible adsorption sites and a low intraparticle diffusion distance. Metal nanoparticles show better results than activated carbon in the removal of heavy metal pollutants [15]. TiO_2 nanoparticles and nanosized magnetite are commonly used for arsenic adsorption [16, 17]. It is found that the maximum efficiency for the removal of arsenic is seen at neutral pH by nanocrystalline TiO_2 as compared to that of commercially available TiO_2 nanoparticles. In nature, chromium is found to exist in two oxidation states, i.e., Cr(VI) and Cr(III). Cr(VI) is more hazardous than Cr(III). The dominant Cr(VI) species is $HCrO^{4-}$ at low pH, although it transforms to $Cr_2O_7^{2-}$ or CrO_4^{2-} at higher pH [20]. In sunlight, a nanocomposite of graphene sheet embedded with TiO_2 and palladium nanoparticles can be used to reduce Cr(VI) to Cr(III).
2. **Adsorption of Cationic Pollutants:** Cationic pollutants include ions of heavy metals Pb, Hg, Cd, Cu, Ni, Co, and Zn, etc. Metal ion adsorption is based on three mechanisms, i.e., formation of complexes, electrostatic interaction, and ion exchange [21]. The protruding oxygen groups attached with graphene oxide (GO) sheets can be used to bind ions electrostatically using an ion exchange or coordinate approach. Adsorption of cationic contaminants is increased as the pH rises. In low pH conditions, proton and metal ion competition on –COO– and –O– sites reduce adsorption capacity. This electrostatic interaction helps adsorb cationic species when the solution's pH is high.

4.2.2 CARBON-BASED NANO-ADSORBENTS

Carbon nanotubes with single and multi-walls are good materials for adsorbing heavy metals like Pb^{2+}, Cd^{2+}, Ni^{2+}, and Zn^{2+} [22]. It is found that even after several cycles of adsorption/desorption, the adsorption capacity of carbon nanotubes remains stable. Graphene and its compound graphene oxide are another key carbon-based nanomaterial that can be used for removing the heavy metals pollutants from water/wastewater. They are widely used for their qualities of good electrical, mechanical, and thermal conductivity [23].

4.2.3 METAL-BASED NANO-ADSORBENTS

For the treatment of contaminated water, nFe_3O_4, cFe_2O_3, and aFe_2O_3 are the most commonly used iron-based nanomaterials. The iron oxide-based nanomaterials could bind heavy metals like Pb^{2+}, Cu^{2+}, $nTiO_2$, nZnO, and nAl_2O_3 are also effective adsorbents for heavy metals and other metallic pollutants [15].

4.2.4 REDUCING NANOMATERIALS

Since the 1980s, nanoscale zero-valent metals have been attracting attention as a form of effective reductant for pollutant removal from water. Nanoscale zero-valent iron (nZVI) has now been effectively used to decontaminate water and wastewater from heavy metals like Cr, Cd, Cu, Ag, and Zn as iron is a strong reductant and has a redox potential of –0.44 V (E0 = –0.44 V) when it reacts with oxidized pollutants in water [24]. The removal efficiency using nZVI of eight metal ions is 36.5%, 71.0%, 92.5%, 97.5%, 99.7%, 99.7%, and 99.8% for Cd^{2+}, Ni^{2+}, Zn^{2+}, Cr^{6+}, Cu^{2+}, Pb^{2+}, and Ag^{+} respectively. Sorption and complex surface are the mechanisms used for the removal of metals like Zn^{2+} and Cd^{2+}, whose standard potential (E_0) is very close to or more negative than that of iron. The removal method for metals whose E_0 is significantly higher than that of iron, such as Cu^{2+}, Ag^{+}, and Hg^{2+}, is mostly reduction. Metals with somewhat higher E_0 than iron, such as Ni^{2+} and Pb^{2+}, can be immobilized through both sorption and reduction at the surface of nZVI.

4.3 ORGANIC CONTAMINANT REMOVAL

Organic contaminants are important water pollutants. Organic contaminants include dyes, pesticides, etc. Industrial, domestic, and agricultural wastes are the major anthropogenic sources of these toxic organic contaminants. Consuming the organic complex material contaminated water can result in serious health problems, including nervous system disorders, hormonal disruptions, and cancer [25]. Various organic pollutants and nanomaterials working as adsorbents are given in Table 4.1.

TABLE 4.1 Different Nanoparticles Efficient in Removing Various Organic Contaminants

Sl. No.	Organic Contaminants	Nanoparticles	References
1.	Phenanthrene/Pentachlorophenol	Natural soil nanoparticle (Inceptisol-nanoparticle, Oxisol-nanoparticle, Ultisol-nanoparticle)	[67]
2.	PCB (tri and tetrachlorobiphenyls)	nZVI combined with thermal desorption at 300–600°C	[68]
3.	Polychlorinated biphenyls (PCBs)	nZVI combined with anaerobic composting.	[69]
4.	Tetrachlorophenol (TCP), 2,4-diclorophenol (2,4-DCP)	nZVI	[70]
5.	Benzene, nitrates	TiO_2	[71]
6.	Pesticides, pharmaceuticals	Carbon nanotubes	[72]
7.	Trichloroethene	PVP-stabilized nZVI	[73]
8.	Ibuprofen (anti-inflammatory drug)	nZVI 15–45 nm	[74]
9.	Chlorinated hydrocarbons	CeO_2 NPs/activated carbon composite.	[75]
10.	Rhodamine B, tetracycline hydrochloride	Carbon quantum dots attached on graphene-C_3N_4 nanosheets.	[76]
11.	Reactive Black 5	Graphite oxide/magnetic chitosan.	[77]
12.	P-nitrophenol, diethyl phthalate, and ciprofloxacin.	Graphene quantum dots/Mn–N–TiO_2/g-C_3N_4, Graphene quantum dots /TCN.	[78]
13.	Polychlorinated biphenyls	G-nZVI	[79]
14.	Bisphenol A	N–F–TiO_2	[80]
15.	Degradation of γ-HCH	Pd/FeO bimetallic NPs	[81]

4.3.1 PESTICIDES REMOVAL

Leaching of various organic pollutants like pesticides, chemical fertilizers, etc., used on forest and agricultural land is one of the major causes of organic pollution in various water resources. Because of their toxic nature and potential to cause mutation, all pesticides and fertilizers are harmful to living beings. Various nanomaterial-based techniques can be utilized to reduce them from water bodies:

1. **Photocatalysts:** Heterogeneous photocatalysis uses light energy and semiconductors to clean up organic pollutants. Any wavelength of light that strikes a semiconductor creates positive holes by moving valence electrons into the conduction band. Organic pollutants are converted to harmless compounds by the arranged movement of electron-hole pairs toward semiconductor surfaces. Ag, TiO_2, and ZnO_2 are widely used as photocatalysts for pesticide decomposition [26]. Burbano et al. [27] evaluated the ZnO catalyst for photodecomposition of 2,4-D AMINA 4® (dimethylamine salt of 2,4-dichlorophenoxyacetic acid). After 8 hours of treatment, 38.1% photocatalytic mineralization was achieved, whereas photolytic degradation occurs slowly without the catalyst (6% only).
2. **Nanocomposites:** These are created by adding nanofillers to improve their optical, mechanical, magnetic, catalytic, and thermal characteristics. Nanocomposites exhibit significant changes in characteristics depending on the materials used. Titanium dioxide, also known as titania (TiO_2) based materials supported on zeolitic matrices, were produced *in situ* with the target of degrading DDVP (Dichlorvos). Around 360 minutes of interaction with TiO_2/Zeolite with DDVP (20%) results in its full breakdown [27].
3. **Nano-Bio-Composites:** Nanofillers are included in the biopolymer matrix of nano-bio-composites. This alteration imparts specific features to a variety of ancient and contemporary applications. Because biopolymers are hydrophilic, they may be used to adsorb pollutants from water bodies. These are contemporary efforts to replace non-biodegradable conventional polymers with biodegradable alternatives. One of the potentially known nano-bio-composites is copper-coated chitosan nanocomposite (CuCH), which was used in a study that removed malathion by creating a dithionate combination with it. It also effectively removes organophosphate insecticides [27].

4.4 XENOBIOTICS AND PERSISTENT ORGANIC POLLUTANTS (POPS) REMOVAL

Xenobiotics are a large group of man-made chemicals, which, when reached a toxic concentration, have the potential to cause an environmental hazard. The majority of xenobiotics are produced as volatilized pollutants that are released from industrial emissions, which then settle on soil and are later added to water bodies through precipitation and runoff. POPs are xenobiotics containing a halogen to which humans are frequently exposed *via* various point and nonpoint sources. These pollutants are highly bio-accumulative and can alter the microbiome's food chain [94]. The various types of nanomaterials used in the treatment of xenobiotics are described in subsections.

4.4.1 *NANO-ADSORBENTS*

Nano-adsorbents made of carbon commonly interact with pollutants *via* hydrophobic, electrostatic, hydrogen bonding, and covalent bonding. Because of flexibility and physical chemistry, each form contains many adsorption sites that can absorb organic contaminants. Modification of the surface chemistry of single-walled and multi-walled carbon nanotubes has recently generated high-energy sites [29]. Magnetic nanoparticles effectively remove harmful chemicals from suitable sorption sites as porosity rises. When the surface of carbon dots is altered with magnetic carbon nanotubes, they show a high adsorption rate with great potential for reuse [30]. CNTs that have negatively charged surfaces are used to remove cationic pollution by sol-gel methods [31].

4.4.2 *NANO-FILTERS*

Nano-filters reduce the rigidity of organic pollutants by lowering the ionic strength of the solution. Polyvinyl fluoride cellulose acetate (PFCA), polypropylene, polyacrylonitrile, and other natural and manmade polymers are utilized to make nanoporous membranes [31]. This process is based on a pressure-driven approach for eliminating components with molecular weights between 10 nm and 1 nm. It filters components using hydrodynamics on the surface of the membrane and nanopores. Filtration efficiency is largely dependent on filter membrane charge, porous nature, and surface concentration polarization [32]. Nanofibers effectively reject monovalent, bivalent, and polyvalent ions, making them ideal for eliminating arsenic from drinking water. Nano-filtration is efficient in removing many dissolved solids with minimal energy and operating pressure.

4.4.3 NANOFIBERS

Various natural and man-made polymers are utilized to create nanopore membranes, including polyvinyl fluoride, polypropylene, and polyacrylonitrile. Nanofibers, unlike tubes and particles, exhibit stable adsorption structures. Nanofibers of titanium dioxide and graphene polymers, when combined, degrades dye by photocatalysis. Pharmacological antibiotics (ciprofloxin and bisphenol) are adsorbed by using electrical spinning processes and produce carbon nanofibers with reduced molecular dimensions. Cellulose extracted from bacterial cells is used to make ultra-thin biofilms with a coating of cyclodextrin and makes it efficient for cleaning water. This film excellently adsorbs POPs like phenol, bisphenol A (BPA), and glyphosate (2,4-DCP) from water. The product was highly reusable and showed optimal adsorption across a wide pH range [33].

4.4.4 NANOCOMPOSITES

Nanocomposites are organic matrices that strongly influence contaminant adsorption [34]. The new hybrid matrices combine electrical, mechanical, and magnet properties to efficiently sequester contaminants and release the toxic load. Because of their recycling ability, nanocomposites are attractive pollutant removal materials [35]. Techniques like co-precipitation, hydrothermal thermal deposition synthesis, sol-gel synthesis, microwave synthesis, chemical vapor deposition, surface modification, and energy-efficient ball milling can produce nanocomposites [94]. Although xenobiotic removal using nanocomposite has many benefits, its stability is still a concern. The possibility of leaching of metal ions from nanocomposites limits its commercial production as it can cause environmental instability and increase production costs [36–40]:

1. **Graphene-based Nanocomposites:** Graphene is an allotrope of carbon that is used for the removal of inorganic contaminants like oil, hydrocarbons, etc. It has many structural and adsorption functions. Graphene nanocomposites are used as sorption materials to remove aromatic pollutants [41]. Graphene oxide-functionalized iron is more efficient than activated carbon at adsorbing fulvic acid. Amine with graphene functionality nanocomposites remove phenolic substances [42]. Hydrophobic interactions of thiourea with tetracycline molecules result in a higher adsorption capacity of thiourea than magnetite. Zhang et al. [43] found that the adsorption of pesticide ametryn by cellulose-graphene oxide was seven times lower than the $Fe_3O_4^-$ graphene oxide when the sample pH was

maintained roughly at 8. It concluded the effectiveness of magnetite nanocomposites over graphene oxide nanocomposites [15].

2. **Magnetic Nanocomposites:** Iron (III) oxide glazed Bi_2O_4 nanorods are successfully utilized for the photocatalytic destruction of the persistent organic contaminant ibuprofen. After recycling magnetically, nanocomposites can be reused five times more efficiently than conventional nanocomposite materials [44]. The photocatalysis by nanocomposites degrades pesticide residues after the UV treatment that are present in the environment. The adsorption efficiency of co-polymers of polycarrageenine, polyvinyl acetate, and Fe_3O_4 composite was found to adsorb crystal violet dye, and the adsorption capacity increased with an increase in carrageenin content [45].

4.5 POLYCHLORINATED BIPHENYL REMOVAL

PCBs are synthetically made aromatic compounds that have a general formula $C_{12}H_{10}$-xClx. It is comprised of two benzene rings connected at the C-1 carbon. There are a total of 209 different structural arrangements that differ with change in position and number of chlorine molecule, and it exhibits unique chemical properties [46]. These pollutants are synthetic organic compounds resistant to environmental degradation through chemical, biological, and photolytic processes due to which they persist in the environment, are transported through long-range distances, and reach remote areas where they have never been used and get bioaccumulated in human and animal tissue, and biomagnify in food chains [47, 48].

4.5.1 NANO-REMEDIATION

Zero valent iron nanoparticles (nZVI) are promising compounds for the degradation of PCBs in aqueous solutions as they can promote reductive de-chlorination [49, 50]. Under ambient conditions, nZVI dechlorinates PCBs to form lower-chlorinated byproducts [51]. nZVI and bimetallic nanoparticles are effective in removing PCBs, especially for highly chlorinated biphenyls that are biodegradable [52, 53]. For the application of nZVI, they can be directly injected into contaminated aquifers [49].

4.6 BIOLOGICAL CONTAMINANTS REMOVAL: BIOLOGICAL CONTAMINANTS

This includes human pathogens and free-living microbes, natural organic matter, biological toxins, etc. Many adsorbents have good removal efficiencies against these biological contaminants, which are influenced by various factors. Disinfecting aquatic disease-causing bacteria from water bodies is possible with nanomaterials like Ag, Ti, and Zn. Antimicrobial nanomaterials include TiO_2 photocatalysts, metals, and metal-oxides due to their charge capacity [13]. Different types of biological contaminants and their potential nano-sorbents are given in Table 4.2.

TABLE 4.2 The Nanoparticles Efficient in Removal of Different Bio-Contaminants

Sl. No.	Nanoparticles	Type of Bio-Pollutant	References
1.	Ag	*Escherichia coli, Pseudomonas aeruginosa, Staphylococcus aureus*	[82]
2.	TiO_2	Influenza A virus (IAV)	[83]
3.	Si–NH_2 Nanocomposites	Influenza A virus (IAV)	[84]
4.	Nano-Ag, nano-cobalt ferrite, silver coated cobalt ferrite, silver ferrite	*Escherichia coli, Pseudomonas aeruginosa, Staphylococcus aureus, Bacillus subtilis*	[85]
5.	Graphene manganese ferrite ($MnFe_2O_4$-G) nanocomposites	*Escherichia coli*	[86]
6.	Si–NH_2 nanocomposites	Influenza A virus (IAV)	[84]
7.	ZnO	*Campylobacter jejuni*	[87]
8.	TiO_2, SiO_2, and ZnO	*Escherichia coli, Bacillus subtilis*	[88]
9.	Ag–TiO_2	*Escherichia coli*	[89]
10.	Ag–Au nanoparticles	Influenza A virus (IAV)	[90]

1. **Silver Nanoparticles:** Silver inactivates bacterial and algal growth in water, making it the most extensively utilized material. Silver nanoparticles are reported biocides that are made from various salts of silver, like silver nitrate and silver chloride. The effectiveness of their antibacterial activity is dependent on their size, most effective Ag nanoparticles are of size about 8 nm, on the other hand, bigger-sized nanoparticles (11–23 nm) have shown reduced bactericidal activity [53]. Furthermore, the antibacterial activity of nanoparticles is affected by their shape, as short, trimmed triangular silver nanomaterials show better bactericidal effects than nanoparticles

that are spherical or rod in shape [54]. Ag nanoparticles cause damage to bacterial membranes and DNA interactions, modifying cell membrane characteristics by adhering to the surface of the cell and causing enzyme destruction [55]. Nanofibers that are made by embedding Ag nanoparticles in cellulose acetate fibers work effectively against both gram-positive and gram-negative bacteria [56, 57]. Water filters made of polyurethane foam containing Ag nanofibers and polysulfone membranes are antibacterial against *E. coli, Pseudomonas*, etc. [58, 59].

2. **TiO_2 Nanoparticles:** These are a promising new photocatalyst for the decontamination of water resources as they are low cost and produce highly reactive hydroxyl (-OH) radicals, which help in the disinfection of various microbes like bacteria, fungi, algae, viruses, and so on [60]. After high solar exposure for about 8 hrs., TiO_2 reduces the survival of various pathogens present in water, such as protozoa, fungi, and *E. coli*. The association of Ag and TiO_2 nanoparticles increases the bacterial inactivation capacity by decreasing the time required to inactivate *E. coli* and increasing disinfection under UV and solar light [61]. Transition metal oxide nanostructures were added to TiO_2. They have promising potential for disinfection of water by using photocatalytic oxidation to disinfect *E. coli*, *P. aeruginosa*, and *B. subtilis* spores under visible light, researchers have improved disinfection efficacy against these bacteria. Due to the stability of TiO_2 in water, it can be used in membrane filters to filter water [62].

3. **Carbon Nanotubes (CNTs):** These have shown great success in killing bacterial pathogens from water. CNTs are antimicrobial against bacteria like *E. coli*, *Salmonella*, and viruses [63]. The adsorbing capacity of CNTs is higher than carbon-based adsorbents for cyanobacterial toxins. This is due to the large surface area and wide external diameter. Multiwalled CNTs are much more harmful to microorganisms than single-walled CNTs. Dispersion of CNTs more important parameter as compared to length. Many researchers found that single-walled CNTs are low-cost and have high bacterial adsorption rates and sorption capacities [64]. A nanocomposite membrane made of graphene nanotubes and polyvinyl-N-carbazole is effective in inactivating the bacteria [65].

4.7 RADIOACTIVE POLLUTANT REMOVAL

Radioactive elements produce radiological pollutants. Ore processing and mining are the possible sources of radioactive pollutants like 32P, 131I, 60Co, 35S, 45Ca, 14C, etc., that are used in research, agriculture, industrial activities, and medicine. A nuclear reactor, power plant radiation, and nuclear weapon testing and deployment are all potential sources of radioactive pollution contamination of the environment by harmful radioactive elements such as 90Sr, 238U, 137Cs, 248Pu, and 235U. These isotopes are harmful to humans because they cause major health problems by accumulating in the bones and teeth and, thus, increasing the risk of cancer. Nanoparticles have been seen to be a good adsorbent or ion exchange for selectively extracting radioactive species from contaminated water resources.

1. **Synthetic Nanoparticles:** Carbon nanotubes are used to quickly remove Uranium (U) (VI) through the exchange of Europium (Eu) (III) and adsorption-mediated surface complexation and Thorium (Th) (IV), while they are also utilized to remove Americium (Am) (III) [92]. Toxic and long-lived radionuclides (Am(III), Th(IV), Plutonium (Pu) (IV), U(VI), Strontium (Sr) (II), Technetium (Tc) (VII), and Neptunium (Np) (V)) are swiftly removed from polluted water using graphene oxide synthesized using the modified Hummers' method [66]. nZVI has been widely used as an effective adsorbent for removing various radionuclides like U(VI) and Co(II) from wastewaters. Radioactive Zn(II) could be removed by nZVI-NH_2 modified with 3-aminopropyl trimethoxysilane. Zn(II) and Co(II) can be removed effectively by nZVI coated with diethylenetriamine/2 pyridinecarboxaldehyde (nZVI-DETAPY) [66]. Many natural materials (e.g., clays, zeolites, birnessite, etc.) can exchange ions and adsorb radioactive nuclides when manufactured to the nanoscale [91]. Magnetic nano-sized zeolite A removed 95.2% Sr(II) and 81.4% Cs(I) from 0.01 N aqueous solutions, while magnetic nano-sized zeolite Y removed 91% Sr(II) at pH 8. A triple-layered Fe-amino clay (FeAC) composite made by layering FeAC/carboxymethyl cellulose (CMC)/polyhedral oligomeric silsesquioxane (POSS) (FeAC/CMC/POSS) was used as an effective adsorbent for extracting Cs(I) from contaminated water [92]. It is found that ultrafine CNFs could effectively remove radioactive UO22+ ions from water. The ultrafine CNFs were

made by oxidizing wood pulp with TEMPO/NaBr/NaClO and then mechanically disintegrating it [66].

2. **Biogenic Nanocomposites:** Plant, bacteria, fungi, yeast, actinomycetes, and algae cellular extracts were utilized as green reagents to make low-cost and non-toxic bio-nanomaterials. The use of these bio-nanocomposites has shown good results in decontaminating the water contaminated by radioactive compounds. Adsorption tests showed the saturated adsorption capacity of bio-nanocomposite to be 120 mg/g for B(II) ions. Researchers used the bacterium *D. radiodurans which is resistant to radiation*, to create bio-Au-NPs for the removal of radioactive I from radioactive effluent. Handley-Sidhu et al. used *Serratia* sp. NCIMB 40259 strain to make bio-HAP for adsorbing the Sr(II) from simulated groundwater [93].

4.8 CONCLUSION AND FUTURE PERSPECTIVES

This chapter discusses briefly about the use of nanotechnology for the remediation of water contaminated with various pollutants, including suspended particles, heavy metals, pesticides, pathogenic microbes, radioactive materials, and various organic materials like POPs, PCBs, and other emerging micropollutants. Among various nanomaterials, nanosorbents such as carbon nanotubes (CNTs), hydrated aluminum silicate, and dendrimers have shown promising results in removing the various micropollutants from the water body due to their low-cost effectiveness and simple design having large active surface area and a large number of functional groups. The key issues with the usage of nanoparticles that need attention are their high cost, production scalability, structural stability, and toxicity. Industrial production of nanoparticles with effective size and functionality is a complicated and multistep procedure that requires high facilities and reagent-grade chemicals. This increases their production cost, thereby increasing the cost of water purification through nanomaterials-based techniques. Since nanomaterials are less stable and due to their leaching into the water bodies, they are directly polluting them. This harms the health of humans and fauna by entering into the food chain. In the future, the removal of nanomaterials used for cleaning water resources could be a potential cause of concern.

ACKNOWLEDGEMENTS

The authors are thankful to the department of botanical and Environmental Sciences, GND University, Amritsar, for the necessary facilities. The authors would also like to acknowledge R.U.S.A. 2.0 for providing financial assistance.

KEYWORDS

- **decontamination**
- **heavy metals**
- **nanomaterials**
- **organic pollutants**
- **radioactive pollutants**
- **water resources**

REFERENCES

1. Yang, D., Yang, Y., & Xia, J., (2021). Hydrological cycle and water resources in a changing world: A review. *Geography and Sustainability, 2*(2), 115–122. https://doi.org/10.1016/j.geosus.2021.05.003.
2. Gittins, J. R., Hemingway, J. R., & Dajka, J. C., (2021). How a water-resources crisis highlights social-ecological disconnects. *Water Research, 194*, 116937. https://doi.org/10.1016/j.watres.2021.116937.
3. Sarma, G. K., Sen, G. S., & Bhattacharyya, K. G. (2019). Nanomaterials as versatile adsorbents for heavy metal ions in water: A review. *Environmental Science and Pollution Research, 26*(7), 6245–6278.
4. Hlongwane, G. N., Sekoai, P. T., Meyyappan, M., & Moothi, K., (2019). Simultaneous removal of pollutants from water using nanoparticles: A shift from single pollutant control to multiple pollutant control. *Science of the Total Environment, 656*, 808–833. https://doi.org/10.1016/j.scitotenv.2018.11.257.
5. Abdelbasir, S. M., & Shalan, A. E., (2019). An overview of nanomaterials for industrial wastewater treatment. *Korean Journal of Chemical Engineering, 36*(8), 1209–1225. https://doi.org/10.1007/s11814-019-0306-y.
6. Khan, S., Naushad, M., Al-Gheethi, A., & Iqbal, J., (2021). Engineered nanoparticles for removal of pollutants from wastewater: Current status and future prospects of nanotechnology for remediation strategies. *Journal of Environmental Chemical Engineering, 9*(5), 106160. https://doi.org/10.1016/j.jece.2021.106160.

7. Abdulhameed, A. S., Jawad, A. H., & Mohammad, A. K. T. (2019). Synthesis of chitosan-ethylene glycol diglycidyl ether/TiO_2 nanoparticles for adsorption of reactive orange 16 dye using a response surface methodology approach. *Bioresource Technology, 293*, 122071. https://doi.org/10.1016/j.biortech.2019.122071.
8. Krishnaraj, C., Ramachandran, R., Mohan, K., & Kalaichelvan, P. T., (2012). Optimization for rapid synthesis of silver nanoparticles and its effect on phytopathogenic fungi. *Spectrochimica Acta – Part A: Molecular and Biomolecular Spectroscopy, 93*, 95–99. https://doi.org/10.1016/j.saa.2012.03.002.
9. Chen, W., Duan, L., & Zhu, D., (2007). Adsorption of polar and nonpolar organic chemicals to carbon nanotubes. *Environmental Science & Technology, 41*(24), 8295–8300.
10. Bankole, M. T., Abdulkareem, A. S., Mohammed, I. A., Ochigbo, S. S., Tijani, J. O., Abubakre, O. K., & Roos, W. D., (2019). Selected heavy metals removal from electroplating wastewater by purified and polyhydroxylbutyrate functionalized carbon nanotubes adsorbents. *Scientific Reports, 9*(1), 1–19. https://doi.org/10.1038/s41598-018-37899-4.
11. Qadri, S., Ganoe, A., & Haik, Y., (2009). Removal and recovery of acridine orange from solutions by use of magnetic nanoparticles. *Journal of Hazardous Materials, 169*(1–3), 318–323.
12. Zahid, M., Rashid, A., Akram, S., Rehan, Z. A., & Razzaq, W. (2018). A comprehensive review on polymeric nano-composite membranes for water treatment. *Journal of Membrane Science and Technology, 8*(2), 1–20.
13. Amin, M. T., Alazba, A. A., & Manzoor, U., (2014). A review of removal of pollutants from water/wastewater using different types of nanomaterials. *Advances in Materials Science and Engineering, 2014*.
14. Lu, C., Chung, Y. L., & Chang, K. F., (2005). Adsorption of trihalomethanes from water with carbon nanotubes. *Water Research, 39*(6), 1183–1189.
15. Sharma, Y. C., Srivastava, V., Singh, V. K., Kaul, S. N., & Weng, C. H., (2009). Nano-adsorbents for the removal of metallic pollutants from water and wastewater. *Environ. Technol., 30*, 583–609.
16. Deliyanni, E. A., Bakoyannakis, D. N., Zouboulis, A. I., & Matis, K. A., (2003). Sorption of As (V) ions by akaganeite-type nanocrystals. *Chemosphere, 50*, 155–163.
17. Mayo, J. T., Yavuz, C., Yean, S., Cong, L., Shipley, H., Yu, W., & Colvin, V. L., (2007). The effect of nanocrystalline magnetite size on arsenic removal. *Science and Technology of Advanced Materials, 8*(1, 2), 71.
18. Cullen, W. R., & Reimer, K. J., (1989). Arsenic speciation in the environment. *Chemical Reviews, 89*, 713–764.
19. Zhu, H., Jia, Y., Wu, X., & Wang, H., (2009). Removal of arsenic from water by supported nano zero-valent iron on activated carbon. *Journal of Hazardous Material, 172*(2), 1591–1596.
20. Liu, S. S., Chen, Y. Z., De, Z. L., Hua, G. M., Xu, W., Li, N., & Zhang, Y., (2011). Enhanced removal of trace Cr(VI) ions from aqueous solution by titanium oxide–Ag composite adsorbents. *Journal of Hazardous Material, 190*(1), 723–728.
21. Moreno-Castilla, C., Alvarez-Merino, M. A., Pastrana-Martı´nez, L. M., & Lo´pez-Ramo´n, M. V., (2010). Adsorption mechanisms of metal cations from water on an oxidized carbon surface. *Journal of Colloid Interface Science, 345*(2), 461–466.

22. Su, F., & Lu, C., (2007). Adsorption kinetics, thermodynamics, and desorption of natural dissolved organic matter by multi-walled carbon nanotubes. *Journal of Environmental Science and Health, 42*, 1543–1552.
23. Avouris, P., & Dimitrakopoulos, C., (2012). Graphene: Synthesis and applications. *Materials Today, 15*(3), 86–97.
24. Scott, T. B., Popescu, I. C., Crane, R. A., & Noubactep, C., (2011). Nano-scale metallic iron for the treatment of solutions containing multiple inorganic contaminants. *Journal of Hazardous Materials, 186*(1), 280–287.
25. Li, X., & Zhang, W., (2007). Sequestration of metal cations with zerovalent iron nanoparticles a study with high resolution x-ray photoelectron spectroscopy (HR-XPS). *Journal of Physical Chemistry C, 111*, 6939–6946.
26. Harvey, R. W., Smith, R. L., & George, L., (1984). Effect of organic contamination upon microbial distributions and heterotrophic uptake in a Cape Cod, Mass., aquifer. *Applied and Environmental Microbiology, 48*(6), 1197–1202.
27. Burbano, J., Cruz, I., Colina–Márquez, J., López– Vásquez, A., & Machuca, F., (2008). Evaluation of zinc oxide-based photocatalytic degradation of a commercial pesticide. *Journal of Advanced Oxidation Technologies, 11*(1), 49–55.
28. Jaiswal, M., Chauhan, D., & Sankararamakrishnan, N., (2012). Copper chitosan nanocomposite: Synthesis, characterization, and application in removal of organophosphorus pesticide from agricultural runoff. *Environmental Science and Pollution Research, 19*(6), 2055–2062.
29. Rushi, A. D., (2019). In: Datta, K. P., (ed.), *Functionalized Carbon Nanotubes for Detection of Volatile Organic Pollutant.* IntechOpen, Rijeka, p. Ch. 6. doi: 10.5772/intechopen.85045.
30. Deng, Y., Ok, Y. S., Mohan, D., Pittman, Jr. C. U., & Dou, X., (2019). Carbamazepine removal from water by carbon dot-modified magnetic carbon nanotubes. *Environmental Research, 169*, 434–444.
31. Konicki, W., & Pełech, I., (2019). Removing cationic dye from aqueous solutions using as grown and modified multi-walled carbon nanotubes. *Polish Journal of Environmental Studies, 28*(2), 717–727.
32. Abdel-Fatah, M. A., (2018). Nanofiltration systems and applications in wastewater treatment: Review article. *Ain Shams Engineering Journal, 9*(4), 3077–3092.
33. Khalil, A. M., & Schäfer, A. I., (2021). Cross-linked β-cyclodextrin nanofiber composite membrane for steroid hormone micropollutant removal from water. *Journal of Membrane Science, 618*, 118228.
34. Srivastava, V., Zare, E. N., Makvandi, P., Xuan-Qi, Z., Iftekhar, S., Wu, A., Padil, V. V. T., et al., (2020). Cytotoxic aquatic pollutants and their removal by nanocomposite-based sorbents. *Chemosphere, 258*, 127324.
35. Guerra-Rodríguez, S., Rodríguez, E., Singh, D. N., & Rodríguez-Chueca, J., (2018). Assessment of sulfate radical-based advanced oxidation processes for water and wastewater treatment: A review. *Water, 10.*
36. Sahu, S., Sahu, U. K., & Patel, R. K., (2018). Synthesis of thorium–ethanolamine nanocomposite by the coprecipitation method and its application for Cr(vi) removal. *New Journal of Chemistry, 42*, 5556–5569.
37. Jaramillo-Fierro, X., P′erez, S. G., Jaramillo, H. A., & Cabello, F. M., (2020). Synthesis of the zntio3 /tio2 nanocomposite supported in Ecuadorian clays for the adsorption and photocatalytic removal of methylene blue dye. *Nanomaterials, 10*, 1–20.

38. Omo-Okoro, P. N., Maepa, C. E., Daso, A. P., & Okonkwo, J. O., (2020). Microwave-assisted synthesis and characterization of an agriculturally derived silver nanocomposite and its derivatives. *Waste Biomass Valor*, *11*(5), 2247–2259.
39. Kumar, D., Pandey, J., Raj, V., & Kumar, P., (2017). A review of the modification of polysaccharide through graft copolymerization for various potential applications. *The Open Medicinal Chemistry Journal, 11*(1), 109–126.
40. Wei, X., Wang, X., Gao, B., Zou, W., & Dong, L., (2020). Facile ball-milling synthesis of CuO/biochar nanocomposites for efficient removal of reactive red 120. *ACS Omega, 5*(11), 5748–5755.
41. Zhou, Y., Cao, S., Xi, C., Li, X., Zhang, L., Wang, G., & Chen, Z. (2019). A novel Fe_3O_4/graphene oxide/citrus peel-derived bio-char based nanocomposite with enhanced adsorption affinity and sensitivity of ciprofloxacin and sparfloxacin. *Bioresources Technology, 292*, 121951.
42. Jebaranjitham, J. N., Mageshwari, C., Saravanan, R., & Mu, N. (2019). Fabrication of amine functionalized graphene oxide–AgNPs nanocomposite with improved dispersibility for reduction of 4-nitrophenol. *Composites Part B: Engineering, 171*, 302–309.
43. Zhang, Q., Teng, J., Zhang, Z., Nie, G., Zhao, H., Peng, Q., & Jiao, T., (2015). Unique and outstanding cadmium sequestration by polystyrene-supported nanosized zirconium hydroxides: A case study. *RSC Advances*, *5*(68), 55445–55452. doi: 10.1039/C5RA09628F.
44. Xia, D., & Lo, I. M., (2016). Synthesis of magnetically separable Bi_2O_4/Fe_3O_4 hybrid nanocomposites with enhanced photocatalytic removal of ibuprofen under visible light irradiation. *Water Research*, *100*, 393–404.
45. Mansoori, Y., Khodayari, A., Banaei, A., Mirzaeinejad, M., Azizian-Kalandaragh, Y., & Pooresmaeil, M., (2016). Fe_3O_4–PVAc nanocomposites: Surface modification of sonochemically prepared magnetite nanoparticles via chemical grafting of poly (vinyl acetate). *RSC Advances*, *6*(54), 48676–48683.
46. Maervoet, J., Covaci, A., Schepens, P., Sandau, C. D., & Letcher, R. J., (2004). A reassessment of the nomenclature of polychlorinated biphenyl (PCB) metabolites. *Environmental Health Perspectives*, *112*(3), 291–294.
47. Koblizkova, M., Ruzickova, P., Cupr, P., Komprda, J., Holoubek, I., & Klánová, J., (2009). Soil burdens of persistent organic pollutants: Their levels, fate, and risks. Part IV. Quantification of volatilization fluxes of organochlorine pesticides and polychlorinated biphenyls from contaminated soil surfaces. *Environmental Science & Technology, 43*(10), 3588–3595.
48. Donaldson, S. G., Van, O. J., Tikhonov, C., Feeley, M., Armstrong, B., Ayotte, P., Boucher, O., et al., (2010). Environmental contaminants and human health in the Canadian arctic. *Science of the Total Environment*, *408*(22), 5165–5234.
49. Wang, C. B., & Zhang, W. X., (1997). Synthesizing nanoscale iron particles for rapid and complete dechlorination of TCE and PCBs. *Environmental Science & Technology, 31*(7), 2154–2156.
50. Lowry, G. V., & Johnson, K. M., (2004). Congener-specific dechlorination of dissolved PCBs by microscale and nanoscale zerovalent iron in a water/methanol solution. *Environmental Science & Technology*, *38*(19), 5208–5216.
51. Varanasi, P., Fullana, A., & Sidhu, S., (2007). Remediation of PCB-contaminated soils using iron nanoparticles. *Chemosphere, 66*, 1031–1038.

52. Chen, X., Yao, X., Yu, C., Su, X., Shen, C., Chen, C., & Xu, X., (2014). Hydrodechlorination of polychlorinated biphenyls in contaminated soil from an e-waste recycling area, using nanoscale zerovalent iron and Pd/Fe bimetallic nanoparticles. *Environmental Science and Pollution Research, 21*(7), 5201–5210.
53. Morones, J. R., Elechiguerra, J. L., Camacho, A., Holt, K., Kouri, J. B., Ramírez, J. T., & Yacaman, M. J., (2005). The bactericidal effect of silver nanoparticles. *Nanotechnology, 16*(10), 2346.
54. Pal, S., Tak, Y. K., & Song, J. M., (2007). Does the antibacterial activity of silver nanoparticles depend on the shape of the nanoparticle? A study of the gram-negative bacterium *Escherichia coli. Applied and Environmental Microbiology, 73*(6), 1712–1720.
55. Xiu, Z. M., Ma, J., & Alvarez, P. J., (2011). Differential effect of common ligands and molecular oxygen on antimicrobial activity of silver nanoparticles versus silver ions. *Environmental Science & Technology, 45*(20), 9003–9008.
56. Balogh, L., Swanson, D. R., Tomalia, D. A., Hagnauer, G. L., & McManus, A. T., (2001). Dendrimer-silver complexes and nanocomposites as antimicrobial agents. *Nano Letters, 1*(1), 18–21.
57. Botes, M., & Eugene, C. T., (2010). The potential of nanofibers and nanobiocides in water purification. *Critical Reviews in Microbiology, 36*(1), 68–81.
58. Chou, W. L., Yu, D. G., & Yang, M. C., (2005). The preparation and characterization of silver-loading cellulose acetate hollow fiber membrane for water treatment. *Polymers for Advanced Technologies, 16*(8), 600–607.
59. Lee, S. Y., Kim, H. J., Patel, R., Im, S. J., Kim, J. H., & Min, B. R., (2007). Silver nanoparticles immobilized on thin film composite polyamide membrane: Characterization, nanofiltration, antifouling properties. *Polymers for Advanced Technologies, 18*(7), 562–568.
60. Adesina, A. A., (2004). Industrial exploitation of photocatalysis: Progress, perspectives, and prospects. *Catalysis Surveys from Asia, 8*(4), 265–273.
61. Liga, M. V., Bryant, E. L., Colvin, V. L., & Li, Q., (2011). Virus inactivation by silver-doped titanium dioxide nanoparticles for drinking water treatment. *Water Research, 45*(2), 535–544.
62. Kwak, S. Y., Kim, S. H., & Kim, S. S., (2001). Hybrid organic/inorganic reverse osmosis (RO) membrane for bactericidal anti-fouling. 1. Preparation and characterization of TiO_2 nanoparticle self-assembled aromatic polyamide thin-film-composite (TFC) membrane. *Environmental Science & Technology, 35*(11), 2388–2394.
63. Brady-Estévez, A. S., Kang, S., & Elimelech, M., (2008). A single-walled-carbon-nanotube filter for removal of viral and bacterial pathogens. *Small, 4*(4), 481–484.
64. Rahaman, M. S., Vecitis, C. D., & Elimelech, M., (2012). Electrochemical carbon-nanotube filter performance toward virus removal and inactivation in the presence of natural organic matter. *Environmental Science & Technology, 46*(3), 1556–1564.
65. Ahmed, F., Santos, C. M., Vergara, R. A. M. V., Tria, M. C. R., Advincula, R., & Rodrigues, D. F., (2012). Antimicrobial applications of electroactive PVK-SWNT nanocomposites. *Environmental Science & Technology, 46*(3), 1804–1810.
66. Zhang, X., & Liu, Y. (2020). Nanomaterials for radioactive wastewater decontamination. *Environmental Science: Nano, 7*(4), 1008–1040.

67. Liu, F., Xu, B., He, Y., Brookes, P. C., & Xu, J. (2019). Co-transport of phenanthrene and pentachlorophenol by natural soil nanoparticles through saturated sand columns. *Environmental Pollution*, *249*, 406–413.
68. Liu, J., Chen, T., Qi, Z., Yan, J., Buekens, A., & Li, X., (2014). Thermal desorption of PCBs from contaminated soil using nano zerovalent iron. *Environmental Science and Pollution Research*, *21*(22), 12739–12746.
69. Long, Y. Y., Zhang, C., Du, Y., Tao, X. Q., & Shen, D. S., (2014). Enhanced reductive dechlorination of polychlorinated biphenyl-contaminated soil by in-vessel anaerobic composting with zero-valent iron. *Environmental Science and Pollution Research*, *21*(6), 4783–4792.
70. Kadar, E., Dyson, O., Handy, R. D., & Al-Subiai, S. N., (2013). Are reproduction impairments of free-spawning marine invertebrates exposed to zero-valent nano-iron associated with the dissolution of nanoparticles? *Nanotoxicology*, *7*(2), 135–143.
71. Baun, A., Hartmann, N. B., Grieger, K., & Kusk, K. O., (2008). Ecotoxicity of engineered nanoparticles to aquatic invertebrates: A brief review and recommendations for future toxicity testing. *Ecotoxicology*, *17*(5), 387–395.
72. Minetto, D., Ghirardini, A. V., & Libralato, G., (2016). Saltwater ecotoxicology of Ag, Au, CuO, TiO_2, ZnO and C60 engineered nanoparticles: An overview. *Environment International*, *92*, 189–201.
73. Tian, H., Liang, Y., Yang, D., & Sun, Y., (2020). Characteristics of PVP–stabilized NZVI and application to dechlorination of soil–sorbed TCE with ionic surfactant. *Chemosphere, 239*, 124807.
74. Machado, S., Stawiński, W., Slonina, P., Pinto, A. R., Grosso, J. P., Nouws, H. P. A., & Delerue-Matos, C., (2013). Application of green zero-valent iron nanoparticles to the remediation of soils contaminated with ibuprofen. *Science of the Total Environment*, *461*, 323–329.
75. Alhooshani, K. R., (2019). Adsorption of chlorinated organic compounds from water with cerium oxide-activated carbon composite. *Arabian Journal of Chemistry*, *12*(8), 2585–2596.
76. Hong, Y., Meng, Y., Zhang, G., Yin, B., Zhao, Y., Shi, W., & Li, C., (2016). Facile fabrication of stable metal-free CQDs/g-C_3N_4 heterojunctions with efficiently enhanced visible-light photocatalytic activity. *Separation and Purification Technology*, *171*, 229–237.
77. Travlou, N. A., Kyzas, G. Z., Lazaridis, N. K., & Deliyanni, E. A., (2013). Functionalization of graphite oxide with magnetic chitosan for the preparation of a nanocomposite dye adsorbent. *Langmuir*, *29*(5), 1657–1668.
78. Nie, Y. C., Yu, F., Wang, L. C., Xing, Q. J., Liu, X., Pei, Y., & Suib, S. L., (2018). Photocatalytic degradation of organic pollutants coupled with simultaneous photocatalytic H2 evolution over graphene quantum dots/Mn-N-TiO_2/g-C_3N_4 composite catalysts: Performance and mechanism. *Applied Catalysis B: Environmental*, *227*, 312–321.
79. Jabeen, H., Kemp, K. C., & Chandra, V., (2013). Synthesis of nano zerovalent iron nanoparticles–graphene composite for the treatment of lead-contaminated water. *Journal of Environmental Management*, *130*, 429–435.
80. He, F., Zhao, D., & Paul, C., (2010). Field assessment of carboxymethyl cellulose stabilized iron nanoparticles for in situ destruction of chlorinated solvents in source zones. *Water Resources*, *44*, 2360–2370.

81. Singh, R., Misra, V., & Singh, R. P., (2012). Removal of hexavalent chromium from contaminated groundwater using zero-valent iron nanoparticles. *Environmental Monitoring and Assessment, 184*(6), 3643–3651.
82. Pacheco, S., Tapia, J., Medina, M., & Rodriguez, R., (2006). Cadmium ions adsorption in simulated wastewater using structured alumina–silica nanoparticles. *Journal of Non-crystalline Solids, 352*(52–54), 5475–5481.
83. Hao, Y. M., Man, C., & Hu, Z. B., (2010). Effective removal of Cu(II) ions from aqueous solution by amino-functionalized magnetic nanoparticles. *Journal of Hazardous Materials, 184*(1–3), 392–399. https://doi.org/10.1016/j.jhazmat.2010.08.048.
84. Chang, Y. C., Chang, S. W., & Chen, D. H., (2006). Magnetic chitosan nanoparticles: Studies on chitosan binding and adsorption of Co(II) ions. *Reactive and Functional Polymers, 66*(3), 335–341. https://doi.org/10.1016/j.reactfunctpolym.2005.08.006.
85. Salehzadeh, J., (2013). Removal of heavy metals Pb^2, Cu^2, Zn^2, Cd^2, Ni^2, Co^2, and Fe^3 from aqueous solutions by using *Xanthium Pensylvanicum*. *Leonardo Journal of Sciences, 23*, 97.
86. Huang, D. L., Chen, G. M., Zeng, G. M., Xu, P., Yan, M., Lai, C., & He, Y., (2015). Synthesis and application of modified zero-valent iron nanoparticles for removal of hexavalent chromium from wastewater. *Water, Air, & Soil Pollution, 226*(11), 1–14.
87. Dionysiou, D. D., Khodadoust, A. P., Kern, A. M., Suidan, M. T., Baudin, I., & Laîné, J. M., (2000). Continuous-mode photocatalytic degradation of chlorinated phenols and pesticides in water using a bench-scale TiO_2 rotating disk reactor. *Applied Catalysis B: Environmental, 24*(3, 4), 139–155.
88. Aydin, S., (2016). Removal of organophosphorus pesticides from aqueous solution by magnetic Fe_3O_4/red mud-nanoparticles. *Water Environment Research, 88*(12), 2275–2284.
89. Ng, Q. H., Lim, J. K., Ahmad, A. L., Ooi, B. S., & Low, S. C., (2015). Magnetic nanoparticles augmented composite membranes in the removal of organic foulant through magnetic actuation. *Journal of Membrane Science, 493*, 134–146.
90. Awual, M. R., Hasan, M. M., & Khaleque, M. A., (2015). Efficient selenium (IV) detection and removal from water by tailor-made novel conjugate adsorbent. *Sensors and Actuators B: Chemical, 209*, 194–202.
91. Omidi, M. H., Azad, F. N., Ghaedi, M., Asfaram, A., Azqhandi, M. H. A., & Tayebi, L., (2017). Synthesis and characterization of Au-NPs supported on carbon nanotubes: Application for the ultrasound-assisted removal of radioactive UO_2^{2+} ions following complexation with Arsenazo III: Spectrophotometric detection, optimization, isotherm and kinetic study. *Journal of Colloid International Sciences, 504*, 68–77.
92. Mertz, J. L., Fard, Z. H., Malliakas, C. D., Manos, M. J., & Kanatzidis, M. G., (2013). Selective removal of Cs^+, Sr^{2+}, and Ni^{2+} by $K_{2x}Mg_xSn_{3-x}S_6$ (x = 0.5–1) (KMS-2) relevant to nuclear waste remediation. *Chemistry of Materials, 25*(10), 2116–2127.
93. Lee, K. Y., Kim, K. W., Park, M., Kim, J., Oh, M., Lee, E. H., Chung, D. Y., & Moon, J. K., (2016). Novel application of nanozeolite for radioactive cesium removal from high-salt wastewater. *Water Resources, 95*, 134–141.
94. Karthigadevi, G., Manikandan, S., Karmegam, N., Subbaiya, R., Chozhavendhan, S., Ravindran, B., & Awasthi, M. K., (2021). Chemico-nanotreatment methods for the removal of persistent organic pollutants and xenobiotics in water–A review. *Bioresource Technology, 324*, 124678.

CHAPTER 5

Nanomaterials-Assisted Decontamination of Heavy Metal from Water Resources

PRIYA SHRIVASTAVA,[1] RUPESH KUMAR BASNIWAL,[1] ABHISHEK CHAUHAN,[2] ANUJ RANJAN,[3] and V. K. JAIN[1]

[1]*Amity Institute of Advanced Research and Studies (M&D), Amity University, Noida, Uttar Pradesh, India*

[2]*Amity Institute of Environmental Toxicology Safety and Management, Amity University, Noida, Uttar Pradesh, India*

[3]*Academy of Biology and Biotechnology, Southern Federal University, Stachki, Rostov-on-Don, Russia*

ABSTRACT

Heavy metals (HMs) pollution in streams is a huge danger to global water security, as excessive doses of these metallic agents may cause poisoning and can affect all living organisms significantly. These contaminants in wastewater have become a serious global problem. The study was performed, and few have access to unpolluted and potable water. The contamination from the metals HMs, like As^{3+}, Pb^{2+}, Cd^{2+}, Ni^{2+}, Cu^{2+}, and several other metals, are causing water pollution, and all of us are connected via the food chain, which is accumulating into our atmosphere at a great speed, and these metals have the potential to bio-accumulate and cause toxicity. To address this issue, several technologies have been created. Nanotechnology has grown in popularity as an emerging technology, and numerous nanomaterials have been designed to eradicate HMs from polluted wastewater because of their superior properties arising from the nanoscale effect. Conventional procedures are employed

Nano-Bioremediation for Water and Soil Treatment: An Eco-Friendly Approach.
Vishnu D. Rajput, Arpna Kumari, and Tatiana M. Minkina (Eds.)

for wastewater treatment and purification, but they are unsuccessful, and the pollutants are only partially removed. The nanotechnology-based industry has recently developed the strategy of eliminating the HMs from water as an emerging technology. To present, various nanotechnologies for HMs remediation from water bodies have been developed, including Carbon Nanotubes, Nanoscale Metals, oxide, Nanofibers, Nano-zeolites, Nano-Ag, Nano-magnetite, and dye-doped silica nanoparticles. Nanotechnology-based HMs removal from water sources has advantages such as good linearity range, low detection limit, and quantification limits with great sensitivity and selectivity. Field use of nanomaterial-based techniques must be simple and safe. The chapter discusses all the latest technology developed so far based on nanomaterial for the removal of HMs from wastewater.

5.1 INTRODUCTION

In today's time, water scarcity is an issue which is haunting the human civilization throughout the world. Despite the abundance of water, the availability of potable water is declining because of an increase in anthropogenic pollution [1]. Every day, 2 million tons of water is discharged due to residential (77%) and industrial (33%) activities. Around 780 million people around the world are without potable water as a result of these actions [2].

The massive increase in the world's population, along with rapid technological advancements, has resulted in a widespread threat to human existence [3]. Water contamination is most commonly caused by wastewater discharge from industries such as chemicals, mining, electroplating, agriculture waste, and many others [4, 5]. Contamination of water by HMs has become a big concern in nations like India and China due to their massive populations [6]. Domestic sewage, industrial water pollution, population increase, pesticides and fertilizers, plastics and polythene bags, urbanization, soil pollution, and agricultural water pollution all accelerated weathering processes, resulting in further freshwater quantity and quality losses. As per a survey, clean and safe drinking water is inaccessible to approximately 780 million people worldwide. Arsenic (As), lead (Pb), cadmium (Cd), nickel (Ni), and copper (Cu) are HMs that build in contaminated water. Their ions can accumulate in the body and prove to be fatal in humans and animals through the food chain [7, 8]. All HMs are non-biodegradable and have a negative impact because they can accumulate in the food chain [9]. Kidney failure, anemia, hypertension, lung damage, diseases pertaining to the nervous system, and memory loss are just a few of the major health problems that could be caused by

HMs in water systems [10, 11]. The United States Environmental Protection Agency (USEPA) and the Bureau of Indian Standards (BIS) have defined the maximum permissible limit of HM in drinking water as shown in Table 5.1 [12]. Over the last 20 years, multiple efforts have been carried out to remove HMs from water sources.

TABLE 5.1 The Standard Limits Set by WHO, EPA, and ICMR in Drinking Water for HM

Heavy Metals	WHO (mgL–1)	EPA (mgL–1)	ICMR (mgL–1)
Zn^{2+}	3	5	0.1
Fe^{2+}	0.1	0.3	1
Ni^{2+}	0.07	0.04	0.02
As^{3+}	0.010	0.010	0.05
$Cr^{3+/6+}$	0.05	0.1	0.05
Se^{6+}	0.01	0.05	–
Cu^{2+}	2	1.3	1.5
Pb^{2+}	0.010	0.015	0.05
Hg^{+}	0.001	0.002	0.001
Cd^{2+}	0.003	0.005	0.01
Ba^{2+}	1.3	2	–
Tl^{6+}	0.5	0.8	–
Mn^{4+}	0.4	0.3	–

Source: Adapted from Ref. [12]

Heavy metal detoxification has been accomplished using a variety of techniques, including electrochemical, chemical, and biological procedures. Chemical coagulants include photocatalytic oxidation, membrane filtration, ion exchange resin, and reverse osmosis [13]. Each method has inherent advantages and limitations in its implementation. Several standard protocols have been established by the World Health Organization (WHO), including ion exchange, precipitation, and reverse ion exchange. Membrane filtration removes effluents from wastewater using osmosis, oxidation, and membrane filtration [14–18]. Despite this, traditional treatments continue to be used. Some solutions have drawbacks, such as corrosion issues, sludge buildup, and so on, which have significant sensitivity to solution pH, and some techniques that aren't regarded as economically viable make it difficult to remove HMs from the water source. Although HMs can be successfully removed from wastewater using traditional sorbents, only in concentrated solutions do they work. As a result, advances are necessary for the effective removal of HMs, and adsorbents must be developed. The use of nanomaterials to remove HMs is the future. The most recent innovation addresses all of the problems that traditional sorbents face. Carbon, metal oxides, polymers, and other Nano-adsorbents can effectively operate to prevent pollution in environmental and industrial process applications. Iron oxide, titanium oxide, cerium oxide, and zinc oxide are among the metal-oxide nanomaterials that have been researched thus far. They have some limitations due to their inability to adsorb HM from wastewater, as they are unstable at such a small size [19]. Carbon-based nanomaterials, on the other hand, have wide use in several applications as functional adsorbents. They have a lot of high-energy binding sites in them, which is an important and valuable adsorption feature [20].

Traditional procedures (distillation, UV therapy, ultrafiltration, chemical transformation, coagulation, flocculation, and biological treatment) have been employed in treating wastewater and purification, but they are ineffective in complete pollutant removal [21]. The most promising upcoming water treatment technique is nanotechnology-based treatment [22]. With an emerging method for eliminating HMs from water, nanotechnology is currently in development. To date, various techniques in nanotechnology have been developed to remove these HMs from water bodies. These techniques comprise carbon nanotubes, nanoscale metal oxide, core-shell nanofibers, nano-zeolites, nano-Ag, nano-magnetite, and dye-doped silica nanoparticles [23, 24]. This chapter covers the updated nanotechnological techniques for removing HMs from water.

5.2 SOURCES OF HEAVY METALS IN WASTE OR CONTAMINATED WATER

The leading source of HMs contamination in water is the mining industry, apart from several other anthropogenic processes, such as industrial waste, agricultural waste, metallurgical operations, etc. Nickel, mercury, lead, cadmium, arsenic, cobalt, and chromium are a few HMs that may be harmful to terrestrial and aquatic animals as well as humans [27–29].

To some extent, natural sources like volcanic eruptions, rock weathering, forest fires, and biogenic sources have also contributed to excess water contamination by HMs, which may be present in the form of oxides, sulfides, hydroxides, sulfate, and phosphates. Water contamination is also caused by chemical fertilizers, lead-acid batteries, paint and varnish products, and microplastics [30].

5.3 NANOPARTICLE FOR REMOVAL OF HEAVY METAL FROM WATER SOURCES

Nanomaterials have unique qualities, including tiny size, huge surface area, tailored functions, etc., all of which are the reasons for their high adsorption capacity and reactivity. Using nanoparticles for the removal of HMs from wastewater is possible because of their exceptional adsorption properties. To assess the efficacy of nanomaterials for removing HMs from contaminated water, many methodologies have been devised. Conventional methodologies of detecting HMs in water have been superseded by the nanotechnology paradigm. Liquid chromatography, UV-Vis spectroscopy, X-ray fluorescence spectroscopy (XFS), capillary electrophoresis (CE), microprobes (MPs), anodic stripping voltammetry (ASV), atomic absorption spectroscopy (AAS), ICP-MS, and ICP-OES are being used [31, 32]. The detection of HM in water has been limited owing to the high cost of instrumentation, sample processing, and other detection limits. Several electrochemical, green nanosensors and nanomaterials (metal and metal oxide nanoparticles, polymeric nanomaterials, silicone, and carbonized nanomaterials) have been developed that can detect HMs in water bodies with more efficient detection limits, robustness, detection of location, and other characteristics [33]. Because nanotechnology combines all of the aforementioned characteristics, it is the best option for detecting metals in polluted water. The role of numerous nanoparticles in metal removal is explored in-depth in this chapter, including metal-based nanomaterials, carbon-based nanomaterials, dendrimers, and nanocomposites.

5.4 TYPE OF NANOMATERIALS USED FOR REMOVING HEAVY METAL

Several types of nanomaterial have been used for HMs removal from water sources, including carbon-based, metal-based, and nanocomposites, which are further categorized into different sections as presented in Figure 5.1.

5.4.1 CARBON-BASED NANOMATERIALS

Carbon-based NMs are the ones with tools that have improved mechanical strength, optical properties, conductivity, chemical versatility, etc. Owing to their versatility, they are the most used NMs for various purposes [34]. Fullerenes and quantum dots among carbon-based NMs are known as

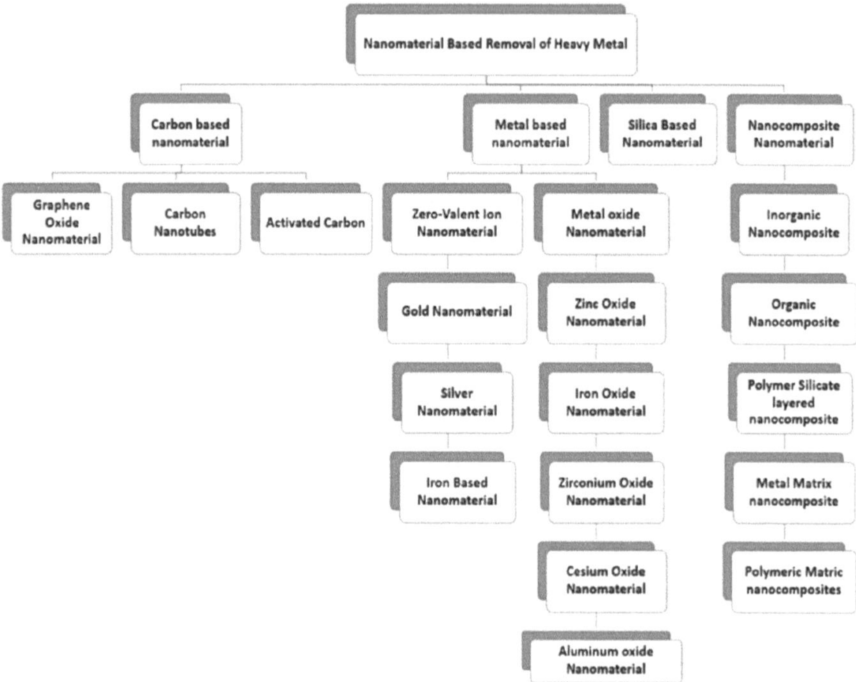

FIGURE 5.1 Schematic representation of nanomaterial-based removal of HM from water resources.

zero-dimensional (0-D) NMs; however, CNTs and graphene are categorized as one (1-D) and two-dimensional (2-D) NMs, respectively [36].

Carbon-based NMs are easy for any chemical or physical manipulation, and they have the capacity to withstand extreme temperatures and desired surface modifications. They have the potential to remove organic as well as inorganic contaminants. Based on shape and size, tubes are found as hollow spheres and ellipsoids.

The CNTs can generate electrostatic contacts. Hence, they have the potential to effectively remove HMs from wastewater [35]. In this chapter, the two most studied carbon-based NMs, CNTs and graphene, are discussed in subsections.

5.4.1.1 CARBON NANOTUBES

In recent decades, carbon nanotubes (CNTs) have been extensively studied, and they have been claimed to have a variety of unique features, including optical, electrical, vibrational, and magnetic capabilities characteristics mechanical and thermal. Due to their unique capabilities, they were employed to remove HMs from wastewater [37]. CNTs have a far higher adsorption potential than activated carbon, owing to their large surface area, which assures a robust contact between CNTs and metal ions. CNTs can effectively remove HMs such as Pb^{2+}, Zn^{2+}, Cd^{2+}, Cu^{2+}, and Ni^{2+}. Based on the number of layers present, single-walled carbon nanotubes (SWCNTs) and multi-walled carbon nanotubes (MWCNTs) are the two varieties of CNT tubes. SWCNTs are single-layer graphene sheets rolled with dozens of concentric tubes, whereas multilayer graphene sheets coiled with dozens of concentric tubes are known as MWCNTs. Due to their potential to create electrostatic interactions with the metal ion and large surface area, CNTs are better adsorbents for divalent metal ions in water. With the exception of Hg, Pb^{2+}, Cd^{2+}, Cr^{6+}, and Cu^{2+} can be efficiently removed from wastewater using CNTs. The possible adsorption active sites of CNTs mainly consist of the outside surface, interstitial channels, internal sites, and exterior groove sites. Functional groups such as OH, NH_2, and COOH have been routinely added to the surface of CNTs to boost their adsorption potential. Few studies suggest that oxidants such as H_2SO_4, HNO_3, $KMnO_4$, and NaOCl significantly improve CNTs adsorption by changing their surface. To make the unique functionalized CNTs, the pre-oxidized CNTs were sonicated using a deep eutectic solvent consisting of glycerol and allyltriphenylphosphonium to produce the unique functionalized CNTs. Hg^{2+} has been removed from wastewater by using this modified CNT [39]. CNTs are also mixed with other substances to generate superior nanocomposites, such as the manufacture of an amino-functionalized Fe_3O_4/carboxylic MWCNTs hybrid with a high Cu^{2+} adsorption capacity [40]. Even while CNTs may successfully remove HMs from wastewater, they come with several drawbacks, including high costs, their removal from wastewater after adsorption, and their potential for increased contamination [41]. There is still more to be done to create CNTs that are both useful and cost-effective. Following adsorption, CNTs would have to be extracted from effluent, increasing treatment costs and time. However, there is a great demand for toxicological research on carbon nanotubes [42].

5.4.2 GRAPHENE OXIDE NANOMATERIAL

Graphene was the first 2-D material known to humans and can now be used as a very efficient tool in removing HMs from water. Graphene comprises a dense stacked hexagonal lattice of carbon atoms and is the thinnest substance known. Because it is 200 times stronger than steel, we find its use in almost every field. Graphene is described as having a huge surface area, thermal mobility, electrical properties, great mechanical strength, and altered surface chemistry [43]. Owing to their unique and fascinating properties, graphene-based nanomaterials have been great precursors in water treatment. Tabish et al. [44] recently developed porous graphene that removed As^{3+} from water with an efficiency of 80% [44]. The breaking of the conjugation network that takes place during the production of graphene oxide (GO) with other compounds causes it to become extremely water-dispersible. By minimizing aggregation between individual GO sheets, their adsorption potential can be boosted. Graphene sheets were mixed with a variety of metal oxides and polymers, resulting in nanocomposites with multifunctional features and excellent performance, as presented in Figure 5.2 [45].

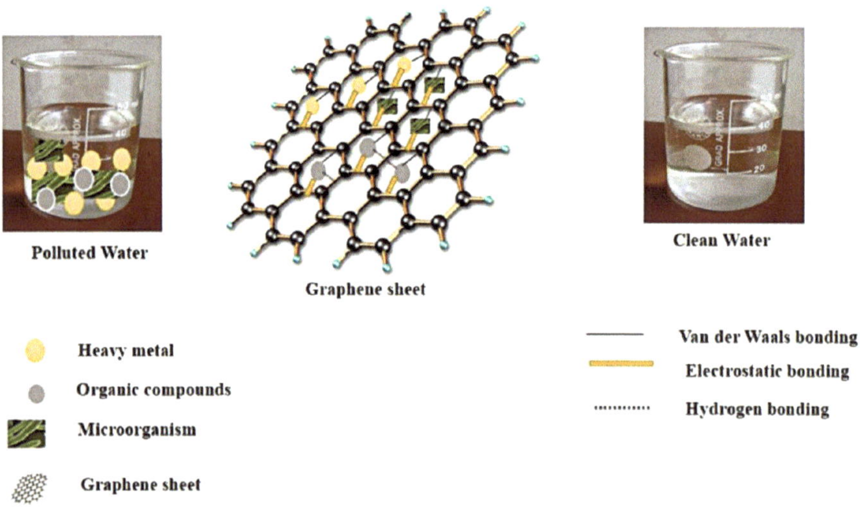

FIGURE 5.2 Schematic representation of remediation of polluted water using graphene sheet.
Source: Reprinted/Adapted from Ref. [45]

Reduced GO (rGO) has more flaws than graphene and can be readily altered by functional groups such as COOH, OH, and other graphene-based nanomaterials that have been widely employed to remove HMs from wastewater.

A nanocomposite made by partially combining rGO and Fe_3O_4 shows great promise in eliminating Pb^{2+} from water. Pb^{2+}, Ni^{2+}, Cr^{3+}, Cd^{2+}, and Cu^{2+} are also removed successfully by the use of modified GO, made by functionalizing rGO with 4-sulfophenylazo compound [46]. A self-propelled graphene-based microbot system with nano-sized multilayers of GO, Ni, and Pt was developed, which successfully removed 80% of Pb^{2+} from water [47]. Although graphene-based NMs are widely used in the removal of HMs, their high water dispersibility makes it challenging to separate them from an aqueous solution [48]. To address this issue, researchers created a modified graphene-based adsorbent wherein calcium alginate beads were embedded inside GO, which was later reduced by polyethyleneimine. The adsorption capacity for HM was improved even further with this modified adsorbent [49]. Arshad et al. [49] have developed a graphene-modified absorbent that can fix this problem. To improve the adsorption capability towards HMs, calcium alginate (CA) beads were initially implanted into graphene oxide and then further lowered by polyethylenimine to increase the adsorption performance towards heavy metals. This adsorption experiment done in batches in a shaker bath. The adsorption capabilities are the highest at 602, 374, and 181 mgg^{-1} of this GO-based nanocomposite towards Pb^{2+}, Hg^{2+}, and Cd^{2+}.

The Langmuir isotherm shows a material has a significant advantage in removing with great superiority for eliminating Pb^{2+}, Hg^{2+}, and Cd^{2+}. The adsorption kinetics was found to be of pseudo-second-order, and thermodynamic adsorption features suggested that the adsorption process was physicochemical. Because of the synergetic effect, the polyethyleneimine-reduced functionalized beads demonstrated higher adsorption; the functionalized beads' ability exceeds that of the non-functionalized beads.

Fullerenes contain distinct physicochemical characteristics, like high electron affinity, hydrophobicity, a large surface-to-volume ratio, and surface defects. Adsorbates enter the defects/spaces contained in fullerenes and are adsorbed. Due to their particular features, fullerenes are successfully used for HMs adsorption [36]. When compared to nanocomposite polystyrene, they are more efficient at removing Cu^{2+} [50]. The adsorption ability of fullerenes is assumed to be due to their surface defects. The Cu^{2+} ion is thought to adsorb in the lattice of fullerenes. Cu^{2+} considerably adsorbs on the C_{60} surface via a charge transfer from Cu (4s) to C (2p) of C_{60}, according to spectroscopic and X-ray absorption studies by Rowe et al. in [51]. Cu_{2+} atoms also change the chemical environment of the C_{60} molecule's interface, according to researchers [51].

5.4.3 *ACTIVATED CARBON*

Activated carbon is widely used in capacitors, catalyst support, gas storage, electrodes, and, most notably, as an adsorbent for removing HMs from water due to its high mechanical strength. Carbon, particularly activated carbon, has an adsorption potential that is influenced by its source. Depending on the source material and activation process, activated carbon has different adsorption properties. Its quality is influenced by biomass type, final activation temperature, chemical agent impregnation rate, and heating rate. As a result, the conclusion was drawn that the type of raw material utilized in the manufacturing of activated carbon, as well as the experimental settings used, constitute the source of its adsorption potential and removal effectiveness [52]. An incident of removal from the water of Al^{3+} and Zn^{2+} ions by using activated carbon made from apricot stone was shown to have a 92% effectiveness [53]. Sewage sludge has also been a source of activated carbon that was used to remove Cu^{2+} and demonstrated a 50% adsorption capability. Several HM ions were removed using sulfonated activated carbon, including Ni^{2+}, Cd^{2+}, Pb^{2+}, and Cu^{2+} are all metals [54]. Agricultural multipore activated carbon (MPAC) which has a greater surface area was made from the agricultural waste of long-root Eichhornia crassipes. It is also a modified type that was found to be successful in removing HMs ions such as Zn^{2+}, Pb^{2+}, Cu^{2+}, Cd^{2+}, and Ni^{2+}. This type is made and modified in a simple, basic, and low-cost manner. They also have a higher ability for the adsorption of HM ions.

5.4.4 SILICA-BASED NANOMATERIALS

Advanced features like non-toxic nature and good surface features, another important type of nanomaterial for eliminating HMs are the silica-based ones. Nano-silica can be modified with surface groups like $–NH_2$, –SH, and others, and it can also be used as a foundation for nanocomposites. Kotsyuda et al. [56] made silica Nanospheres having biofunctionalized 3-aminopropyl and phenyl groups and tested their ability to remove Cu^{2+} and cationic thiazine dye in a static mode [56]. It was observed that when compared to the amino-functionalized nano-silica, these silica nano-spheres exhibited greater adsorption capabilities for Cu^{2+} and methylene blue. For methylene blue, biofunctionalized nano-silica has approximately twice the adsorption capacity of similar amino silica nanoparticles. Antibacterial characteristics were also present in this modified nanomaterial. Najafi et al. [57] experimented and assessed the removal effects for Cd^{2+}, Ni^{2+}, and Pb^{2+} in a batch method by using three silica-based nanomaterials: amino-functionalized silica gel (NH_2-SG), amino-functionalized silica nano-hollow sphere (NH_2-SNHS), and non-functionalized silica nano-hollow sphere. The metal adsorption trend was Pb^{2+} > Cd^{2+} > Ni^{2+}, and the adsorption capabilities of these three nanomaterials were determined and were found to be in the order as follows: NH_2-SNHS > NH_2-SG > SNHS. Pb^{2+}, Cd^{2+}, and Ni^{2+} adsorption capabilities utilizing NH_2–SNHS were 96.79, 40.73, and 31.29 mgg^{-1}, respectively. The kinetic data fit the pseudo-second-order model well, and the adsorption isotherms stood linked to the Langmuir-Freundlich (Sips) isotherm. In a test tube, he used the Stöber reaction to coat magnetite particles with silica layers and established its industrial application [57]. Silica has also been used and extensively reported for the synthesis of nano-composites, with magnetic silica compounds receiving particular interest. Pogorilyi et al. [58] employed the Stöber reaction to coat magnetite particles with silica layers in a test tube and established its industrial use potential [58]. Nano-polyaniline was immobilized onto nano-silica to form Sil-Phy-N PANI nanocomposites and the effects of Cu^{2+}, Cd^{2+}, Hg^{2+}, and Pb^{2+} on Cu^{2+}, Cd^{2+}, Hg^{2+}, and Pb^{2+} were compared utilizing a batch approach. For Cu^{2+}, Cd^{2+}, Hg^{2+}, and Pb^{2+}, the best Sil-Phy-NPANI had adsorption capacities of 1,700, 800, 600, and 600, respectively, whereas the adsorption capacities of Sil-Phy-Cross N PANI for similar ions are 900 $molg^{-1}$ and according to the Langmuir isotherm, the values were 1,650, 1,050, 1,350, and 1,450 $molg^{-1}$ correspondingly. The Sil-Phy-Cross N PANI was found to be an effective adsorbent for Cd^{2+}, Hg^{2+}, and other HMs [59].

5.4.5 ZERO-VALENT METAL-BASED NANOMATERIALS

The Zero-valent metal-based NPs have shown promising capabilities in water treatment and cleanup. Due to their antibacterial properties, Ag nanoparticles have been used to treat wastewater. Dioxins have been found to have a high susceptibility to the degradation of zero-valent zinc nanoparticles [60]. It has been observed to have outstanding sensitivity to dioxins [60]. When it comes to HMs ions, the most common was zero-valent iron. This section has been thoroughly examined and is the subject of the majority of the discussion. Gold, Silver, and Iron NPs are a few more nano-sized particles used for the removal of HMs from wastewater.

1. **Gold Nanomaterial (AuNPs):** These have a diameter of fewer than 100 nanometers, making them an excellent choice of NMs for water treatment. To enhance the stability of gold nanoparticles in biological fluids, agents such as polyethylene glycol (PEG), lipid bilayers, and protein coatings have been used [61]. Because of their

effective adsorption potential, AuNPs are used to remove Hg^{2+} from water. It has been well established that when supported on Al_2O_3, AuNPs in the 10–20 nm range are unable to adsorb Hg^{2+} due to a lack of electrostatic contact. However, by using a reducing agent like $NaBH_4$ and converting Hg(II) to Hg(0), the Hg metal's adsorption may be improved [41]. For water treatment, the filter membrane is extensively in use, and research is still in progress to improve its performance and make it more cost-effective. Metallic NPs deposition on the membrane has proven to be an effective technology, and it is now being applied [62]. The expenses of wastewater treatment with AuNPs following metal ion reduction are relatively less, as is the possibility for Au nanoparticle recovery. Ojea-Jimenez and colleagues conducted another investigation in which they employed citrate-coated AuNPs to adsorb Hg^{2+} from water [63]. The Hg metal, which was in a zero-oxidation state, was adsorbed on the surface of AuNPs as Au_3Hg and was readily extracted from the sample by centrifugation. Membrane-coated AuNPs have been discovered to have substantially better effectiveness than ordinary membranes. The flow efficiency has also been greatly improved by utilizing membrane-coated AuNPs [64].

2. **Silver Nanomaterial (Ag):** Even though the reactivity of Hg^{2+} and bulk silver is low, Ag NPs can have a better reactivity due to Ag's reduction potential being reduced as the parity fell. Different detection approaches have been described using modified Ag NPs as HMs sensors, including changes in color because of particle aggregation, enhanced intensity of fluorescence, and production of silver chloride precipitate.

By combining silver nanoparticles with mercaptosuccinic acid (MSA), Sumesh et al. [65] created a unique silver nanoparticle-based adsorbent [65]. Alteration in the ratio of Ag to MSA, two distinct materials were created and examined. When compared to other adsorbents, 1:6 Ag@MSA exhibited a better removal capacity for Hg^{2+} (800 mgg^{-1}) than the other adsorbents. Additionally, the authors also claimed the expenses of removing Hg^{2+} using Ag@MSA were economical, implying that Ag@MSA can prove to be a viable alternative for the removal of Hg^{2+}. At an optimum pH of 6.0 and 7.0, silver nanoparticles were dispersed on the surface of CNTs for effective removal of Cu^{2+} and Cd^{2+}. The remarkable features of Ag NPs have been investigated, including easy removal, catalytic activity, higher adsorption capacity, reusability, and surface modification [66].

5.4.6 ZERO VALENT IRON

Made up of Fe(0) with a ferric oxide coating makeup nanoscale zero-valent iron (nZVI) composites have received a lot of interest as a novel adsorbent for the treatment of HMs [67, 68]. In essence, Fe(0) brings forth a reducing capability, whereas the ferric oxide shell provides reactive and electrostatic interaction sites. Furthermore, the particle size of nZVI can also be adjusted, and the surface also possesses a greater number of reactive sites. Because of these reasons, nZVI is more effective at removing HMs from wastewater. To further enhance the performance of the nZVI, various modification processes have been devised, like modifying surface chemistry or doping nZVI with other metals. By mixing nZVI with sodium dodecyl sulfate (SDS), an anionic surfactant with good migration and dispersion properties [69], Huang et al. [69] generated a unique nZVI-modified material. The greatest removal performance of this unique nZVI material towards Cr^{6+} was 253.68 mgg^{-1} in a batch adsorption method, suggesting a viable adsorbent with enhanced adsorption potential and less aggregation. The Freundlich model and the pseudo-second-order kinetic model were finely obeyed by the adsorption course. The pH, contact time, volume, and concentration were all researched, and under ideal conditions, a maximum removal efficiency of 98.919% could be reached. Su et al. [70] looked into the elimination of both Cd^{2+} and nitrate from drinking water, where nZVI and Au-doped nZVI NPs were used to purify water in a batch process [70]. With the use of Au-doped in comparison to nZVI, the nitrite yield ratio lowered from nitrate may be reduced greatly. While the elimination ratio of Cd^{2+} remained high, the bare nZVI remained unchanged. This outcome revealed indicated the Au-doped nZVI may be used to remediate Cd^{2+} and nitrate-containing effluent.

5.4.7 METAL OXIDE-BASED NANOMATERIALS

Owing to their distinctive physical and chemical properties, metal oxide NPs are widely used to remove hazardous HMs ions from polluted water. Green chemical approaches, like natural biopolymers and biological wastes, have gained popularity in recent years. Because of their inexpensive cost and abundant availability, they have been designed to make magnetic nano-adsorbents biodegradability, as well as a high preference for metal capping.

To name a few, metal oxide-based nanomaterials include nanosized as HMs adsorbents. Nano-sized iron oxides, manganese oxides, zinc oxides, titanium oxides, aluminum oxides, magnesium oxides, cerium oxides, and zirconium oxides.

1. **Manganese Oxides-based Nanomaterials:** The increased surface area of Nanocrystalline manganese oxide has been discovered to contribute to its superior adsorption capability. M-O+ and M-O units on the surface of manganese oxide assist the metal ion sorption. Wang et al. [71] developed a manganese dioxide/gelatin dumbbell and studied its adsorption properties for Pb^{2+} and Cd^{2+}. As per the Langmuir model, the maximum adsorption capabilities for Pb^{2+} and Cd^{2+} in a batch adsorption investigation were 318.7 and 105.1 mgg^{-1}, respectively. The adsorption kinetics have been explained using a pseudo-second-order model. Hydrous manganese oxide (HMO) is effective in removing HMs due to its characterization of the adsorption kinetics. Because of its wide surface area, porous structure, and large number of adsorption sites, HMO has shown greater efficacy in removing HMs. Coordination chemistry is crucial in HMO adsorption due to the pairing of hydroxyl groups with HM ions on the surface of HMO. On HMO, heavy metal ion adsorption normally took two steps: quick HMs adsorption on the outer surface and sluggish intra-particle diffusion along the micro-pore walls. An HMO-BC nanocomposite was developed by Wan et al. [72] by combining HMO NPs and biochar (BC). Over a wide pH range, the HMO-BC nanocomposite effectively removed Pb^{2+} and Cd^{2+}. According to fixed-bed column adsorption tests, the effective treatment capability of HMO-BC for simulated Pb^{2+} or Cd^{2+} containing wastewater was roughly 4–6 times higher than that of the BC host. As a result, it appears that HMO-BC could be a viable choice for eliminating HM from polluted water.
2. **Iron Oxide Nanomaterial:** Iron oxide (FeO) NPs have been effective as adsorbents in the removal of hazardous HMs from water samples [73]. In comparison to other water treatment processes, this method is quite inexpensive. The adsorbents' simple removal from the solution is owing only to the magnetic properties of the FeONPs, which allow them to be reused. Due to their larger surface

area, tiny size, and magnetic characteristics, FeONPs are traditionally a popular alternative for water treatment. FeONP synthesis has already been thoroughly examined. The surface modification of FeONPs could improve the nanomaterial's efficacy and stability in water decontamination. The hydroxyl group on the surface of FeONPs provides a versatile synthetic tool for attaching various functional groups [74]. For example, altering FeONPs with fly ash could be used to generate cementitious (cement-based composites) composites for removing HM from water samples [75]. Wang et al. created MoS_2/Fe_3O_4, FeONPs modified by molybdenum disulfide that effectively removed Pb^{2+} and Hg^{2+} from water and soil samples [76].

3. **Aluminum Oxide Nanomaterial:** Due to their large surface area to volume ratio, cheaper raw material, and greater decontamination power, aluminum oxide (Al_2O_3) NPs have emerged as significant adsorbents for HM ions such as Cd^{2+}, Zn^{2+}, and many others [77]. These are available in a variety of crystalline forms, the most popular of which is Al_2O_3 for the removal of contaminants from water. The standard size of Al_2O_3 NPs was revealed to be 6–13 nm, and their HMs removal efficacy was reported to be approximately 97% for Pb^{2+} and 87% for Cd^{2+}. The adsorptive capacity of Al_2O_3 NPs has been significantly improved by chemical treatment by means of several functional groups, which also include donor atoms. HMs are also removed from water by chemical interaction after a reagent is added to the alumina nanoparticles. The efficiency of Al_2O_3 NPs has been dramatically enhanced in the presence of phosphate and humic acid. Nano-Al_2O_3 NPs were produced in another work, and their efficacy for removing HM as Cr^{6+}, Ni^{2+}, Cd^{2+}, and Pb^{2+} was studied. The sol-gel method was used to synthesize Al_2O_3, and HMs adsorption was investigated for Pb^{2+} and Cd^{2+} metal ions [78].

4. **Titanium Oxide Nanomaterial:** Chemical behavior and catalytic activity have been discovered to be influenced by the varied surface planes [79] in the case of titanium oxide (TiO_2). On combining TiO_2 with zirconium, the surface characteristics changes and the adsorbent capacity is increased effectively [80]. By hydrothermally synthesizing TiO_2 nanowires, few researchers undertook a relative examination of HMs removal from wastewater samples. Titanium nanowires are a preferred choice due to their huge surface area and

good magnetic characteristics. Lead adsorption was determined to be the highest, while zinc adsorption was the lowest [81]. The removal of HMs using TiO_2 and the wetland technique was compared to find that TiO_2 nanoparticles could reduce HMs contamination by 90–96% [82]. Some scientists have recently developed mesoporous phosphate TiO_2 (PTO) for the removal of chromium ions from polluted water. They created 8-PTO (using phosphoric acid at an 8% molar ratio) with a surface area of 278 m^2g^{-1} and a distribution size of less than 5 nm. In dirty water, modified TiO_2 showed a remarkable adsorption capability of up to 92 mgg^{-1}, which was significantly higher than TiO_2 alone (10–83 mg/g).

5. **Zinc Oxide Nanomaterial:** Due to their greater surface area, cheaper expense, and better removal capacity, zinc oxide NPs (ZnO) have become popular as HMs adsorbents [83]. Cr^{6+}, Cu^{2+}, Ni^{2+}, and other metals have been treated using Nano-sized zinc oxides. Sheela et al. investigated the removal efficacy of ZnO nanoparticles from Zn^{2+}, Cd^{2+}, and Hg^{2+} using a batch approach [84]. From the Langmuir model, the maximal adsorption capacities for these HM are 357, 387, and 714 mgg^{-1}, indicating a highly competitive adsorbent. To synthesize zinc oxide nanoparticles, casein was used as a reducing and capping agent by Somu et al. [85]. Casein-capped ZnO NPs with an average size of 10 nm were used for the remediation of wastewater having three metals and two colors. The experiment was carried out in a batch process. Cd^{2+}, Pd^{2+}, and Co^{2+} adsorption capacities were 156.74, 194.93, and 67.93 mgg^{-1}, respectively, and methylene blue and Congo red absorption capacities were 115.47 and 62.19 mgg^{-1}, respectively, and Cd^{2+}, Pd^{2+}, and Co^{2+} adsorption capabilities were 156.74, 194.93, and 67.93 mgg^{-1}, respectively. In addition, the casein-capped ZnO NPs had outstanding antibacterial activity, indicating that they could be employed to absorb wastewater in real-world circumstances.

6. **Cerium and Zirconium Oxide Nanomaterial:** Cerium oxide (CeO_2) is a rare-earth oxide that is non-toxic, has Ce quadrivalve and has been used in a wide range of applications, including photocatalysis sen,sing, and water purification, among many others [86]. Some scientists studied the effect of CeO_2 nanoparticles generated averaged 12 nm in diameter and had a surface BET area of 65 m^2g^{-1}. According to the Langmuir isotherm, the adsorption potential for

these two ions was 71.9 and 36.8 mgg^{-1}, respectively, and a mere 10 minutes were required for the entire adsorption process [87]. It is worth mentioning that anions like $H_2PO_4^{-1}$, SO_4^{-1}, and HCO_3^{-1} diminish adsorption capacity when they coexist. In another study, the elimination impact of Samaria-doped ceria nanopowder (SDC) on Pb^{2+}, Cu^{2+}, and Zn^{2+} was assessed. It was observed that spherical SDC Nanopowder (SDC-F) has better adsorption potential than similarly made cluster plate SDC Nanopowder (SDC-I).

Nano-sized zirconium oxides are also potential metallic oxide adsorbent that has been utilized for the extraction of HMs from wastewater. Their benefits can be attributed to a large amount of hydroxide on their surfaces and a larger area. Nano-sized zirconium oxides also have good chemical stability and adsorption affinities for HMs like Pb^{2+}, Zn^{2+}, and Cd^{2+}. Gulaim et al. studied the effects of TiO_2, ZrO_2, HfO_2, Nb_2O_5, and Ta_2O_5 in solutions, as well as other transition metals with mesoporous structures [88]. The metal oxides were made of partially integrated homogenous NPs with shells that are amorphous and crystalline cores, each with a large surface area. In the nanoscale, hydrous zirconia (HZO) in the nanoscale has also shown potential as an HMs adsorbent. Zhang et al. produced polystyrene-supported Nano-sized zirconium hydroxide HZO-PS and researched on its ability to remove Cd^{2+} [89]. The results revealed that this nanocomposite could remove Cd^{2+} in a pH range of 2.5–7.0 with negligible Zr^{3+} emissions. The results of a fixed-bed column adsorption experiment revealed better applicability of such a nanocomposite with a treated capacity of 750-bed volume per run.

5.4.8 NANOCOMPOSITE NANOMATERIAL FOR HEAVY METAL REMOVAL

A nanocomposite is a mixture of more than two materials that are separated by an interface with different properties, both physical and chemical. Generally, there are two phases to the composite: a phase with a matrix (continuous phase) and another dispersed phase in which one of the two phases is dispersed, whereas a nanocomposite makes up tiny particles [90]. The distinctiveness is generally found in the material utilized, stoichiometry, and other factors, all of which play an important part in defining the properties of nanocomposites along with the particle size and shape in various phases. Stiffness, strength, stability, corrosion resistance, low density, and thermal insulation can be included in the properties of a nanocomposite.

Water filtration membranes made of polymeric-based nanocomposite materials are currently being researched. Water treatment membranes made of polyether sulfone, polysulfone, polyvinylidene fluoride, and polypropylene are effective for the treatment of wastewater [91]. Regarded as one of the most progressive technologies for water treatment techniques, membrane technology, with the addition of NMs, becomes additionally encouraging. Polymeric membranes have been made using NPs, nanotubes, or fiber with diameters ranging from 4–100 nm [92]. Nano-zeolites are often utilized in the membrane of thin-film nanocomposites due to their superior molecular filter, more membrane permeability, active layer thickness, hydrophilic nature, and enhanced negative charge.

Although each of the NMs described above has its own set of advantages, it also has its own set of drawbacks. CNTs, for example, are difficult to suspend in a variety of solvents reliably, whereas nZVI is prone to oxidation. Furthermore, when used in fixed-bed and flow-through systems, NPs usually induce aggregation, poor separation, and an excessive pressure drop [93]. A usually utilized method for overcoming such challenges is by creating hybrid nanocomposites wherein the advantages of multiple nanocomposites are combined.

On the basis of chemical composition, the material of the matrix used, fillers, microstructures, and usage, polymeric matrix nanocomposites, metal-based matrix nanocomposites, polymer/layered silicate (PLS) nanocomposites, and polymer nanocomposites are the four types of nanocomposites. This section delves further into nanocomposites using inorganic and organic polymer supports, as well as magnetic nanocomposites.

1. **Inorganic-based Nanocomposite:** The inorganic supports of nanocomposites for HMs removal are activated carbon (AC), CNTs, and natural minerals like bentonite, montmorillonite, zeolite, and others. The most simple and cost-effective among these is AC, which removes pollutants from aqueous solutions. Several researchers have published regarding the AC-backed program to remove HMs from water using nanocomposites, and these composites have shown to be capable of removing Cr^{6+}, Pb^{2+}, Cd^{2+}, and other HMs [94, 95]. A type of nanomaterial that is predominantly supported by carbon nanotubes is a nanocomposite. Salam et al. sonicated chitosan and CNTs suspensions before crosslinking them with glutaraldehyde to create an MWCNTs/chitosan nanocomposite, which was packed onto a glass column to efficiently extract Cu^{2+}, Zn^{2+}, and Cd^{2+}, and Ni^{2+} from aqueous solutions. Other compounds that have been

reported to produce nanocomposites with CNTs include polyethylenimine, 3-mercaptopropyltriethoxysilane, 8-hydroxyquinoline, cyclodextrin, and others. Hayati et al. revealed the synthesis of a CNT-coated polyamidoamine dendrimer (PAMAM) nanocomposite (CNT/PAMAM) in a fixed-bed technique and explored its adsorption efficacy toward As^{3+}, Co^{2+}, and Zn^{2+} [96]. In column adsorption tests, the maximum absorption capacities for these three ions were 432, 494, and 470 mgg^{-1}, respectively. Because of its advantages, such as a larger specific area, the capacity of cationic exchange, and adsorptive affinity, bentonite is one of the most promising solutions for dealing with contamination at higher concentrations. NZVI, magnetite, hexadecyltrimethylammonium bromide (CTMAB), ethylene diamine tetra acetic acid (EDTA), 2-mercaptobenzothiazole (MBT), cellulose, and other materials have been used with bentonite to extract HMs from aqueous solution [22, 97]. The hydroxyapatite/zeolite nanocomposite (HAp/NaP) has been synthesized using a batch process to treat Pb^{2+} and Cd^{2+} in water, with maximum adsorption capacities of 55.55 mgg^{-1} for Pb^{2+} and 40.16 mgg^{-1} for Cd^{2+}, respectively. The kinetics of adsorption might be properly connected with the pseudo-second model. The HAp/zeolite nanocomposite is also known to possess antibacterial activity against the majority of Gram-positive and negative bacteria, indicating that it might be used in water treatment. Many different inorganic nanocomposites support, such as GO, sand, clay, and others, have been discovered to be potential alternatives for removing HMs from wastewater.

2. **Organic Polymer Nanocomposite:** Two major types of organic polymer-supported nanocomposites are synthetic organic polymer-supported nanocomposites and biopolymer-supported nanocomposites [98]. Polymer nanocomposites are prepared by two means: direct compounding and in situ synthesis. The creation of nanocomposites for the treatment of HMs has been widely described using synthetic organic polymers like polystyrene (PS) and polyaniline, nanocomposites for the treatment of HMs. Biopolymers, in addition to synthetic organic polymers, are often employed. Since HMs have plenty of coordination sites on the hydroxyl groups on the glucose ring of cellulose, one of the most prevalent biopolymers, they can be used as an adsorbent precursor [99].

For eliminating Cr^{6+} from wastewater, few scientists created a chitosan/alginate nanocomposite [100]. A batch adsorption experiment was performed where Cr^{6+} had a maximum adsorption capacity of 108.8 mgg^{-1}, and the adsorbent favored multilayer adsorption. The thorough analysis of polymer functionalized nanocomposites (PFNCs) for metal removal from the water was performed by Lofrano et al. [101], where they also examined the manufacturing, characterization, toxicity, removal capacities, and interactions between the polymer hosts and NPs. Although polymer-based nanocomposites have shown promise in the HMs removal of heavy metals, additional research into their synthesis, prices, recovery techniques, and safety is needed.

3. **Polymer Silicate Layered Nanocomposite:** Layered silicates are mixed into polymer matrices to create PLS nanocomposites [102]. In comparison with the traditional macro-composites, they have a significant improvement in their properties. In a study, the direct integration of polymer on clay in a paper was analyzed [103]. The removal of numerous HMs such as Hg^{2+}, Ag^{+}, Au^{3+}, Pb^{2+}, Zn^{2+}, Cu^{2+}, Ni^{2+}, Mn^{2+} was examined using a nanocomposite made up of silica gel microspheres sum up in imidazole functionalized polystyrene (SG-PS-azo-IM) which was found to have the best adsorption potential for Au^{3+} out of all the HMs [104]. A silica-sphere-poly(catechol hexamethylenediamine) (PCHA-SiO_2) nanocomposite has also been developed for the selective removal of Cd^{2+}, Pb^{2+}, and Cu^{2+} metal even at very low concentrations [105]. The selective removal of HM such as Pb^{2+}, Cu^{2+}, and Zn^{2+} was achieved by using microporous layered silicate AMH-3 as a sorbent. The lattice d-spacing of zeolite, which is a manufactured PLS nanocomposite, increased, resulting in enhanced adsorption of Pb^{2+} and Cd^{2+}, 97.20% and 85.06%, respectively, from the solution [106].
4. **Polymeric Matrix Nanocomposites:** In polymeric matrix nanocomposites, a polymeric matrix is reinforced with small NPs with a high aspect ratio [107]. In this, the filler NPs must form a good interaction with the matrix and should be well disseminated inside the matrix. Polypyrrole/polyacrylonitrile (PANI/PPy) polymer nanofibers conveniently absorbed Co^{2+} ions, resulting in a 99.68% removal efficiency. A hybrid nanocomposite combining poly(3,4-ethylenedioxythiopheneethylene dioxythiophene)/polystyrene

sulfonate (PEDOT/PSS) and the biopolymer lignin was employed for the removal of Pb^{2+} from wastewater [105].

5. **Metal Matrix Nanocomposites:** Metal matrix-based nanocomposites are created when Nano-dimension fillers are added to a matrix of metal or an alloy that is larger. These nanocomposites have properties that are similar to both metals and ceramics. They have a high degree of ductility property, as well as high mechanical and thermal stability. When properties like the thermal and mechanical stability of CNTs are combined with metal matrices, a wider range of applications can be realized.

To absorb Sb^{3+} and Sb^{5+} from polluted water, polyvinyl alcohol (PVA) was synthesized with nanoscale Fe^0. When compared to metal-based NPs, the PVA-Fe^0 nanocomposite showed quick antimony adsorption [108]. Cr^{6+} and As^{5+} metals were rapidly and efficiently released from wastewater using Fe^0 NPs supported by PANI nanofibers [109]. Cu-NPs were recently diffused efficiently on chitosan-tripolyphosphate beads to build nanocomposites, which was employed for the improvement in Cr^{6+} adsorption over chitosan-polyphosphate beads alone [110].

5.5 CONCLUSION AND FUTURE PERSPECTIVES

Removal of HMs from polluted water is urgently needed to protect both water quality and human health. A number of commercial solutions for the removal of HMs are now available on the market, however, for the removal of HMs, the application of NMs has its advantages due to their unique properties. This chapter explains how nanotechnology has made substantial advances in the analysis and removal of HMs using nanotechnology. NMs have different roles in the analysis; nanomaterials play a variety of roles, including HMs like: (i) adsorbent, (ii) filter membrane, (iii) reducing agent, (iv) peroxide catalyst, and (v) conjugator. This chapter also covers carbon-based, magnetic, zero-valent metal ion NPs and metal oxide NMs, among other topics. Many challenges come in the way during the treatment of water with NMs, like instability and aggregation, thus, limiting their efficiency. This can be resolved by using nanocomposite materials instead of a single type of NM. These materials exhibit a large linear range, a lower limit of detection, higher sensitivity, and better selectivity as HMs nano-sensors. Many industrialists use magnetic NP-based composites for the removal of

HMs to save time and money because of their easy recoverability with the external magnet.

Future research should focus on the following points for efficient removal of HMs with the application of nanotechnology:

- In-depth research is required on the stability of nanocomposite materials.
- By changing one or two parameters, can we reduce the complexity of the process?
- For addressing the concern of environmental safety, a toxicity study of NMs is mandatory.
- Scale-up study for industrial application purposes.
- For a viable and economical process, reducing the application cost of materials as much is required.
- Research for new efficient NMs should continue.

ACKNOWLEDGEMENTS

The research was financially supported by the Strategic Academic Leadership Program of the Southern Federal University ("Priority 2030").

KEYWORDS

- **carbon nanotubes**
- **graphene oxide**
- **heavy metal**
- **nanomaterials**
- **nanoparticle**
- **zero-valent metal**

REFERENCES

1. Cheraghi, M., Lorestani, B., & Yousefi, N., (2009). Effect of wastewater on heavy metal accumulation in Hamedan province vegetables. *International Journal of Botany, 5*(2), 109–193.
2. Reglero, M. M., Taggart, M. A., Monsalve-Gonzalez, L., & Mateo, R., (2009). Heavy metal exposure in large game from a lead mining area: Effects on oxidative stress and fatty acid composition in the liver. *Environmental Pollution, 157*(4), 1388–1395.

3. Luvita, V., Darmayanti, N. E., Zaid, G., & Setyarini, D., (2019). Testing instrument for water quality and drinking water using oxidation and electromagnetic methods (case study: Local water company at Bangka barat). *IOP Conference Series: Materials Science and Engineering* (p. 012048). IOP Publishing.
4. Wu, C., Chen, X., Cai, Y., Zhou, X., Xu, S., Huang, H., Zhang, L., et al., (2020). Risk factors associated with acute respiratory distress syndrome and death in patients with coronavirus disease 2019 pneumonia in Wuhan, China. *JAMA Internal Medicine, 180*(7), 934–943.
5. Shrivastava, P., Jain, V., & Nagpal, S., (2022). Nanoparticle intervention for heavy metal detection: A review. *Environmental Nanotechnology, Monitoring & Management*, 100667.
6. Stamatakis, E., Gale, J., Bauman, A., Ekelund, U., Hamer, M., & Ding, D., (2019). Sitting time, physical activity, and risk of mortality in adults. *Journal of the American College of Cardiology, 73*(16), 2062–2072.
7. Afroze, S., & Sen, T. K., (2018). A review on heavy metal ions and dye adsorption from water by agricultural solid waste adsorbents.*Water, Air, & Soil Pollution, 229*(7), 1–50.
8. Cocârţă, D., Neamţu, S., & Deac, A. R., (2016). Carcinogenic risk evaluation for human health risk assessment from soils contaminated with heavy metals. *International Journal of Environmental Science and Technology, 13*(8), 2025–2036.
9. Azimi, A., Azari, A., Rezakazemi, M., & Ansarpour, M., (2017). Removal of heavy metal rom industrial wastewaters: A review. *ChemBioEng Reviews, 4*(1), 37–59.
10. Engwa, G. A., Ferdinand, P. U., Nwalo, F. N., & Unachukwu, M. N., (2019). Mechanism and health effects of heavy metal toxicity in humans. *Poisoning in the Modern World Tricks for an Old Dog, 10*(2019).
11. Laks, E., McPherson, A., Zahn, H., Lai, D., Steif, A., Brimhall, J., Biele, J., et al., (2019). Clonal decomposition and DNA replication states are defined by scaled single-cell genome sequencing. *Cell, 179*(5), 1207–1221. e22.
12. Carolin, C. F., Kumar, P. S., Saravanan, A., Joshiba, G. J., & Naushad, M. (2017). Efficient techniques for the removal of toxic heavy metal from aquatic environment: A review. *Journal of Environmental Chemical Engineering, 5*(3), 2782–2799.
13. Elkacmi, R., & Bennajah, M., (2019). Advanced oxidation technologies for the treatment and detoxification of olive mill wastewater: A general review. *Journal of Water Reuse and Desalination, 9*(4), 463–505.
14. Gadysz-paska, A., Skwarek, E., Budnyak, T., & Koodyska, D., (2017). Metal ions removal using nano oxide pyrolox (TM) material. *Nanoscale Research Letters, 12*(1), 1.
15. Fu, Y., Freymueller, J. T., & Jensen, T., (2012). Seasonal hydrological loading in southern alaska observed by GPS and GRACE. *Geophysical Research Letters, 39*(15).
16. Thaçi, B. S., & Gashi, S. T., (2019). Reverse osmosis removal of heavy metal from wastewater effluents using biowaste materials pretreatment. *Pol. J. Environ. Stud, 28*(1), 337–341.
17. Gupta, V. K., Kumar, R., Nayak, A., Saleh, T. A., & Barakat, M., (2013). Adsorptive removal of dyes from aqueous solution onto carbon nanotubes: A review. *Advances in Colloid and Interface Science, 193*, 24–34.
18. Qdais, H. A., & Moussa, H., (2004). Removal of heavy metal from wastewater by membrane processes: A comparative study. *Desalination, 164*(2), 105–110.

19. Mantel, A., Newsome, A., Thekkudan, T., Frazier, R., & Katdare, M. (2016). The role of aldo-keto reductase 1C3 (AKR1C3)-mediated prostaglandin D2 (PGD2) metabolism in keloids. *Experimental Dermatology, 25*(1), 38–43.
20. Tian, F., Dong, X., Zhao, Z., He, J., & Wang, H. T., (2012). Super hard F-carbon predicted by ab initio particle-swarm optimization methodology. *Journal of Physics: Condensed Matter, 24*(16), 165504.
21. Namasivayam, C., & Sangeetha, D., (2006). Recycling of agricultural solid waste, coir pith: Removal of anions, heavy metals, organics, and dyes from water by adsorption onto $ZnCl_2$ activated coir pith carbon. *Journal of Hazardous Materials, 135*(1–3), 449–452.
22. Sudilovskiy, P., Kagramanov, G., Trushin, A., & Kolesnikov, V., (2007). Use of membranes for heavy metal cationic wastewater treatment: Flotation and membrane filtration. *Clean Technologies and Environmental Policy, 9*(3), 189–198.
23. Tran, T. K., Leu, H. J., Chiu, K. F., & Lin, C. Y., (2017). Electrochemical treatment of heavy metal-containing wastewater with the removal of COD and heavy metal ions. *Journal of the Chinese Chemical Society, 64*(5), 493–502.
24. Mavrov, V., Erwe, T., Blöcher, C., & Chmiel, H., (2003). Study of new integrated processes combining adsorption, membrane separation, and flotation for heavy metal removal from wastewater. *Desalination, 157*(1–3), 97–104.
25. Goutam, S. P., Saxena, G., Singh, V., Yadav, A. K., Bharagava, R. N., & Thapa, K. B., (2018). Green synthesis of TiO_2 nanoparticles using leaf extract of *Jatropha curcas* L. for photocatalytic degradation of tannery wastewater. *Chemical Engineering Journal, 336*, 386–396.
26. Bora, T., & Dutta, J., (2014). Applications of nanotechnology in wastewater treatment—A review. *Journal of Nanoscience and Nanotechnology, 14*(1), 613–626.
27. Enfrin, M., Lee, J., Gibert, Y., Basheer, F., Kong, L., & Dumée, L. F. (2020). Release of hazardous nanoplastic contaminants due to microplastic fragmentation under shear stress forces. *Journal of Hazardous Materials, 384*, 121393.
28. Shrivastava, P., Jain, S., Kumar, N., Jain, V. K., & Nagpal, S., (2021). Handheld device for rapid detection of lead (Pb^{2+}) in gunshot residue for forensic application. *Microchemical Journal, 165*, 106186.
29. Shrivastava, P., Singh, B. P., Jain, S. K., Jain, V. K., & Nagpal, S., (2021). A novel approach to detect barium in gunshot residue using a handheld device: A forensic application. *Analytical Methods, 13*(38), 4379–4389.
30. Sharma, S., & Bhattacharya, A., (2017). Drinking water contamination and treatment techniques. *Applied Water Science, 7*(3), 1043–1067.
31. Shrivastava, P., Jain, V., & Nagpal, S., (2021). Gunshot residue detection technologies—A review. *Egyptian Journal of Forensic Sciences, 11*(1), 1–21.
32. Kumar, S., Ahlawat, W., Kumar, R., & Dilbaghi, N., (2015). Graphene, carbon nanotubes, zinc oxide, and gold are elite nanomaterials for the fabrication of biosensors for healthcare. *Biosensors and Bioelectronics, 70*, 498–503.
33. Buzea, C., & Pacheco, I., (2017). Nanomaterial and nanoparticle: Origin and activity. *Nanoscience and Plant–Soil Systems* (pp. 71–112). Springer.
34. Dresselhaus, M. S., & Terrones, M., (2013). Carbon-based nanomaterials from a historical perspective. *Proceedings of the IEEE, 101*(7), 1522–1535.
35. Gangadhar, G., Maheshwari, U., & Gupta, S., (2012). Application of nanomaterials for the removal of pollutants from effluent streams. *Nanoscience & Nanotechnology-Asia, 2*(2), 140–150.

36. Baby, R., Saifullah, B., & Hussein, M. Z. (2019). Carbon nanomaterials for the treatment of heavy metal-contaminated water and environmental remediation. *Nanoscale Research Letters, 14*(1), 1–17.
37. Gupta, D. K., Palma, J. M., & Corpas, F. J., (2016). *Redox State as a Central Regulator of Plant-Cell Stress Responses*. Springer.
38. Kumar, V., Kumar, V., Som, S., Neethling, J., Olivier, E., Ntwaeaborwa, O., & Swart, H., (2014). The role of surface and deep-level defects on the emission of tin oxide quantum dots. *Nanotechnology, 25*(13), 135701.
39. He, P., Zhang, G., Liao, X., Yan, M., Xu, X., An, Q., Liu, J., & Mai, L., (2018). Sodium ion stabilized vanadium oxide nanowire cathode for high-performance zinc-ion batteries. *Advanced Energy Materials, 8*(10), 1702463.
40. Fritsche, L. G., Igl, W., Bailey, J. N. C., Grassmann, F., Sengupta, S., Bragg-Gresham, J. L., Burdon, K. P., et al., (2016). A large genome-wide association study of age-related macular degeneration highlights the contributions of rare and common variants. *Nature Genetics, 48*(2), 134–143.
41. Ye, S., Yan, M., Tan, X., Liang, J., Zeng, G., Wu, H., Song, B., et al., (2019). Facile-assembled biochar-based nanocomposite with improved graphitization for efficient photocatalytic activity driven by visible light. *Applied Catalysis B: Environmental, 250*, 78–88.
42. Kumar, R., Khan, M. A., & Haq, N., (2014). Application of carbon nanotubes in heavy metal remediation. *Critical Reviews in Environmental Science and Technology, 44*(9), 1000–1035.
43. Mohan, V. B., Lau, K. T., Hui, D., & Bhattacharyya, D. (2018). Graphene-based materials and their composites: A review on production, applications, and product limitations. *Composites Part B: Engineering, 142*, 200–220.
44. Tabish, T. A., Lin, L., Ali, M., Jabeen, F., Ali, M., Iqbal, R., Horsell, D. W., et al., (2018). Investigating the bioavailability of graphene quantum dots in lung tissues via Fourier transform infrared spectroscopy. *Interface Focus, 8*(3), 20170054.
45. Ren, H., Tang, M., Guan, B., Wang, K., Yang, J., Wang, F., Wang, M., et al., (2017). Hierarchical graphene foam for efficient omnidirectional solar–thermal energy conversion. *Advanced Materials, 29*(38), 1702590.
46. Qu, R., Zhang, W., Liu, N., Zhang, Q., Liu, Y., Li, X., Wei, Y., & Feng, L., (2018). Antioil Ag_3PO_4 nanoparticle/polydopamine/Al_2O_3 sandwich structure for complex wastewater treatment: Dynamic catalysis under natural light. *ACS Sustainable Chemistry & Engineering, 6*(6), 8019–8028.
47. Vilela, D., Parmar, J., Zeng, Y., Zhao, Y., & Sánchez, S., (2016). Graphene-based microbots for toxic heavy metal removal and recovery from water. *Nano Letters, 16*(4), 2860–2866.
48. Zhang, C., Wang, W., Duan, A., Zeng, G., Huang, D., Lai, C., Tan, X., et al., (2019). Adsorption behavior of engineered carbons and carbon nanomaterials for metal endocrine disruptors: Experiments and theoretical calculation. *Chemosphere, 222*, 184–194.
49. Arshad, A., Jabbal, M., Yan, Y., & Reay, D., (2019). A review on graphene based nanofluids: Preparation, characterization, and applications. *Journal of Molecular Liquids, 279*, 444–484.
50. Alekseeva, A., Shaikhiev, I., Stepanova, S., Mavrin, G., Fazullin, D., Kharlyamov, D., & Shaimardanova, A., (2016). *The Use of Leaves of Different Tree Species as a Sorption*

Material for Extraction of Heavy Metal Ions from Aqueous Media. https://dspace.kpfu.ru/xmlui/handle/net/145648.

51. Wren, S. A., & Rowe, R. C., (1992). Theoretical aspects of chiral separation in capillary electrophoresis: I. Initial evaluation of a model. *Journal of Chromatography A, 603*(1–2), 235–241.
52. Yeganeh, H., Atai, M., Hojati, T. P., & Jamshidi, S., (2006). Synthesis, characterization, and properties of novel poly (urethane-imide) networks as electrical insulators with improved thermal stability. *Macromolecular Materials and Engineering, 291*(7), 883–894.
53. El-Saharty, A., Mahmoud, S. N., Manjood, A. H., Nassar, A. A. H., & Ahmed, A. M., (2018). Effect of apricot stone activated carbon adsorbent on the removal of toxic heavy metal ions from aqueous solutions. *Int. J. Ecotoxicol. Ecobiol., 3*(2), 13.
54. Li, J., Wang, X., Zhao, G., Chen, C., Chai, Z., Alsaedi, A., Hayat, T., & Wang, X., (2018). Metal–organic framework-based materials: Superior adsorbents for the capture of toxic and radioactive metal ions. *Chemical Society Reviews, 47*(7), 2322–2356.
55. Yuan, J., Zhang, Y., Zhou, L., Zhang, G., Yip, H. L., Lau, T. K., Lu, X., et al., (2019). Single-junction organic solar cell with over 15% efficiency using fused-ring acceptor with electron-deficient core. *Joule, 3*(4), 1140–1151.
56. Kotsyuda, S. S., Tomina, V. V., Zub, Y. L., Furtat, I. M., Lebed, A. P., Vaclavikova, M., & Melnyk, I. V., (2017). Bifunctional silica nanospheres with 3-aminopropyl and phenyl groups. Synthesis approach and prospects of their applications. *Applied Surface Science, 420*, 782–791.
57. Najafi, M., Yousefi, Y., & Rafati, A., (2012). Synthesis, characterization, and adsorption studies of several heavy metal ions on amino-functionalized silica nano hollow sphere and silica gel. *Separation and Purification Technology, 85*, 193–205.
58. Pogorilyi, R., Melnyk, I., Zub, Y., Carlson, S., Daniel, G., Svedlindh, P., Seisenbaeva, G., & Kessler, V., (2014). New product from the old reaction: Uniform magnetite nanoparticles from iron-mediated synthesis of alkali iodides and their protection from leaching in acidic media. *RSC Advances, 4*(43), 22606–22612.
59. Mahmoud, M. E., Fekry, N. A., & El-Latif, M. M., (2016). Nanocomposites of nanosilica-immobilized-nanopolyaniline and crosslinked nanopolyaniline for removal of heavy metals. *Chemical Engineering Journal, 304*, 679–691.
60. Bokare, V., Jung, J. L., Chang, Y. Y., & Chang, Y. S., (2013). Reductive dechlorination of octachlorodibenzo-p-dioxin by nanosized zero-valent zinc: Modeling of rate kinetics and congener profile. *Journal of Hazardous Materials, 250*, 397–402.
61. Waszak, S. M., Northcott, P. A., Buchhalter, I., Robinson, G. W., Sutter, C., Groebner, S., Grund, K. B., et al., (2018). Spectrum and prevalence of genetic predisposition in medulloblastoma: A retrospective genetic study and prospective validation in a clinical trial cohort. *The Lancet Oncology, 19*(6), 785–798.
62. Ouyang, T., Ye, Y. Q., Wu, C. Y., Xiao, K., & Liu, Z. Q., (2019). Heterostructures composed of N-doped carbon nanotubes encapsulating cobalt and β-Mo2C nanoparticles as bifunctional electrodes for water splitting. *Angewandte Chemie, 131*(15), 4977–4982.
63. Ojea-Jiménez, I., García-Fernández, L., Lorenzo, J., & Puntes, V. F., (2012). Facile preparation of cationic gold nanoparticle-bioconjugates for cell penetration and nuclear targeting. *ACS Nano, 6*(9), 7692–7702.
64. Mueller, S., Ahmed, K., Singh, A., Davis, A., Mukhopadyay, S., Swaminathan, M., Mano, Y., et al., (2016). Design of high-efficiency integrated voltage regulators with

embedded magnetic core inductors. In: *2016 IEEE 66th Electronic Components and Technology Conference (ECTC)* (pp. 566–573). IEEE.
65. Sumesh, E., Bootharaju, M., & Pradeep, T., (2011). A practical silver nanoparticle-based adsorbent for the removal of Hg^{2+} from water. *Journal of Hazardous Materials, 189*(1, 2), 450–457.
66. Ali, S., Ganai, B. A., Kamili, A. N., Bhat, A. A., Mir, Z. A., Bhat, J. A., Tyagi, A., et al., (2018). Pathogenesis-related proteins and peptides as promising tools for engineering plants with multiple stress tolerance. *Microbiological Research, 212*, 29–37.
67. Liu, T., Wang, Z. L., & Sun, Y., (2015). Manipulating the morphology of nanoscale zero-valent iron on pumice for removal of heavy metal from wastewater, *Chemical Engineering Journal, 263*, 55–61.
68. Zhang, Z., Hao, Z. W., Liu, W. L., & Xu, X. H., (2009). Synchronous treatment of heavy metal ions and nitrate by zero-valent iron. *Huan Jing Ke Xue= Huanjing Kexue, 30*(3), 775–779.
69. Huang, D. L., Chen, G. M., Zeng, G. M., Xu, P., Yan, M., Lai, C., Zhang, C., et al., (2015). Synthesis and application of modified zero-valent iron nanoparticles for removal of hexavalent chromium from wastewater. *Water, Air, & Soil Pollution, 226*(11), 1–14.
70. Su, Y., Adeleye, A. S., Huang, Y., Sun, X., Dai, C., Zhou, X., Zhang, Y., & Keller, A. A., (2014). Simultaneous removal of cadmium and nitrate in aqueous media by nanoscale zerovalent iron (nZVI) and Au-doped nZVI particles. *Water Research, 63*, 102–111.
71. Wang, X., Huang, K., Chen, Y., Liu, J., Chen, S., Cao, J., Mei, S., et al., (2018). Preparation of dumbbell manganese dioxide/gelatin composites and their application in the removal of lead and cadmium ions. *Journal of Hazardous Materials, 350*, 46–54.
72. Wan, S., Wu, J., Zhou, S., Wang, R., Gao, B., & He, F. (2018). Enhanced lead and cadmium removal using biochar-supported hydrated manganese oxide (HMO) nanoparticles: Behavior and mechanism. *Science of the Total Environment, 616*, 1298–1306.
73. Martínez-Cabanas, M., López-García, M., Barriada, J. L., Herrero, R., & De Vicente, M. E. S., (2016). Green synthesis of iron oxide nanoparticles. Development of magnetic hybrid materials for efficient As (V) removal. *Chemical Engineering Journal, 301*, 83–91.
74. Wu, S., Mickley, L. J., Jacob, D. J., Rind, D., & Streets, D. G., (2008). Effects of 2000–2050 changes in climate and emissions on global tropospheric ozone and the policy-relevant background surface ozone in the United States. *Journal of Geophysical Research: Atmospheres, 113*(D18).
75. Gautam, J., Thanh, T. D., Maiti, K., Kim, N. H., & Lee, J. H., (2018). Highly efficient electrocatalyst of N-doped graphene-encapsulated cobalt-iron carbides towards oxygen reduction reaction. *Carbon, 137*, 358–367.
76. Yang, J., Zheng, Y., Gou, X., Pu, K., Chen, Z., Guo, Q., Ji, R., et al., (2020). Prevalence of comorbidities and its effects in patients infected with SARS-CoV-2: A systematic review and meta-analysis. *International Journal of Infectious Diseases, 94*, 91–95.
77. Mahdavi, M., Ahmad, M. B., Haron, M. J., Namvar, F., Nadi, B., Rahman, M. Z. A., & Amin, J., (2013). Synthesis, surface modification, and characterization of biocompatible magnetic iron oxide nanoparticles for biomedical applications. *Molecules, 18*(7), 7533–7548.
78. Tabesh, S., Davar, F., & Loghman-Estarki, M. R., (2018). Preparation of γ-Al_2O_3 nanoparticles using the modified sol-gel method and its use for the adsorption of lead and cadmium ions. *Journal of Alloys and Compounds, 730*, 441–449.

79. Zhang, W., & Xiong, R. G., (2012). Ferroelectric metal–organic frameworks. *Chemical Reviews, 112*(2), 1163–1195.
80. Dong, X., Chen, J., Ma, Y., Wang, J., Chan-Park, M. B., Liu, X., Wang, L., et al., (2012). Superhydrophobic and superoleophilic hybrid foam of graphene and carbon nanotube for selective removal of oils or organic solvents from the surface of water. *Chemical Communications, 48*(86), 10660–10662.
81. Youssef, A. M., & Malhat, F. M., (2015). Selective removal of heavy metal from drinking water using titanium dioxide nanowire. *Macromolecular Symposia* (pp. 96–101). Wiley Online Library.
82. Kalpana, T., Brik, M., Sudarsan, V., Naresh, P., Kumar, V. R., Kityk, I., & Veeraiah, N., (2015). Influence of Al^{3+} ions on luminescence efficiency of Eu^{3+} ions in barium boro-phosphate glasses. *Journal of Non-Crystalline Solids, 419*, 75–81.
83. Kumar, K. Y., Muralidhara, H., Nayaka, Y. A., Balasubramanyam, J., & Hanumanthappa, H., (2013). Low-cost synthesis of metal oxide nanoparticles and their application in adsorption of commercial dye and heavy metal ion in aqueous solution. *Powder Technology, 246*, 125–136.
84. Sheela, T., Nayaka, Y. A., Viswanatha, R., Basavanna, S., & Venkatesha, T., (2012). Kinetics and thermodynamics studies on the adsorption of Zn(II), Cd(II), and Hg(II) from aqueous solution using zinc oxide nanoparticles. *Powder Technology, 217*, 163–170.
85. Somu, P., & Paul, S., (2018). Casein-based biogenic-synthesized zinc oxide nanoparticles simultaneously decontaminate heavy metals, dyes, and pathogenic microbes: A rational strategy for wastewater treatment. *Journal of Chemical Technology & Biotechnology, 93*(10), 2962–2976.
86. Umar, A., Kumar, R., Akhtar, M., Kumar, G., & Kim, S., (2015). Growth and properties of well-crystalline cerium oxide (CeO_2) nanoflakes for environmental and sensor applications. *Journal of Colloid and Interface Science, 454*, 61–68.
87. Recillas, S., Colón, J., Casals, E., González, E., Puntes, V., Sánchez, A., & Font, X., (2010). Chromium VI adsorption on cerium oxide nanoparticles and morphology changes during the process. *Journal of Hazardous Materials, 184*(1–3), 425–431.
88. Seisenbaeva, G. A., Daniel, G., Kessler, V. G., & Nedelec, J. M., (2014). A general facile approach to transition-metal oxides with highly uniform mesoporosity and their application as adsorbents for heavy-metal-ion sequestration. *Chemistry–A European Journal, 20*(34), 10732–10736.
89. Zhang, Q., Teng, J., Zhang, Z., Nie, G., Zhao, H., Peng, Q., & Jiao, T., (2015). Unique and outstanding cadmium sequestration by polystyrene-supported nanosized zirconium hydroxides: A case study. *RSC Advances, 5*(68), 55445–55452.
90. Abdel-Aty, H., Boyé, P., Zagrosek, A., Wassmuth, R., Kumar, A., Messroghli, D., Bock, P., et al., (2005). Diagnostic performance of cardiovascular magnetic resonance in patients with suspected acute myocarditis: Comparison of different approaches. *Journal of the American College of Cardiology, 45*(11), 1815–1822.
91. Oki, E., Tokunaga, S., Emi, Y., Kusumoto, T., Yamamoto, M., Fukuzawa, K., Takahashi, I., et al., (2016). Surgical treatment of liver metastasis of gastric cancer: A retrospective multicenter cohort study (KSCC1302). *Gastric Cancer, 19*(3), 968–976.
92. Mehwish, N., Kausar, A., & Siddiq, M., (2014). Advances in polymer-based nanostructured membranes for water treatment. *Polymer-Plastics Technology and Engineering, 53*(12), 1290–1316.

93. Hotze, E. M., Phenrat, T., & Lowry, G. V., (2010). Nanoparticle aggregation: Challenges to understanding transport and reactivity in the environment. *Journal of Environmental Quality, 39*(6), 1909–1924.
94. Parlayici, S., Eskizeybek, V., Avcı, A., & Pehlivan, E., (2015). Removal of chromium (VI) using activated carbon-supported-functionalized carbon nanotubes. *Journal of Nanostructure in Chemistry, 5*(3), 255–263.
95. Jayaweera, H. C., Siriwardane, I., De Silva, K. N., & De Silva, R. M., (2018). Synthesis of multifunctional activated carbon nanocomposite comprising biocompatible flake nano hydroxyapatite and natural turmeric extract for the removal of bacteria and lead ions from aqueous solution. *Chemistry Central Journal, 12*(1), 1–14.
96. Hayati, B., Maleki, A., Najafi, F., Gharibi, F., McKay, G., Gupta, V. K., Puttaiah, S. H., & Marzban, N., (2018). Heavy metal adsorption using PAMAM/CNT nanocomposite from aqueous solution in batch and continuous fixed bed systems. *Chemical Engineering Journal, 346*, 258–270.
97. Eskandari, M., Khatir, M. Z., Darban, A. K., & Meshkini, M., (2018). Decreasing Ni, Cu, Cd, and Zn heavy metal using magnetite-bentonite nanocomposites and adsorption isotherm study. *Materials Research Express, 5*(4), 045030.
98. Lu, F., & Astruc, D., (2018). Nanomaterials for removal of toxic elements from water. *Coordination Chemistry Reviews, 356*, 147–164.
99. Cai, J., Lei, M., Zhang, Q., He, J. R., Chen, T., Liu, S., Fu, S. H., et al., (2017). Electrospun composite nanofiber mats of Cellulose@ Organically modified montmorillonite for heavy metal ion removal: Design, characterization, evaluation of absorption performance. *Composites Part A: Applied Science and Manufacturing, 92*, 10–16.
100. Gokila, S., Gomathi, T., Sudha, P., & Anil, S., (2017). Removal of the heavy metal ion chromium (VI) using chitosan and alginate nanocomposites. *International Journal of Biological Macromolecules, 104*, 1459–1468.
101. Lofrano, G., Carotenuto, M., Libralato, G., Domingos, R. F., Markus, A., Dini, L., Gautam, R. K., et al., (2016). Polymer functionalized nanocomposites for metals removal from water and wastewater: An overview. *Water Research, 92*, 22–37.
102. Pavlidou, S., & Papaspyrides, C., (2008). A review on polymer–layered silicate nanocomposites. *Progress in Polymer Science, 33*(12), 1119–1198.
103. Fornes, T., Yoon, P., Keskkula, H., & Paul, D., (2001). Nylon 6 nanocomposites: The effect of matrix molecular weight. *Polymer, 42*(25), 09929–09940.
104. He, C., Hu, Y., Yin, L., Tang, C., & Yin, C., (2010). Effects of particle size and surface charge on cellular uptake and biodistribution of polymeric nanoparticles. *Biomaterials, 31*(13), 3657–3666.
105. Zhao, G., Huang, X., Tang, Z., Huang, Q., Niu, F., & Wang, X., (2018). Polymer-based nanocomposites for heavy metal ions removal from aqueous solution: A review. *Polymer Chemistry, 9*(26), 3562–3582.
106. Mbui, D. W., Andala, D. M., Abong'o, D. A., & Mmbaga, J. N., (2018). Design and fabrication of polymer layered silicate nanocomposites for water purification. *MRS Advances, 3*(36), 2099–2107.
107. Saleh, T. A., & Al-Absi, A. A., (2017). Kinetics, isotherms, and thermodynamic evaluation of amine functionalized magnetic carbon for methyl red removal from aqueous solutions. *Journal of Molecular Liquids, 248*, 577–585.

108. Tan, H., Zhao, Z., Zhu, W. B., Coker, E. N., Li, B., Zheng, M., Yu, W., Fan, H., & Sun, Z., (2014). Oxygen vacancy enhanced the photocatalytic activity of pervoskite $SrTiO_3$. *ACS Applied Materials & Interfaces, 6*(21), 19184–19190.
109. Roy, A. S., Mondal, J., Banerjee, B., Mondal, P., Bhaumik, A., & Islam, S. M., (2014). Pd-grafted porous metal–organic framework material as an efficient and reusable heterogeneous catalyst for C–C coupling reactions in water. *Applied Catalysis A: General, 469*, 320–327.
110. Bauernfeind, F. G., Horvath, G., Stutz, A., Alnemri, E. S., MacDonald, K., Speert, D., Fernandes-Alnemri, T., et al., (2009). Cutting edge: NF-κB activating pattern recognition and cytokine receptors license NLRP3 inflammasome activation by regulating NLRP3 expression. *The Journal of Immunology, 183*(2), 787–791.

CHAPTER 6

Nanomaterials for Inorganic Pollutants Removal from Contaminated Water

SHIV VENDRA SINGH,[1] RASHMI SHARMA,[1] PRIYANKA BALAN,[2] SHUBHAM DURGUDE,[3] and SUKANYA GHOSH[1]

[1]*School of Agriculture, Graphic Era Hill University, Dehradun, Uttarakhand, India*

[2]*Department of Environmental Science, Dr. Y.S. Parmar University of Horticulture and Forestry, Solan, Himachal Pradesh, India*

[3]*School of Agriculture, Graphic Era Hill University, Bhimtal, Uttarakhand, India*

ABSTRACT

Water is an essential natural resource for living beings' survival; hence, the quality becomes a major concern. The increasing industrial development is depleting freshwater availability and elevating contamination in the ecosystem, which is expected to worsen in the future. Water reservoirs are getting contaminated with pollutants like hydrocarbons, pesticides, pharmaceutics, cosmetics, heavy metals, etc. The toxicity of heavy metals may have adverse effects on the liver, kidney, blood, nervous, circulatory, gastrointestinal system, bones, and skin. Considering their adverse effects on living beings, the environment, and the ecosystem, water remediation technologies are of utmost importance. Extensive studies are being carried out to find an eco-friendly and economically sustainable water treatment approach. Several mechanisms, such as coagulation, adsorption, membrane filtration, dialysis, electrocoagulation, and nanoremediation, are proving their effectiveness. The adsorption and membrane filtration processes for the removal of pollutants with nanomaterials, due to their cheaper, easy to use,

Nano-Bioremediation for Water and Soil Treatment: An Eco-Friendly Approach.
Vishnu D. Rajput, Arpna Kumari, and Tatiana M. Minkina (Eds.)

lesser treatment time, and higher efficacy, have been widely explored by the scientific community. The present chapter emphasizes the potential utilization of nanomaterials for inorganic pollutants removal from contaminated water, highlighting their applicability, advantages, and limitations.

6.1 INTRODUCTION

Rapid industrialization and globalization are imposing threats to human health and the ecosystem by water contamination or pollution, which is expected to get worse in the future [1]. Not only industrialization and urbanization but faulty savage and slug management and heavy pesticide inclusion in crop production systems have added plenty of pollutants to the groundwater and water bodies [2, 3]. Nonbiodegradable organic and inorganic pollutants in underwater ecosystems known to be persistent are causing a plethora of problems associated with human beings, water bodies, as well as environment [4]. The inorganic pollutants are showing harmful mutagenic and carcinogenic in aquatic creatures and traveling into human food chains as biproducts besides the other direct impacts [5]. In many parts of the world, heavy metals and other toxicants, like fluoride and nitrate, are being reported beyond the threshold limit in the groundwater, which is making it undesirable [6]. The rising levels of these metals/heavy metals are becoming a concern worldwide as these are being widely distributed into the water ecosystem with potential threats to human health and the atmosphere (Figure 6.1) [7].

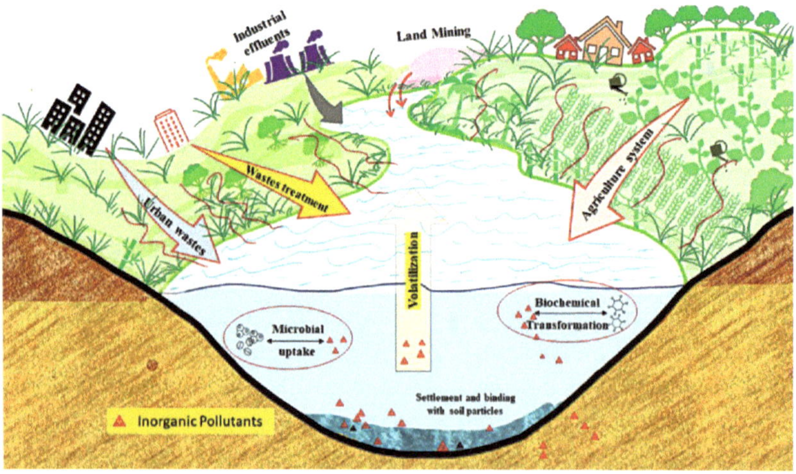

FIGURE 6.1 Heavy metals get into water bodies from the potential sources vi,z., municipal wastewater treatment plants, manufacturing industries, mining, and rural agricultural cultivation and fertilization.

6.2 INORGANIC WATER POLLUTANTS, THEIR SOURCES, AND IMPACTS

In the last few decades, industrialization, urbanization, faulty savage and slug management, and heavy pesticide inclusion in crop production systems have added plenty of pollutants like inorganic chemicals, heavy metals, hydrocarbons, insecticides, pharmaceutics, cosmetics, radionuclide microbes, synthesized organic reactants into the groundwater and water bodies [3, 8]. Out of numerous pollutants, frequently observed inorganic contaminants include heavy metals arsenic (As), fluoride (F), iron (Fe), nitrate (NO_3^-), oxyanions, cations, radioactive materials, etc. [9, 10]. Effluents from mines, smelters, sewage, agro-industries, urban wastes, alloys, and electronic factories are the major sources of metal contaminants like As, Cd, Cr, Cu, Hg, Pb, and Zn [10]. Among the inorganic pollutants, heavy metals and metalloids are the primary constituents associated with anthropogenic activities like fossil fuel burning, mining, faulty industrial and municipal waste management, pesticide, and fertilizer inclusion in agricultural production systems, and so on. Grouping of inorganic pollutants as: (i) acids and bases: industrial discharges/effluents and coal burning power plants; (ii) chemical wastes: manufacturing process and agricultural inputs; (iii) heavy metals: occurs from motor vehicles, mine, drainage; and (iv) sediments: powdery constituents, typically arises through creation sites, land clearing sites [11].

Metallic ions like Cd, Cu, Cr, As, Hg, Ni, Pb, Sn, etc., having elemental density of $>4 \pm 1$ g cc^{-1}, are considered heavy metals [12]). These are widely distributed in the ecosystem due to increasing involvement in numerous domestic, pharmaceutical, pesticide, and fertilizers industries [13]. Besides these, corrosion of metals, soil erosion, metals leaching and runoff into water bodies and soil systems, weathering, and volcanic eruptions also lead to metal contamination. Some of the heavy metals like Zn, Cu, Se, Cr, Co, Mn, Mo, Al, and Ni are considered essential trace elements for plant and microbial physicochemical and biological activities [14]. Therefore, metals or heavy metal pollutants, like cadmium, lead, and mercury, are considered significantly hazardous. Acids are basically the effluents of different industries and chemical research laboratories that can alter the pH. Acids are pollutants with low hazardous potential as they can be comparatively easily neutralized but can excessively impose significant hazardous impacts on the environment and human health. Some of the potential inorganic water pollutants and their sources are listed in Table 6.1.

TABLE 6.1 Inorganic Pollutants, Their Sources and Permissible Limits in Drinking Water and Their Impact on Living Beings

Sl. No.	Inorganic Pollutant	Sources	Allowable Limits	Harmful Effects	References (effects reported by)
1.	Arsenic	Industrial discharge	0.05*	Cancer (skin, bladder, and lung), infant mortality, epigenetic changes, hepatitis, and cardiovascular disorders.	[15]
2.	Lead	Heavy industry	0.05**	Neurological disorders, skeletal, endocrine, and immune system damage, toxicity in the kidney, hypertension, and stroke.	[16]
3.	Mercury	Industry	0.002*	Neurotoxicity, genotoxicity, cardiovascular toxicity, disrupting endocrine systems, impaired growth and development, induced liver and kidney damage, sensory disturbance, visual field constriction, ataxia, dysarthria, and immunomodulation.	[16]
4.	Nickel	Battery, electronics, and pigments manufacture, electroplating.	0.1*	Dermatitis, lung fibrosis, lung cancer, nasal cancer, severe headaches, gastrointestinal and cardiovascular infestations.	[17]
5.	Selenium	Discharge from mines and refineries, natural deposits, and agricultural runoff.	0.05*	Dermatitis, renal failure, infertility, myocardial infarction, damage of the liver, kidney, and spleen, nervousness, and even death.	[18]
6.	Cadmium	Geogenic, metallurgical process, discarded batteries.	0.005*	Emphysema, chronic inflammation of the pharynx, renal tubular dysfunction, bone osteomalacia, and cancer.	[19]
7.	Chromium	Geogenic, steel factories, paper, and pulp mills.	0.1*	Apoptosis, inflammatory cytokines, hepatotoxicity, respiratory problems, reproductive system damage, cancer (lung, stomach, intestinal tract).	[20]
8.	Uranium	Naturally occurring mineral	0.02**	Kidney toxicity, increased risk of cancer.	[21]
9.	Antimony	Mining, smelting waste, tailings dam, underground tunnel wastewater.	0.006**	Affects the skin, irritates the eyes, causes kidney disorders, damages the respiratory system, and disrupts the gastrointestinal tract.	[22]
10.	Zinc	Industrial effluents	5**	Lethargy, neuronal deficits, metal fume fever, nausea, vomiting, epigastric pain, diarrhea, and prostate cancer.	[23]
11.	Copper	Chemicals used for the preservation of wood and corrosion of pipes.	1.3*	Hepatotoxicity, anemia, oxidative damage, Wilson disease, intravascular coagulation, jaundice, renal failure, low leucocyte counts, and neurological defects.	[24]

TABLE 6.1 *(Continued)*

Sl. No.	Inorganic Pollutant	Sources	Allowable Limits	Harmful Effects	References (effects reported by)
12.	Aluminum	Industrial wastewater effluents.	0.2**	Alzheimer's disease, neurofibril degeneration, lymphopenia, and memory loss.	[25]
13.	Iron	Natural deposits, industrial effluents, refining iron ores, corrosion of Fe-containing metals.	0.1**	Hepatic fibrosis, cirrhosis, osteoarthritis, infertility, and increases the risk of liver diseases.	[26]
14.	Manganese	Industrial emissions, soil erosion, volcanic emissions.	0.5*	Osteoporosis, muscle and joint pain, sexual dysfunction, mental illness, and neurotoxicity with extrapyramidal symptoms.	[26]
15.	Sulfur dioxide	Volcanoes, acid rain, snow, sleet, fog.	0.02	Increased asthma attacks, heart and lung disease, and respiratory problems.	–
16.	Hydrogen sulfide	Industry, anaerobic fermentation	0.05**	Nausea, vomiting, and epigastric pain following ingestion.	[27]
17.	Chlorides	Industrial fumes	250*	Pulmonary edema, shock, circulatory collapse, metabolic acidosis, and respiratory depression.	[28]
18.	Sulfates	Mining, industrial effluents (paper, pulp, and textile industries).	250*	Osmotic diarrhea	–
19.	Hydrogen fluoride	Industry	4*	Fluorosis, dental and skeletal abnormalities, and adverse effects on the gastrointestinal tract, kidneys, and immune systems.	[29]
20.	Ammonia	Industry	0.5*	Lung edema, nervous system dysfunction, acidosis, and kidney damage.	[30]
21.	Nitrate salts	Farms, factories	45**	Methemoglobinemia or blue baby syndrome.	–

**United States Environmental Protection Agency (USEPA);*

***World Health Organization (WHO)*

6.3 APPLICATION OF NANOMATERIALS IN WATER REMEDIATION

Considering the numerous past examples of incidents of water pollution on human health and the ecosystem, governments worldwide introducing numerous laws to maintain the quality of groundwater and drinking water. To monitor the water quality with minimum hazard, the Clean Water Act and Safe Drinking Water Act are being followed worldwide. Besides these acts and lows, several standard water purifying methods, *viz*., osmosis, coagulation, membrane filtration, precipitations, microbial fuel cell, and advanced oxidation process, are being adopted. Recently, nanotechnology has shown a role in the environmental restoration of aquatic ecosystems.

6.4 METAL ORGANIC FRAMEWORK FOR WATER REMEDIATION

In the last few decades, attention has been drawn to the class of porous materials called metal organic frameworks (MOF) that are being widely employed for treatment [31, 32]. These MOFs have metal centers linked with organic compounds to create a crystalline network possessing hydrophobic functionalities, flexible pore size and shape, thermal stability, and huge surface area. These features make these materials versatile for application in sustainable inorganic pollutant removal from aqueous solutions. Furthermore, MOFs are making evolutions in adsorption, catalytic degradation, and membrane separation of heavy metals with well-developed porosity, surface functionality, and surface area. The efficacy of UiO-66-NH_2(Zr) was studied with thiourea, isothiocyanate, and isocyanate framework for metal ions bioremediation from aqueous ecosystem [33]. They reported up to 99% removal efficiency against Pb^{+2}, Cr^{+3}, Cd^{+2}, and Hg^{+2}. Similarly, Audu et al. [34] also reported that UiO-66(Zr) based MOF mediated the arsenate ion's removal increment.

6.5 MECHANISM INVOLVED IN INORGANIC POLLUTANTS REMOVAL

Nanotechnology has emerged with potential mechanism absorption of metals and non-metal oxides, effective photocatalysis, and membranes for filtering, reduction, and oxidation of various organic and inorganic water pollutants using of nanomaterials (Figure 6.2). The various nanomaterials used to perform inorganic pollutant remediation from contaminated water in the form of nanosorbents, nano-photocatalysts, nanomotors, nanomembranes, etc. Properties of nanomaterials such as active surface adsorption sites, adsorption-specific species, activation of specific chemical bonds, and size-dependent customized absorption and emission are some being widely explored nowadays. Various metals and metal oxide-based nanoparticles have been reported for adsorption, photocatalysis, oxidation, disinfection, and sensing [35]. Nanomaterials of metal compounds, titanium dioxide nanoparticles, have displayed wider applicability in arsenic adsorption and nanosized magnetite [36]. Redox reactions primarily initiate the structural transformation of organic and inorganic molecules. The hydroxyl moieties are introduced in the molecular structure or form the keto groups for further continuous reaction by molecular oxidations. Photocatalysis is an advanced process of oxidation where electron pairs (e^-/h^+) get generated as the catalytic particles are exposed to sunlight, UV lamps, or xenon lamps [37]. Photocatalyst based photo oxidation process depends on reactive radicals' formation during the photocatalytic reactions. These materials alter the reaction rate due to greater surface ratio and shape-dependent features without their own involvement in the chemical transformation. Photocatalytic removal of heavy metals from aqueous ecosystems can be achieved by reducing toxic pollutants with high-valence into low or zero-valence ions [38]. Several studies have reported the use of nanomaterials in membrane nanotechnology to get polymer-based multifunction membranous substances. These substances contain porous composite layers allowing nanofiltration, ultrafiltration, reverse osmosis, etc. [39]. The composite layers may consist of carbon-based material dispersed into a polymer matrix, which increases the fouling resistance and aqueous transport. The traditional filtration approach requires primarily calcium, magnesium, and sodium ions as exchangers, while nanomembranes do not require such exchangers [40, 41].

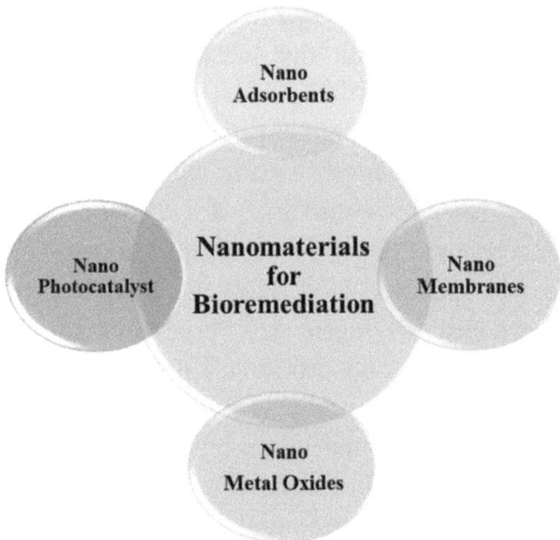

FIGURE 6.2 Potential organic and inorganic pollutant removal mechanisms by nanosorbents.

6.6 NANOMATERIAL FOR INORGANIC POLLUTANTS REMOVAL

Nanomaterials such as iron oxide-based nanoparticles (Fe-NPs), photocatalysts: titanium dioxide (TiO_2), carbon nanotubes (CNTs), graphene-based nanosorbents (GO), metal organic frameworks (MOFs) and layered hydroxides have been explored for water clean-up. Potential nanoparticles frequently reported for removing inorganic pollutants have been listed in Table 6.2.

1. **Nanosorbents:** These are the nanoparticles that have high sorptivity, making them efficient and powerful for water treatment. The majority of explored nanosorbents are carbon-based (CNTs, GO, and rGO), metal oxide-based, and polymeric nanosorbents. Carbon-based nanomaterial like carbon nanotubes (CNTs) having single or multiwalled cylindrical nanotubes hold significant potential with high adsorption sites and surface area. Polymeric nanosorbents (dendrimers) and other adsorbents like copper, zeolites, silver ions, nano-aerogel, etc., are efficient for removing heavy metals from wastewater [42]. The nanomaterials as sorbents have been widely explored with remarkable physiochemical properties. The ideal sorbent to treat the contaminant efficiently must have a superior surface area, size, exceptional adsorptivity, etc. [39]. Some of the widely explored nanomaterials for inorganic pollutants adsorption are activated carbon and carbon nanotubes, MnO, ZnO, FeO, MgO based nanoparticles [43, 44]. Iron-based nanoparticles are low-cost, eco-friendly, and easily synthesized materials for the adsorption of noxious metals with the least secondary contamination. The adsorption efficiency of Fe_2O_3 NPs is pH, temperature, adsorbent, and incubation time dependent [43]. Sorption affinity can be further increased by surface modification with 3-aminopropyltrimethoxysilane [44, 45]. Increased affinity for Cr(III), Co(II), Ni(II), Cu(II), Cd(II), Pb(II), and As(III) due to surface modification of nanoadsorbents was reported [43]. However, MnO-based NPs with huge surface area and polymeric structures have shown great adsorption capacity [46]. Higher adsorption affinity for heavy metals like Pb(II), Cd(II), and Zn(II) on modified hydrous manganese oxide (HMOs) usually occurs due to the inner-sphere development mechanism or ion exchange [43]. Similarly, the morphological modification of ZnO and MgO-based nanomaterials with higher surface area brings out the increased affinity for heavy metals adsorption. Reports on

the removal efficacy of microporous ZnO nano-assemblies for Co^{2+}, Ni^{2+}, Cu^{2+}, Cd^{2+}, Pb^{2+}, Hg^{2+}, and As^{3+} due to electropositive nature were noted [43, 47, 48].

Besides oxides-based nanosorbents, carbon nanotubes (CNTs) and graphene oxide (GO), having a 2D structure developed due to graphite layer oxidation, have also shown the potential to remove heavy metals through adsorption [49]. Many other studies have reported the significant and efficient removal of heavy metals like Pb(II) and Mn(II) [50], Cu(II) [51], and Cr(VI) [42, 52] by using single-walled carbon nanotubes (SWCNTs) and multiwalled-carbon nanotubes (MWCNTs). Surface coating of CNTs by acid treatment, metals grafting, and functional molecules implanting can further increase sorption capacity by altering surface area, functionality, dispersion, hydrophobicity, and functionality. Graphene oxides are carbon-based NPs arranged in 2-D layers of carbon atoms connected in a hexagonal lattice structure [53, 54]. Some of the graphene-based nanosorbents such as reduced-graphene oxide (r-GO), graphene quantum dots (GQD), graphene oxide (GO), graphene nanosheets, and multilayer graphene are graphene-related materials used for water treatments (Table 6.2).

2. **Nanophotocatalyst:** The photocatalysts employ substance catalysis stimulation using light. Most of the nanophotocatalysts (NPCs) are cheaper, stable, and easily accessible, with excellent photoactive properties. NPCs are also known to expand the oxidation ability, producing oxidizing species at the material's surface, which helps in pollutant degradation effectively [55]. Metal oxide-based NPCs, *viz*., SiO_2, ZnO, TiO_2, Al_2O_3, etc., have been reported as the most used nanomaterials for inorganic pollutants removal [12, 56]. TiO_2 is one of the most widely explored photocatalysts due to its easy and cheaper availability and chemical stability. However, ZnO is known for efficiently removing contaminants from wastewater and effective reusability [35, 57] synthesized Synthesis of nanocomposites consisting of carbon quantum dots and precursor CdS nanosheet having photocatalytic reduction efficiency for Cr(VI) by 94% [58]. Multifunctional catalyst TiO_2/Alg/FeNPs magnetic beads have a removal efficiency of nearly 98% for a variety of heavy metal ions, specifically Cr(III), Cu(II), and Pb(II) ions [59]. Photocatalyst and adsorbent properties of nanoscale goethite for heavy metals were

also reported [60]. They have also reported photocatalyst activity of nanoscale goethite for methylene blue (MB) solution under UV light. Khadi et al. [61] reported 4.6 times and 6.8 times higher photocatalysis activity of mesoporous α-Fe_2O_3/g-C_3N_4 nanocomposites than pure α-Fe_2O_3 NPs and g-C_3N_4 nanosheets for Hg(II). Au-decorated TiO_2 nanotubes also exhibited the photocatalytic remediation of Hg(II) from aqueous solutions [62]. Similarly, Bhunia et al. [63] engineered an Ag-doped SnS_2@$InVO_4$ hybrid system for the elimination of arsenic through photodegradation at pH 6 with 97.6% removal efficiency.

TABLE 6.2 Overview of Process and Nanomaterials Application for Inorganic Pollutants Removal from Contaminated Aqueous Solution

Process	Examples of Nanoparticles	Contaminants	Adsorption Capacity (mg/g)/Efficiency (%)	References
Sorption	rGO-Fe(O)	As(III)	37.3 mg/g	[64]
	rGO-Fe_3O_4	Pb(II)	373 mg/g	[65]
	GO-Fe_3O_4	Pb(II)	126.6	[66]
	rGO	Hg(II)	109.49 mg/g	[67]
	RGO/NiO	Cd(II)	–	[68]
	(Graphene oxide naboribbons) GONRs	As(V)	155.61 mg/g	[69]
		Hg(II)	33.02 mg/g	
	MWCNTs	Pb(II)	105 mg/g	[70]
	SWCNTs	Hg(II)	41.6 mg/g	[43]
	SWCNTs-polysulfone	Cr(VI)	96.8%	[71]
	CeO_2-CNTs	Hg(II)	–	[72]
Photocatalytic oxidation/ reduction	CA/TiO_2	Pb(II)	99.7%	[73]
		Cu(II)	98.9%	
	TiO_2 NPs	Cd (II)	70.67%	[74]
		Cr (VI)	76%	[75]
	ZnFe-MMOs	As(III)	95.7%	[57]
	ZnFe-LDHs	As(III)	94.2%	[76]
	Ni-Fe_2O_4-Pd	Pb(II)	98%	[77]
		Cd(II)	97%	
	Au-decorated TiO_2 nanotube	Hg(II)	–	[62]
	TiO_2/Alg/FeNPs	Cu(II)	98%	[59]

TABLE 6.2 *(Continued)*

Process	Examples of Nanoparticles	Contaminants	Adsorption Capacity (mg/g)/Efficiency (%)	References
Micro/ Nanomotors (MNMs)	Fe-Me binary oxide NPs.	Se(VI)	82–92%	[78]
	BC-nZVI	Cr(VI)	98.7%	[79]
	nZVI/Cu	Cr(VI)	94.7%	[80]
	nZVI-HCS	Pb(II)	97.5%	[81]
		Cu(II)	81.0%	
		Zn(II)	54.8%	
		Hg^{2+}	–	[82]
		Pb^{2+}	–	[83]
Nanomembrane filtration	Na-TNB	Sr(II)	97.5%	[84]
	Functionalized MWCNTs	Pb(II)	93%	[85]
		Ni(II)	83%	
		Cu(II)	73%	
		Cd(II)	15%	
	CNT	Br^-, Cl^-, SO_4^{2-}, NO^{-3}	99.99%	[86, 87]
	Nano-filtration membrane bioreactor	NH_3-N	88%	[88]
		NO_3-N	80%	
		PO_4-P	68%	
	Na-TNB	Sr^{2+}	97.5%	[89]
		Cs^+	57.7%	
	Carbon nanofiber membrane	Metal and metal oxide nanoparticles.	95%	[90]

3. **Nanomotors:** The trends in water decontamination and bioremediation throughout the world are growing rapidly due to escalating demand for purity and water superiority regulations. Numerous approaches have been adopted for inorganic pollutants decontamination. The nano/micromotors convert the energy into machine-driven force to achieve the goals of pollutants [91]. Nano/micromotors are proving their remediation potential against organic and inorganic pollutants through adsorption, H_2O_2 assisted, photocatalytic, and biocatalytic degradation [92, 93]. Nanomotors can achieve diffusion by energetic blending owing to self-propulsion capabilities. Self-propelled motors stimulate decontamination efficiency by merging

the microstructure with surface area and activities [94, 95]. Most of the positive implications of nano/micromotors for wastewater decontamination listed in Table 6.2 are fuel-dependent. Though continuous efforts are being made to upscale the photocatalytic, biocatalytic, and magnetically driven nanomotors, challenges like the lifecycle of multi-functional nanomotors restrict the residual materials into its physique, which are consumed in oxidation reactions [96].

4. **Nanomembranes:** These are the membranes created by different nanofibers focused on reclaiming contaminants present in the aqueous phase. Several studies have reported the use of polymer matrix has the multifunctional significance which has been enhanced by the doping of nanoparticles for ultrafiltration membranes [39]. The composite layers may consist of carbon material, i.e., GO, rGO, CNT, etc., dispersed into a matrix to increase the fouling resistance and aqueous transport. Several studies reported metallic NPs immobilization on membrane for dichlorination and degradation of pollutants with properties like hindrance of NPs, reactivity, reduced agglomeration, and surface area [97, 98]. The traditional filtration approach requires primarily calcium, magnesium, and sodium ions as exchangers, while nanomembranes don't require such exchangers [40, 41]. Nanomembranes consist of one-dimensional nanomaterials such as nanotubes, nanoribbons, and nanofibers. The membranes of carbonaceous nanofibers (CNFs) are being widely used with outstanding efficiency.

Similarly, Carbon nanotubes (CNTs) are getting attention as nanomaterials synthesis as polymer composite membranes to maximize the performance with low density, extremely high strength and tensile modulus, high flexibility, and large aspect ratio [99]. Synthesized carbon nanotubes can be single-walled carbon nanotubes (SWCNTs), and multi-walled carbon nanotubes (MWCNTs) consists of singular or multi-walled tubular structure. The CNTs possess anti-microbial properties, which tend to reduce fouling and biofilm formation and reduce mechanical failures [100]. Metal oxide NPs *viz.*, Al_2O_3, TiO_2, antimicrobial NPs *viz.*, nano-Ag, CNTs, and aquaporin-based membranes are deemed material to overwhelm the layer blocking due to high hydrophilicity, permeability, fouling resistance, and homogenous pore distribution [39]. Besides CNFs and CNTs, electrospun nanofiber membranes (ENMs) are developing membranes being used to remove the inorganic and heavy metal pollutants from water [101, 102]. Taha et al. [103] successfully integrated amine cellulose acetate/silica nanofiber membranes having significant Cr(VI) removal capacity from wastewater. Lin et al. [104] reported efficient removal of Cr(IV) to Cr(III) by using PAN/$FeCl_3$ composite membranes. Chitosan nanofiber mats (CNM) showed the removal of metal toxins like Pb and Cu [105].

6.7 CONCLUSION

In summary, nanoparticles and their nanoremediation capacity are basically based on the adsorption/sorption site, surface area, functionality, stability, etc., properties. It is critical to understand the chemical nature of the pollutant to select the optimal adsorbent and modification as nanotubes, nanofillers, and nanomembranes. Aside from porosity and surface area, the synthesis technique is an important issue to consider. Another key aspect to examine is loadings and coatings. Metals and metal oxides offer the greatest flexibility in terms of altering the time and efficacy of an adsorption process. Many of the applications of different metallic or non-metallic nanomaterials, such as nanosorbents, nanophotocatalysts, nanomotors, nanomembranes, etc., have proved their high efficiency in pollutant removal on wider scales. The photocatalysis of the pollutant under varying environmental situations and the potential mechanism responsible should be further investigated in the photocatalytic treatment of inorganic pollutants with improved efficiency by investigating electrostatic communication among catalytic substances and pollutants.

KEYWORDS

- **inorganic pollutants**
- **membrane filtration**
- **nanomaterial**
- **nanoremediation**
- **sorption and desorption**
- **water pollution**
- **water remediation**

REFERENCES

1. Bansal, N., (2018). Industrial development and challenges of water pollution in coastal areas: The case of Surat, India. *IOP Conference Series: Earth and Environmental Science, 2017 5th International Conference on Environment Pollution and Prevention (ICEPP 2017) 14–16 December 2017* (Vol. 120). Singapore.
2. Naz, A., Chowdhury, A., Mishra, B. K., & Gupta, S. K. (2016). Metal pollution in water environment and the associated human health risk from drinking water: A case study of Sukinda chromite mine, India. *Human and Ecological Risk Assessment: An International Journal, 22*(7), 1433–1455.
3. Hsu-Kim, H., Eckley, C. S., Achá, D., Feng, X., Gilmour, C. C., Jonsson, S., & Mitchell, C. P., (2018). Challenges and opportunities for managing aquatic mercury pollution in altered landscapes. *Ambio, 47*(2), 141–169.
4. Ali, I., & Aboul-Enein, H. Y., (2006). *Instrumental Methods in Metal Ion Speciation.* CRC Press.
5. Gautam, R. K., & Chattopadhyay, M. C., (2016). Nanotechnology for water cleanup. *Nanomaterials for Wastewater Remediation,* 1–18. https://doi.org/10.1016/B978-0-12-804609-8.00001-7.
6. Shrivastava, M., Sharma, I. K., & Sharma, D. D., (2013). Groundwater scenario in Indira Gandhi Nahar Pariyojna (IGNP) in parts of Sri Ganganagar, Hanumangarh, Churu, Bikaner, Jaisalmer Jod10hpur, and Barmer Districts, Rajasthan. *Memoir Geological Society of India, 82*, 16-35.
7. Garbarino, J. R., Hayes, H. C., Roth, D. A., Antweiler, R. C., Brinton, T. I., & Taylor, H. E., (1995). *Contaminants in the Mississippi River: Heavy Metals in the Mississippi River* (p. 1133). U.S. Geological Survey Circular, Reston, VA.
8. Grandclément, C., Seyssiecq, I., Piram, A., Wong-Wah-Chung, P., Vanot, G., Tiliacos, N., Roche, N., & Doumenq, P., (2017). From the conventional biological wastewater treatment to hybrid processes, the evaluation of organic micropollutant removal: A review. *Water Research, 111*, 297–317.
9. Mohan, D., Sarswat, A., Ok, Y. S., & Pittman, C., (2014). Organic and inorganic contaminants removal from water with biochar, a renewable, low cost and sustainable adsorbent – A critical review. *Bioresource Technology, 160*, 191–202.
10. Srivastav, A. L., & Ranjan, M., (2020). Inorganic water pollutants. In: *Inorganic Pollutants in Water* (pp. 1–15). Elsevier.
11. Borah, P., Kumar, M., & Devi, P. (2020). Types of inorganic pollutants: Metals/metalloids, acids, and organic forms. In: *Inorganic Pollutants in Water* (pp. 17–31). Elsevier.
12. Ali, I., Basheer, I. A., Mbianda, X. Y., Burakov, A., Galunin, E., Burakova, I., Mkrtchyan, E., et al., (2019). Graphene-based adsorbents for remediation of noxious pollutants from wastewater. *Environment International, 127*, 160–180.
13. Shallari, S., Schwartz, C., Hasko, A., & Morel, J. L., (1998). Heavy metals in soils and plants of serpentine and industrial sites of Albania. *Science of the Total Environment, 192*(09), 133–142.
14. Pacyna, J. M., (1996). Monitoring and assessment of metal contaminants in the air. *Toxicology of Metals, 1*, 9–28.

15. Chikkanna, A., Mehan, L., Sarath, P. K., & Ghosh, D., (2019). Arsenic exposures, poisoning, and threat to human health: Arsenic affecting human health. *Environmental Exposures and Human Health Challenges*, 86–105.
16. Matta, G., & Gjyli, L., (2016). Mercury, lead, and arsenic: Impact on environment and human health. *Journal of Chemical and Pharmaceutical Sciences*, *9*, 718–725.
17. Genchi, G., Carocci, A., Lauria, G., Sinicropi, M. S., & Catalano, A., (2020). Nickel: Human health and environmental toxicology. *International Journal of Environmental Research and Public Health*, *17*(3), 679.
18. Ramady, H. L., Faizy, S. E. D., Abdalla, N., Taha, H., et al., (2020). Selenium and nano-selenium biofortification for human health: Opportunities and challenges. *Soil Systems*, *4*, 57.
19. Nordberg, G. F., Bernard, A., Diamond, G. L., et al., (2018). Risk assessment of effects of cadmium on human health (IUPAC technical report). *Pure and Applied Chemistry*, *90*(4), 755-808.
20. Renu, K., Chakraborty, R., Haritha, M., et al. (2021). Molecular mechanism of heavy metals (lead, chromium, arsenic, mercury, nickel, and cadmium) induced hepatotoxicity: A review. *Chemosphere*, 129735.
21. US EPA, (2013). *Basic Information About Radionuclides in Drinking Water*. Retrieved from http://water.epa.gov/drink/contaminants/basicinformation/radionuclides.cfm (accessed on 23 August 2023).
22. Jiang, J., Wu, Y., sun, G., Zhang, L., Li, Z., Sommar, J., Yao, H., & Feng, X., (2021). Characteristics, accumulation, and potential health risks of antimony in atmospheric particulate matter. *Journal American Chemical Society*, *6*(14), 9460–9470.
23. Plum, L. M., Rink, L., & Haase, H., (2010). The essential toxin: Impact of zinc on human health. *International Journal of Environmental Research and Public Health*, *7*(4), 1342–1365.
24. Karim, N., (2018). Copper and human health: A review. *The Journal of Bahria University Medical and Dental College*, *8*(2), 117–122.
25. Krupińska, I., (2020). Aluminum drinking water treatment residuals and their toxic IMPACT on human health. *Molecule*, *25*(3), 641.
26. Jamshaid, M., Khan, A. A., Ahmed, K., & saleem, M., (2018). Heavy metal in drinking water its effect on human health and its treatment techniques: A review. *International Journal of Biosciences*, *12*(4), 223–240.
27. WHO (World Health Organization), (1981). *Hydrogen Sulfide*. Geneva, (Environmental Health Criteria, No. 19).
28. NRC (National Research Council), (1987). Emergency and continuous exposure guidance levels for selected airborne contaminants. *Ammonia, Hydrogen Chloride, Lithium Bromide, and Toluene* (Vol. 7). Washington, D.C.: National Academy Press.
29. Chen, J., Gao, Y., Qian, H., Ren, W., & Qu, W. (2021). Hydrogeochemical evidence for fluoride behavior in groundwater and the associated risk to human health for a large irrigation plain in the yellow River Basin. *Science of the Total Environment*, *800*, 149428.
30. WHO (World Health Organization), (1986). *Ammonia.* Geneva, (Environmental Health Criteria, No. 54).
31. Daliran, S., Ghazagh-Miri, M., Oveisi, A. R., Khajeh, M., Navalón, S., Âlvaro, M., et al., (2020). A pyridyltriazol functionalized zirconium metal–organic framework

for selective and highly efficient adsorption of palladium. *ACS Applied Materials & Interfaces, 12*, 25221–25232. doi: 10.1021/acsami.0c06672.
32. Hu, C., Hu, X., Li, R., & Xing, Y. (2020). MOF-derived ZnO/C nanocomposite with enhanced adsorption capacity and photocatalytic performance under sunlight. *Journal of Hazardous Materials, 385*, 121599. doi: 10.1016/j.jhazmat.2019.121599.
33. Saleem, H., Rafique, U., & Davies, R. P., (2016). Investigations on post-synthetically modified UiO-66-NH_2 for the adsorptive removal of heavy metal ions from aqueous solution. *Microporous Mesoporous Mater, 221*, 238–244.
34. Audu, C. O., Nguyen, H. G. T., Chang, C. Y., Katz, M. J., Mao, L., Farha, O. K., Hupp, J. T., & Nguyen, S. T., (2016). The dual capture of As^{V} and As^{III} by UiO-66 and analogues. *Chemical Science*, *7*, 6492–6498.
35. Das, R., Vecitis, C. D., Schulze, A., Cao, B., Ismail, A. F., Lu, X., Chen, J., & Ramakrishna, S., (2017). Recent advances in nanomaterials for water protection and monitoring. *Chemical Society Reviews, 46*, 6946–7020.
36. Palani, G., Arputhalatha, A., Kannan, K., Lakkaboyana, S. K., Hanafiah, M. M., Kumar, V., & Marella, R. K., (2021). Current trends in the application of nanomaterials for the removal of pollutants from industrial wastewater treatment—A review. *Molecules*, *26*, 2799.
37. Pedroza-Herrera, G., Medina-Ram´ırez, I. E., Lozano-A´ lvarez, J. A., & Rodil, S. E., (2019). Evaluation of the photocatalytic activity of copper-doped TiO_2 nanoparticles for the purification and/ or disinfection of industrial effluents. *Catalysis Today*. In press.
38. Ren, K., Chakraborty, R., Haritha, M., & Koti, R., (2021). Molecular mechanism of heavy metals (lead, chromium, arsenic, mercury, nickel, and cadmium) induced hepatotoxicity- A review. *Chemosphere*, *271*, 129735.
39. Yaqoob, A. A., Parveen, T., Umar, K., & Ibrahim, M. N. M., (2020). Role of nanomaterials in the treatment of wastewater: A review. *Water*, *12*, 495.
40. Muntha, S. T., Kausar, A., & Siddiq, M., (2017). A review featuring fabrication, properties, and application of polymeric mixed matrix membrane reinforced with different fillers. *Polymer-Plastics Technology and Engineering*, *56*(18), 2043–2064.
41. Kumar, A., Sadhu, S. D., & Singh, R. (2018). Nanotechnology: Key for sustainable future. *Nanotechnology in Environmental Science*, 759–804.
42. Wang, X., Guo, Y., Yang, L., Han, M., Zhao, J., & Cheng, X., (2012). Nanomaterials as sorbents to remove heavy metal ions in wastewater treatment. *J. Environ. Anal. Toxicol*, *2*(7), 154-158.
43. Gupta, V. K., Tyagi, I., Sadegh, H., Shahryari-Ghoshekand, R., Makhlouf, A. S. H., & Maazinejad, B., (2015). Nanoparticles as adsorbent; a positive approach for removal of noxious metal ions: A review. *Science Technology and Development*, *34*, 195.
44. Palani, G., Karthik, K., Lakkaboyana, S. K., Hanafiah, M. M., Kumar, V., & Marella, R. K., (2021). *Advanced Nanomaterial's for Industrial Wastewater Treatment – A Review.* Preprints. 2021020337. https://doi.org/10.20944/preprints202102.0337.v1.
45. Wang, X., Zhan, C., Kong, B., Zhu, X., Liu, J., Xu, W., Cai, W., & Wang, H., (2015). Self-curled coral-like -Al_2O_3 nanoplates for use as an adsorbent. *Journal of Colloid Interface Science*, *453*, 244–251.
46. Luo, T., Cui, J., Hu, S., Huang, Y., & Jing, C., (2010). Arsenic removal and recovery from copper smelting wastewater using TiO_2. *Environmental Science and Technology*, *44*(23), 9094–9098.

47. Singh, S., Barick, K. C., & Bahadur, D., (2013). Fe_3O_4 embedded ZnO nanocomposites for the removal of toxic metal ions, organic dyes, and bacterial pathogens. *Journal of Materials Chemistry A, 1*(10), 3325–3333.
48. Kumar, K. Y., Muralidhara, H. B., Nayaka, Y. A., Balasubramanyam, J., & Hanumanthappa, H., (2013). Low-cost synthesis of metal oxide nanoparticles and their application in adsorption of commercial dye and heavy metal ion in aqueous solution. *Powder Technology. 246*, 125–136.
49. Lingamdinne, L. P., Koduru, J. R., Roh, H., Choi, Y. L., Chang, Y. Y., & Yang, J. K., (2016). Adsorption removal of Co(II) from waste water using graphene oxide. *Hydrometallurgy, 165*, 90–96.
50. Tarigh, G. D., & Shemirani, F., (2013). Magnetic multi-wall carbon nanotube nanocomposite as an adsorbent for preconcentration and determination of lead(II) and manganese(II) in various matrices. *Talanta, 115*, 744–750.
51. Tang, K., Gong, C.,Wang, D., (2016). Reduction potential, shadow prices, and pollution costs of agricultural pollutants in China. *Science of the Total Environment, 541*, 42–50.
52. Dehghani, M. H., Taher, M. M., Bajpai, A. K., Heibati, B., Tyagi, I., Asif, M., Agarwal, S., & Gupta, V. K., (2015). Removal of noxious Cr (VI) ions using single-walled carbon nanotubes and multi-walled carbon nanotubes. *Chemical Engineering Journal, 279*, 344–352.
53. Al-Saydeh, S. A., El-Naas, M. H., & Zaidi, S. J., (2017). Copper removal from industrial wastewater: A comprehensive review. *Journal of Industrial and Engineering Chemistry, 56*, 35-44.
54. Jedla, M. R., Koneru, B., Franco A. Jr., Rangappa, D., & Banerjee, P., (2022). Recent developments in nanomaterials-based adsorbents for water purification techniques. *Biointerface Research in Applied Chemistry, 12*(5), 5821–5835.
55. Gómez-Pastora, J., Dominguez, S., Bringas, E., Rivero, M. J., Ortiz, I., & Dionysiou, D. D., (2017). Review and perspectives on the use of magnetic nanophotocatalysts (MNPCs) in water treatment. *Chemical Engineering Journal, 310*, 407–427.
56. Bhanvase, B. A., Shende, T. P., & Sonawane, S. H., (2017). A review on graphene–TiO_2 and doped graphene–TiO_2 nanocomposite photocatalyst for water and wastewater treatment. *Environmental Technology Reviews, 6*(1), 1–14.
57. Di, G., Zhu, Z., Zhang, H., Zhu, J., Lu, H., Zhang, W., *et al.*, (2017). Simultaneous removal of several pharmaceuticals and arsenic on Zn-Fe mixed metal oxides: Combination of photocatalysis and adsorption. *Chem. Eng. J., 328*, 141–151.
58. Zhang, J., Cao, Y., Zhao, P., Xie, T., Lin, Y., & Mu, Z. (2020). Visible-light-driven pollutants degradation with carbon quantum dots/N-TiO_2 under mild conditions: Facile preparation, dramatic performance, and deep mechanism insight. *Colloids and Surfaces A: Physicochemical and Engineering Aspects, 601*, 125019.
59. Kanakaraju, D., Rusydah Bt, M. S. N., Lim, Y. C., & Pace, A., (2019). Concurrent removal of Cr(III), Cu(II), and Pb(II) ions from water by multifunctional TiO_2/Alg/FeNPs beads. *Sustain. Chem. Pharm., 14*, 100176.
60. Chen, Y. H., & Li, F. A., (2010). Kinetic study on removal of copper(II) using goethite and hematite nano-photocatalysts. *Journal of Colloid Interface Science, 347*, 277–281.
61. Kadi, M. W., Mohamed, R. M., Ismail, A. A., & Bahnemann, D. W., (2020). Performance of mesoporous _-Fe_2O_3/g-C_3N_4 heterojunction for photoreduction of Hg(II) under visible light illumination. *Ceramic International, 46*, 23098–23106.

62. Spanu, D., Bestetti, A., Hildebrand, H., Schmuki, P., Altomare, M., & Recchia, S., (2019). Photocatalytic reduction and scavenging of Hg(II) over templated-dewetted Au on TiO_2 nanotubes. *Photochem. Photobiol. Sci., 18*, 1046–1055.
63. He, S., Yang, Z., Cui, X., Zhang, X., & Niu, X. (2020). Fabrication of the novel Ag-doped SnS_2@$InVO_4$ composite with high adsorption photocatalysis for the removal of uranium (VI). *Chemosphere*, *260*, 127548.
64. Bhunia, P., Kim, G., Baik, C., & Lee, H., (2012). A strategically designed porous iron–iron oxide matrix on graphene for heavy metal adsorption. *Chemical Communications Journal*. *48*(79), 9888–9890.
65. Atkovska, K., Lisichkov, K., Ruseska, G., Dimitrov, A. T., & Grozdanov, A., (2018). Removal of heavy metal ions from wastewater using conventional and nanosorbents: A review. *Journal of Chemical Technology & Metallurgy*, *53*(2).
66. Mousavi, S. M., Hashemi, S. A., Amani, A. M., Esmaeili, H., Ghasemi, Y., Babapoor, A., Mojoudi, F., & Arjomand, O., (2018). Pb(II) removal from synthetic wastewater using kombucha scoby and graphene oxide/Fe_3O_4. *Physical Chemistry Research*, *6*(4), 759–771.
67. Tene, T., Bellucci, S., Guevara, M., Viteri, E., Arias, P. M., Salguero, O., Vera-Guzmán, E., et al., (2022). Cationic pollutant removal from aqueous solution using reduced graphene oxide. *Nanomaterials*, *12*(3), 309.
68. Zhang, K., Li, H., Xu, X., & Yu, H. (2018). Synthesis of reduced graphene oxide/NiO nanocomposites for the removal of Cr(VI) from aqueous water by adsorption. *Microporous Mesoporous Mater.*, *255*, 7–14.
69. Sadeghi, M. H., Tofighy, M. A., & Mohammadi, T. (2020). One-dimensional graphene for efficient aqueous heavy metal adsorption: Rapid removal of arsenic and mercury ions by graphene oxide nanoribbons (GONRs). *Chemosphere*, *253*, 126647.
70. Sui, Z., Meng, Q., Zhang, X., Ma, R., & Cao, B., (2012). Green synthesis of carbon nanotube–graphene hybrid aerogels and their use as versatile agents for water purification. *Journal of Materials Chemistry*, *22*, 8767-8771.
71. Huang, X., Jin, J., Lu, X., Wang, Y., Liu, Y., Pang, S. Y., Cheng, H., Xiang, Z., & Ma, J., (2015). Adsorption and oxidation of thallium(I) by a nanosized manganese dioxide. *Water Air Soil Pollut., 226*, 2272.
72. Hoang, A. T., Sandro, N., Chin, K. C., Rafael, L., Sabu, T., Tien, L. B., Van, V. P., & Xuan, P. N., (2022). Heavy metal removal by biomass-derived carbon nanotubes as a greener environmental remediation: A comprehensive review. *Chemosphere*, *287*(1), 131959.
73. Gebru, K. A., & Das, C., (2017). Removal of Pb(II) and Cu(II) ions from wastewater using composite electrospun cellulose acetate/titanium oxide (TiO_2) adsorbent. *J. Water Process Eng.*, *16*, 1–13.
74. Mahmoud, M. E., Ali, S. A. A., & Elweshahy, S. M., (2019). Microwave functionalization of titanium oxide nanoparticles with chitosan nanolayer for instantaneous microwave sorption of Cu(II) and Cd(II) from water. *Int. J. Biol. Macromol.*, *111*, 393–399.
75. Gautam, S., Kaithwas, G., Bharagava, R. N., & Saxena, G. (2017). Pollutants in tannery wastewater: Pharmacological effects and bioremediation approaches for human health protection and environmental safety. In: *Environmental Pollutants and Their Bioremediation Approaches* (pp. 369–396). CRC Press.

76. Wang, Q., Lin, Q., Li, Q., Li, K., Wu, L., Li, S., & Liu, H., (2021). As(III) removal from wastewater and direct stabilization by in-situ formation of Zn-Fe layered double hydroxides. *J. Hazardous Materials*. https://doi.org/10.1016/j.jhazmat.2020.123920.
77. Kefeni, K. K., & Mamba, B. B., (2020). Photocatalytic application of spinel ferrite nanoparticles and nanocomposites in wastewater treatment. *Sustainable Materials and Technologies*, *23*, e00140.
78. Xie, W., Liang, Q., Qian, T., & Zhao, D. (2015). Immobilization of selenite in soil and groundwater using stabilized Fe–Mn binary oxide nanoparticles. *Water Res*., *70*, 485–494.
79. Khurshid, H., Mustafa, M. R. U., & Isa, M. H., (2022). Adsorption of chromium, copper, lead, and mercury ions from aqueous solution using bio and nano adsorbents: A review of recent trends in the application of AC, BC, nZVI, and MXene. *Environmental Research*, *212*, 113138.
80. Alani, O. A., Ari, H. A., Offiong, N. A. O., Alani, S. O., Li, B., Zeng, Q. R., & Feng, W., (2021). Catalytic removal of selected textile dyes using zero-valent copper nanoparticles loaded on filter paper-chitosan-titanium oxide heterogeneous support. *J. Polym. Environ*., *29*, 2825–2839.
81. Yang, F., Zhang, S., Sun, Y., Cheng, K., Li, J., & Tsang, D. C., (2018). Fabrication and characterization of hydrophilic corn stalk biochar-supported nanoscale zero-valent iron composites for efficient metal removal. *Bioresource Technology*, *265*, 490–497.
82. Wang, S., Zhao, M., Zhou, M., Li, Y. C., Wang, J., Gao, B., Sato, S., et al., (2019). Biochar-supported nZVI (nZVI/BC) for contaminant removal from soil and water: A critical review. *Journal of Hazardous Materials*, *373*, 820–834.
83. Li, J., Chang, H., Ma, L., Hao, J., & Yang, R. T., (2018). Low-temperature selective catalytic reduction of NOx with NH_3 over metal oxide and zeolite catalysts: A review. *Catalysis Today*, 175: 147–156.
84. Maphutha, S., Moothi, K., Meyyappan, M., & Lyuke, S. E., (2013). Pollution remediation-nanoscale materials-technical paper: Water watch. *Inside Mining*, *6*(5), 44–47.
85. Farghali, A. A., et al., (2017). Functionalization of acidified multi-walled carbon nanotubes for removal of heavy metals in aqueous solutions. *J. Nanostructure Chem.*, *7*(2), 101–111.
86. Tijing, L. D., Woo, Y. C., Shim, W. G., He, T., Choi, J. S., Kim, S. H., & Shon, H. K., (2016). Superhydrophobic nanofiber membrane containing carbon nanotubes for high-performance direct contact membrane distillation. *Journal of Membrane Science*, *502*, 158–170.
87. DesOrmeaux, J. P. S., Winans, J. D., Wayson, S. E., Gaborski, T. R., Khire, T. S., Striemer, C. C., & McGrath, J. L., (2014). Nanoporous silicon nitride membranes fabricated from porous nanocrystalline silicon templates. *Nanoscale*, *6*(18), 10798–10805.
88. Kootenaei, F. G., & Rad, H. A., (2013). Treatment of hospital wastewater by a novel nano-filtration membrane bioreactor (NF-MBR) *Iranica J. Energy Environ*., *4*(1). 60–67.
89. Wen, T., Zhao, Z., Shen, C., Li, J., Tan, X., Zeb, A., Wang, X., & Xu, A. W., (2016). Multifunctional flexible free-standing titanate nanobelt membranes as efficient sorbents for the removal of radioactive $90Sr^{2+}$ and $137Cs^{+}$ ions and oils. *Scientific Reports*, *6*(1), 1–10.

90. Faccini, M., Borja, G., Boerrigter, M., Morillo, M. D., Martìnez, C. S., Vázquez-Campos, S., Aubouy, L., & Amantia, D., (2015). Electrospun carbon nanofiber membranes for filtration of nanoparticles from water. *Journal of Nanomaterials, 2015.*
91. Jurado-Sánchez, B., & Wang, J., (2018). Micromotors for environmental applications: A review. *Environmental Science-Nano, 5,* 1530–1544.
92. Parmar, J., Vilela, D., Villa, K., Wang, J., & Sanchez, S., (2018). Micro-and nanomotors as active environmental microcleaners and sensors. *Journal of the American Chemical Society, 140*(30), 9317–9331.
93. Zhang, Y., Wu, B., Xu, H., Liu, H., Wang, M., He, Y., & Pan, B., (2016). Nanomaterials-Enabled water and wastewater treatment. *Nano Impact., 3,* 22–39.
94. Chi, Q., Wang, Z., Tian, F., You, J. A., & Xu, S., (2018). A review of fast bubble-driven micromotors powered by biocompatible fuel: Low-Concentration fuel, bioactive fluid, and enzyme. *Micromachine, 9,* 537.
95. Pacheco, M., López, M. Á., Jurado-Sánchez, B., & Escarpa, A., (2019). Self-Propelled micromachines for analytical sensing: A critical review. *Anal. Bioanal. Chem., 411,* 6561–6573.
96. Ying, Y., & Pumera, M., (2019). Micro/nanomotors for water purification. *Chemistry—A European Journal, 25,* 106–121.
97. Mekaru, H., Lu, J., & Tamanoi, F., (2015). Development of mesoporous silica-based nanoparticles with controlled release capability for cancer therapy. *Advanced Drug Delivery Review, 95,* 40–49.
98. Jawed, A., Saxena, V., & Pandey, L. M., (2020). Engineered nanomaterials and their surface functionalization for the removal of heavy metals: A review. *Journal of Water Process Engineering, 33,* 101009.
99. Pitroda, J., Jethwa, B., & Dave, S. K., (2016). A critical review on carbon nanotubes. *Int. J. Constr. Res. Civ. Eng, 2*(5), 36–42.
100. Hogen-Esch, T., Pirbazari, M., Ravindran, V., Yurdacan, H. M., & Kim, W., (2019). *High Performance Membranes for Water Reclamation Using Polymeric and Nanomaterials.* U.S. Patent No. 20160038885A.
101. Nasreen, S. A. A. N., Sundarrajan, S., Nizar, S. A. S., Balamurugan, R., & Ramakrishna, S., (2013). Advancement in electrospun nanofibrous membranes modification and their application in water treatment. *Membranes, 3*(4), 266–284.
102. Qu, X., Alvarez, P. J., & Li, Q., (2013). Applications of nanotechnology in water and wastewater treatment. *Water Research, 47*(12), 3931–3946.
103. Taha, A. A., Wu, Y. N., Wang, H., & Li, F., (2012). Preparation and application of functionalized cellulose acetate/silica composite nanofibrous membrane via electrospinning for Cr(VI) ion removal from aqueous solution. *J. Environ. Manage, 112,* 10–16.
104. Lin, Y., Cai, W., Tian, X., Liu, X., Wang, G., & Liang, C., (2011). Polyacrylonitrile/ferrous chloride composite porous nanofibers and their strong Cr-removal performance. *Journal of Material Chemistry, 21*(4), 991–997.
105. Teng, M., Wang, H., Li, F., & Zhang, B., (2011). Thioether-functionalized mesoporous fiber membranes: Sol-gel combined electrospun fabrication and their applications for Hg^{2+} removal. *Journal of Colloid Interface Science, 355,* 23–28.

CHAPTER 7

Applications of Nanomaterials in the Restoration of Aquatic Ecosystems

GAURI SHARMA,[1] SNEH RAJPUT,[1] SHUBHAM THAKUR,[2] PREETI RAINA,[1] AKANKSHA JASROTIA,[1] ARPNA KUMARI,[3] AKSHI BHARDWAJ,[1] RINKY KUMARI,[1] SUBHEET KUMAR JAIN,[2] RITU BALA,[4] and RAJINDER KAUR[1]

[1]Department of Botanical and Environmental Sciences, Guru Nanak Dev University, Amritsar, Punjab, India

[2]Department of Pharmaceutical Sciences, Guru Nanak Dev University, Amritsar, Punjab, India

[3]Academy of Biology and Biotechnology, Southern Federal University, Rostov-on-Don, Russia

[4]Department of Chemistry, Guru Nanak Dev University, Amritsar, Punjab, India

ABSTRACT

The demands of the expanding population have directly impacted the upsurge in global industry. Industrialization and urbanization have led to an increase in various pollutants, such as pesticides, heavy metals, hydrocarbons, pharma waste, and explosives, which pose a great threat to all types of ecosystems, especially aquatic ecosystems. The availability of traditional approaches is inadequately successful in dealing with the current challenges of environmental contamination, thus, an alternative strategy like nanoremediation can offer better alternatives. Nanoremediation is a novel technique for environmental remediation that employs synthetically engineered nanomaterials (ENMs). These nano-sized particles have characteristic properties

Nano-Bioremediation for Water and Soil Treatment: An Eco-Friendly Approach.
Vishnu D. Rajput, Arpna Kumari, and Tatiana M. Minkina (Eds.)

that provide the opportunity to recognize, monitor, prevent, reduce, and treat contaminants at much lower cost and energy with greater efficiency. A variety of nanomaterials are being explored, yet commercially, only a few are being used viz inorganic nanomaterials (metal or metal-oxide NMs) and organic nanomaterials (carbon NMs, silica NMs, and polymer-based NMs). Despite many advantages, nanoremediation faces some drawbacks, too. Therefore, the present aim of this chapter is to describe the different uses of nanomaterials, notably in the context of water remediation, and provide current studies and methodologies linked to nanotechnology applications for aquatic remediation.

7.1 INTRODUCTION

Removal of pollutants from various environmental matrixes, such as freshwater, surface water, sediments, air, soil, etc., is known as environmental remediation. Remediation is often subjected to a variety of regulatory criteria, which are based on risk assessment to the environment and human beings. Some of the commonly used remediation methods are ion exchange, solvent extraction, electrochemical treatment, ultrafiltration, and chemical precipitation [9]. However, these methods have several drawbacks, such as high cost and energy requirements and being unsuitable for the aqueous medium. Thereby, such technology that can monitor, identify, and ideally remediate even minute amounts of pollutants in the soil, water, and air is required. In the last decade, a new method based on the adsorption principle has been found to be efficient at a low cost. Moreover, the process is reversible, which means the adsorbents used can be regenerated by the desorption principle. Further, based on the nature of the material used adsorption technique can be divided into two sub-categories: (i) bioremediation (biological adsorbent) and (ii) nanoremediation (synthetic adsorbent) [10]. In recent years, nanoremediation techniques have gained a lot of attention. In this technique, nanoparticles or engineered materials arehas used that act as an adsorbent and help in the removal of pollutants, especially heavy metals. During the nanoremediation process, synthetically synthesized nanoparticles come in contact with the target contaminants under specific conditions. Nanotechnology provides the opportunity to recognize, monitor, prevent, reduce, and treat contaminants at much lower cost and energy with greater efficiency. Nanoremediation is known to detoxify groundwater, wastewater, soil, sediment, and other aquatic contaminants [11]. A variety of nanomaterials are being tested, yet commercially, only a few are being used, such as zero-valent metals (ZVMs), calcium carbonate, graphene, carbon nanotubes, titanium dioxide, and iron oxide [12–14].

Several metals, such as aluminum, iron, magnesium, nickel, palladium, titanium, zinc, etc. that have reducing properties are adopted for the remediation process. These ZVMs have strong chemical reducibility, higher efficiency, and larger specific surface area. Generally, these materials are synthesized via aqueous phase reduction, ball milling process, ion sputtering, chemical vapor deposition, spark discharge, sol-gel, and plasma-enhanced chemical vapor deposition methods [15]. ZVMs are most widely used in groundwater remediation. Nanoparticles such as carbon nanotubes and titanium dioxide are used for the purification, disinfection, and desalination of surface water, which are based on the principles of sorption or membrane filtration acting as nano-filters. Furthermore, detecting the traces of heavy metals (mercury, lead, cadmium, chromate, arsenate, uranium, actinides, etc.) in a field setting can be a tedious job with less sensitivity. For this, nanoremediation methods such as solid-phase micro-extraction or sol-gel processes can be most effective. Moreover, the most widely commercialized nanomaterial "titanium dioxide" is the leading candidate for wastewater treatment. It produces hydroxyl radicals when exposed to ultraviolet light, leading to the oxidization of contaminants. Despite many advantages, nanoremediation faces some drawbacks, too. As a result, the goal of this chapter is to describe the different uses of nanomaterials, notably in the context of water treatment, while also providing current studies and methodologies linked to nanotechnology applications for environmental remediation.

7.2 TYPES OF NANOMATERIALS

Nanoparticles range from 1–100 nm in size and can be divided into the following four material-based categories:

1. **Carbon-based Nanomaterials:** These nanomaterials are made of carbon and come in the form of hollow tubes, ellipsoids, or spheres. Graphene, carbon nanotubes (CNTs), fullerenes, carbon nanofibers, carbon onions, and graphene oxide nanosheets are the several types of carbon-based NMs. Chemical vapor deposition (CVD), arc discharge, and laser ablation are the processes used to create these carbon-based compounds [16].

2. **Inorganic-based Nanomaterials:** Metal nanoparticles are synthesized from metals such as gold or silver, whereas metal oxide nanoparticles are synthesized from titanium oxide (TiO_2) and zinc oxide (ZnO) [16–18].
3. **Organic-based Nanomaterials:** These nanomaterials are derived from organic matter and have non-covalent weak interactions, which help molecules to transform organic NMs such as micelles, liposomes, polymers, and dendrimers [17, 18].
4. **Composite-based Nanomaterials:** These are multi-phased nanoparticles that can be formed either by combining a nanoparticle with another nanoparticle or with larger materials like hybrid nanofibers. Composite NMs can be any combination of metal, carbon, or organic-based nanomaterials, such as biochar-supported zerovalent nanocomposites, grapheme-based nanocomposites, and silica-coated magnetic nanocomposites [16, 17].

7.3 VARIOUS MECHANISMS FOR RESTORATION OF AQUATIC ECOSYSTEMS

Numerous studies have been reported in the literature regarding wastewater treatment. However, removing minute particles from wastewater streams remains a difficult task. Several methods for the removal of suspended particles have been developed, with nanotechnology being the most successful. In this regard, nanomaterials such as carbon, inorganic, organic, or composite-based particles are used for the purification of water bodies [19–21]. These NPs are further synthesized using various methods such as electro-beam lithography, pulse laser ablation, sol-gel process, electrochemical, ultrasonication, pyrolysis, chemical reduction, inert gas condensation, etc. For the restoration of water bodies, these NPs are developed into adsorbents. Furthermore, these adsorbents are incorporated with magnetism to enhance the purification efficiency. Magnetic adsorbents are combined with other adsorbents that contain the oxides of Fe, Co, Ni, and Cu. The occurrence of a magnetic core provides an easy recovery from the medium using an external magnetic field. Hence, nanotechnology has great potential for the treatment of aquatic ecosystems. Due to strong magnetic interaction, bulky materials have a tendency to agglomerate into larger clusters. Therefore, surface coating with suitable agents can be fabricated [21, 22]. Sharma and co-workers suggested a silica coating around the magnetic nanoparticles for the removal of metal ions from wastewater [23]. Furthermore, various ligands are also used to improve the efficiency of magnetic NPs (Figure 7.1).

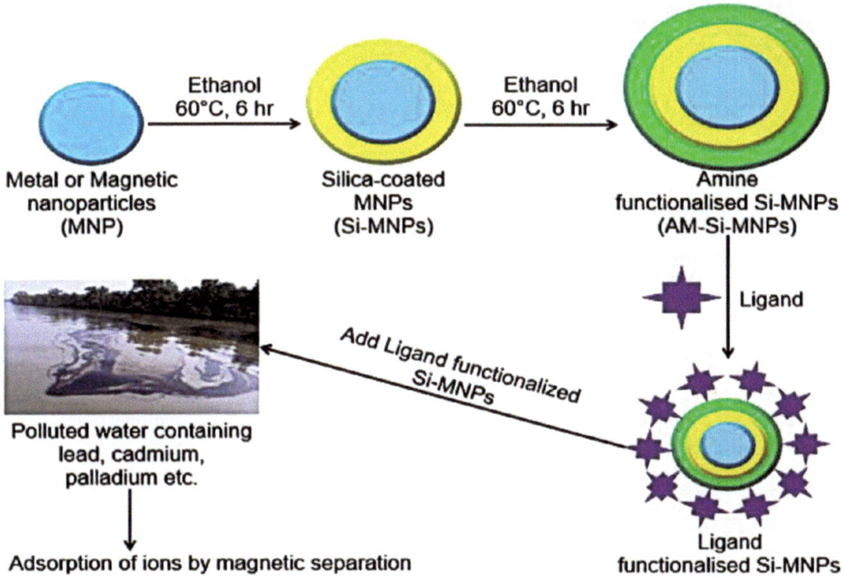

FIGURE 7.1 Schematic representation that illustrates of synthesis of nano-adsorbent and magnetic nanoparticles.

For the removal of various inorganic and organic pollutants, different carbon-based materials, namely carbon nanotubes (CNTs), graphene oxide sheets, etc., have been utilized as these materials have unique properties such as large surface area, high porosity, ability to interact with molecules, and layered structures that enhance their pollutant removal capacity. Gupta et al. [22] combined CNTs with iron oxide to form a composite, which was further used for the removal of Cr(III) [22–25]. An illustration of chemical reaction between zero-valent iron nanoparticles, organic contaminants, and heavy metals in water is shown in Figure 7.2. Whether the catalyst has a reducing, sorption, or oxidizing nature, iron is quickly transformed into ferrous ions and gradually to ferric ions [26–28]. By these thermodynamic mechanisms, aquatic systems are decontaminated as metallic iron has a high capacity to degrade organic compounds [28, 29]. Hence, the process of adsorption is highly beneficial for the removal of contaminants from water. Moreover, adsorbents have good recycling and reusable quality.

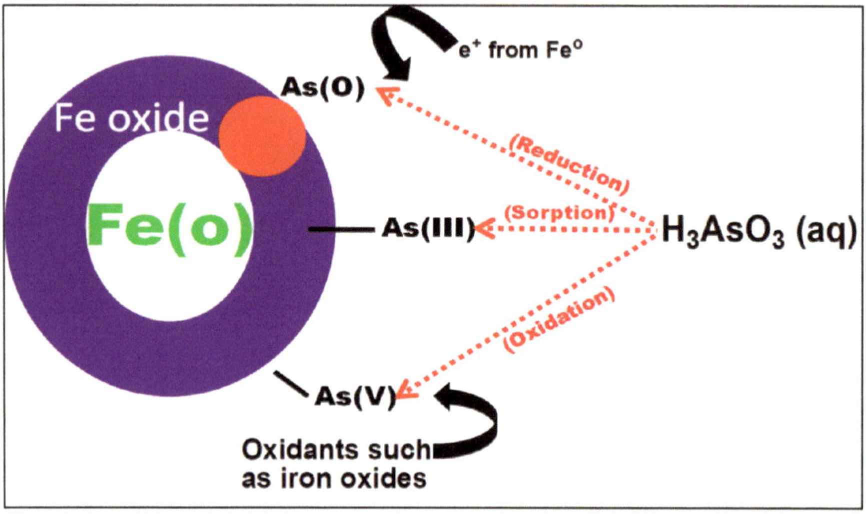

FIGURE 7.2 Illustrates of chemical reaction between zero-valent iron nanoparticles, organic contaminants, and heavy metals in water.

The advantages of nanomaterials for remediation of various environmental sites like contaminated soil, water, groundwater, wastewater treatments, and oil spill decontamination are given in Table 7.1. Nano-zerovalent iron (nZVI) has been used for the removal of chlorinated compounds (PCE, TCE, DCE), heavy metals like chromium, arsenic, copper, polychlorinated biphenyls, and carbon tetrachloride. It has been reported that PCE is reduced to 100–200 μg/L from 20,000–30,000 μg/L, and TCE reduction is 80% from groundwater [30]. nZVI has also been successfully applied for the degradation of several heavy metals from groundwater by 95–99% [31]. Apart from nZVI, agarose stabilized nZVIs and natural or synthetic goethite ZVI have been used for the efficient removal of Chromium [Cr(IV)] from water resources [33, 45]. Several international companies like Toda Kogyo (Japan), Heuer (USA), and Nanoiron (Czech Republic) have been marketing nZVIs in the remediation field because of various advantages like strong reducing power, low cost, high movability in contaminated areas and have great potential for interdependent effect with remediation techniques [50]. Similarly, hexachlorobenzene and methyl orange were detoxified from water bodies by bi-metallic nanoparticles Ag/Fe, Fe/Ni with a 99% removal rate [32, 48]. Carbon-based nanoparticles such as carbon nanotubes and graphite oxides were employed for the removal of dyes, insecticides, medicines, and medications from wastewater and activated sludge due to their flexibility,

TABLE 7.1 Degradation Efficiency of Nanoparticles for Restoration of Aquatic Ecosystems

Sl. No.	Nanomaterials Used	Pollutants	Media	Degradation Efficiency	References
1.	nZVI Flakes	Perchloroethylene, trichloroethylene, 1,2-dichloroethane	Groundwater	Perchloroethylene was decreased from 20,000–30,000 g/L to 100–200 g/L, while Trichloroethylene was lowered by 80%.	[30]
2.	ZVI	Palladium, Chromium, Copper	Groundwater	~95%–99%	[31]
3.	Ag/Fe bimetallic nanoparticles	Hexachlorobenzene	Groundwater	Byproduct's degradation in 2 hours.	[32]
4.	Agarose stabilized nZVI	Hexavalent chromium	Water	100%	[33]
5.	Silica–gold nanocomposite	Metal ions (Cd^{2+}, Pb^{2+}, Hg^{2+}, AS^{3+})	Water	Efficient removal was reported.	[34]
6.	nZVI/Pd	Polychlorinated biphenyls	Contaminated soil	60–80% removal was reported.	[35]
7.	ZVI	Arsenic, phenol	Wastewater	>99% removal was reported.	[36]
8.	Carbon-nanotubes	Dyes, pesticides, pharmaceuticals/drugs	Wastewater	100% removal rate	[37]
9.	Super-hydrophobic nanomembrane and oleophilic-graphite nanoplatelets.	Hydrocarbons	Crude oil spill	Recyclability was observed.	[38]
10.	Modified magnetic hybrid nanoparticles.	Bisphenol A, Cu^+	Soil	Removal of pollutants was efficient and non-toxic.	[39]
11.	Graphite oxide/magnetic chitosan	Reactive black 5	Aqueous solution	Removal efficiency was 391.0 mg/g.	[40]
12.	TiO_2 nanomembrane	Cations, nitrates, biological contaminants, organic matter.	Water	High removal efficiency.	[41]
13.	CdS nanoparticles	Acid Blue-29	Aqueous solution	Efficient removal and photodegradation were reported.	[42]

TABLE 7.1 *(Continued)*

Sl. No.	Nanomaterials Used	Pollutants	Media	Degradation Efficiency	References
14.	Post-sulfidized nZVI	Trichloroethene	Batch experiments in aqueous solutions.	Improvement of S-nZVI electron efficiency for TCE degradation	[43]
15.	TiO_2/WO_3	Diclofenac	Aqueous solution	92% removal was reported.	[44]
16.	Natural and synthetic goethite ZVI	Hexavalent chromium	Water	99% removal was reported.	[45]
17.	CNCs-g-nBA	Pb^{2+}	Batch experiments in aqueous solution.	Removal rate was 140.95 mg/g.	[46]
18.	Fe_3O_4/PANI/MnO_2	Cd^{2+}	Batch experiments in aqueous solution.	99.1% removal was reported.	[47]
19.	Fe/Ni bimetallic NPs 20	Methyl orange	Aqueous solution	99.6% removal was reported.	[48]
20.	nZVI/Pd	Carbon tetrachloride	Water/methanol solution	>60% 24 h degradation was reported.	[49]

high thermal and electrical conductivity, and adsorption capabilities [37]. Metal-based oxides, such as titanium oxide/dioxide nanomembranes, have shown a high success rate in the removal of pharmaceuticals, nitrates, cations, organic debris, and biological pollutants from water bodies [41]. Alvia et al. [38] and Wang et al. [39] reported the graphite nanoplatelets and magnetic nanoparticles for the remediation of crude oil spills and soil, respectively [38, 39].

7.4 APPLICATIONS TO RESTORE AQUATIC ECOSYSTEMS VIA NANOREMEDIATION

Apart from chemical and physical approaches, nano-remediation has gained rapid importance due to its cost effectiveness and fewer bad impacts in restoring the environment contaminated with toxic pollutants. Various nanoparticles have been identified which have potential in the treatment of aquatic ecosystems. Some of the applications to restore groundwater, surface water, wastewater, and aquaculture through nanoremediation have been discussed.

7.4.1 GROUNDWATER TREATMENT

An injection well is used to introduce nanoparticles into the aquifer to start the process. The contaminating source is then reached by these nanoparticles. Nanoparticles then transform toxic pollutants into less harmful chemicals when they come into touch with them. Transformations often involve redox reactions, where the nanoparticle is the reluctant oxidant [51]. Nano-scale iron particles have been shown to be efficient in treating groundwater pollutants [52].

7.4.2 SURFACE WATER TREATMENT

Nanotechnology has emerged as a promising alternative for the treatment of polluted water with improved effectiveness and reduced hazardous chemical production [53]. Ultrafiltration has been driven by membrane technology to create clean water through the utilization of the Binxian Reservoir [54]. Membrane permeability has been recovered by alkaline (NaOH) and oxidizing reagents like Sodium hypochlorite (NaClO), and irreversible contamination has been examined by organic matter (polysaccharide) and metals such as iron and manganese [55]. The successful use of PAC-ultrafiltration (PAC/

UF) and hybrid adsorption methods to remove organic pollutants from Miedwie Lake has been reported. Organic pollutants can also be eliminated more effectively by combining UF with adsorption on PAC [56].

7.4.3 WASTEWATER TREATMENT

Wastewater removal is a growing concern due to its effects on living organisms. Nanotechnology is one of the wastewater treatment strategies used for bioremediation. TiO_2 and ZnO nano-catalyst and paper-pulp production become effective for the degradation of dyes [57]. Photo-catalysis with ozone shows great potential to degrade organic pollutants in wastewater. Photo-catalysis can be used to decontaminate purification systems, and degradation in wastewater treatment has grown in popularity [58]. Semiconductor-sensitized photosynthetic processes have been reported to be effective in the removal of organics, bacteria, viruses, and cyano-toxins [59]. Nanotechnology methods used in the treatment of wastewater are discussed below:

1. **Nano-Adsorption:** Adsorption is primarily generated by physical forces, although weak chemical interactions can also cause it [60]. During the adsorption process, pollutants are adsorbed on a solid surface. In most cases, nano-adsorbents are employed to filter out inorganic and organic pollutants from water and wastewater [61]. Nano-adsorbents are intriguing adsorbent materials for wastewater treatment due to their unique qualities, which facilitate their quick interaction with various contaminants [62].
2. **Carbon-based Nano-Adsorbents:** Carbon nanotubes (CNTs) are cylindrical carbon-based nano-adsorbents which are being researched as a carbon substitute for activated carbon. CNTs have a huge specific surface area and a distinct adsorption site [63]. Because of their hydrophobic surface, CNTs form loose bundles/aggregates in aqueous media, reducing the active surface area. These aggregates are high-energy adsorption sites for organic pollutants in water [64].

7.4.4 DESALINATION

Nanotechnology has significantly improved desalination. These technologies may be commercialized in the near future, but some difficulties like manufacturing, realistic desalinization efficacy, and long-term stability need to be overcome. For the desalination of water, several materials, including zeolites, carbon nanotubes, graphene, and nanostructured materials, have high-flow molecular sieve membranes. The correct operation of these membranes likewise depends on the sub-nanometer regime. Unique transport characteristics have been emphasized by steric, electrostatic, and ultrafast interactions; nevertheless, other nanostructures may be advantageous for desalination [65, 66].

7.4.5 HEAVY METAL REMOVAL

Due to their huge surface area and high reactivity, nanoparticles are increasingly routinely utilized to remove heavy metals from water or wastewater. Heavy metal adsorption in aqueous settings is highly preferred by metal oxide nanoparticles with large surface areas [67].

7.4.6 WIRELESS NANO-SENSORS

Under laboratory conditions, this technique is thought to sense and detect numerous pollutants in water. In this field, work is being done to improve the detection of contaminants, allowing them to be eradicated even more effectively [68].

7.4.7 PURIFICATION OF WATER

In water purification, nanotechnology is employed to use nanoscopic materials such as carbon nanotubes and alumina fibers for nanofiltration. Zeolite filtration membranes containing nanopores, nanocatalysts, and magnetic nanoparticles are also included. For the analytical identification of pollutants in water samples, nanosensors based on palladium or titanium oxide nanowires are utilized [69].

7.4.8 AQUATIC MANAGEMENT

The use of nanotechnology in seawater shrimp farming shown that the nanodevice may lower water exchange rates while also improving water quality. Nannette therapy was the best of numerous nanodevices; the results revealed a 100% improvement in fish survival rate, as well as a decrease in both nitrite and nitrate concentration, with nitrite dropping to 1/4 of the control group. Nanotechnology has also enhanced the efficiency of water for aquatic survival by lowering the concentration of heavy metals, including Al, Cd, Cr, Fe, Zn, Ni, and Pb, and increasing the pH of the water [70].

7.4.9 FISH DISEASE CONTROL

As a result, nanotechnology is seen as a solution for preventing and monitoring illnesses and infections, as well as multiplying the benefits of aquaculture. Nanosensors in aquaculture systems for pathogen detection in water, antibacterial or antifungal surfaces made with porous nanostructures, and nano-delivery of veterinary products and fish treatments via fish meals are a few applications of nanotechnology in fish health. Nano-trace element usage is up to 100 times more than standard inorganic trace element usage, which is often extremely limited because the former enters the animal body by direct penetration [71].

7.4.10 WATER NANOFILTRATION

Using nanotechnology at various phases of the manufacturing process, more effective aquaculture fish feed is being developed to improve the physical, chemical, and nutritional uniformity of feeds and their constituents [72].

7.4.11 POND WATER STERILIZATION

Nanomaterials in the form of activated materials such as carbon or alumina, with chemicals such as zeolite and iron-containing compounds, can be utilized to retain aerobic and anaerobic biofilm for the removal of ammonia, nitrites, and nitrate pollutants in aquaculture applications at various stages and refine the water in order to maintain the good water quality for healthy fish [73, 74].

7.5 SURFACE WATER

Emerging contaminants detected in surface water can be remediated via several nanoparticles:

1. **Nano-Structured Membranes:** Carbon nanotube (CNT) filters are made up of hollow cylinders with radial alignment of carbon nanotube walls, which are used to remove bacterial pathogens (*Escherichia coli and Staphylococcus aureus*) and *Poliovirus sabin 1* from polluted water. These reusable containers may be cleaned using ultrasonication or autoclaving [75].

2. **Nano-Reactive Membranes:** Water filtering techniques have also employed nanostructured membranes in conjunction with other reactive and functionalized membranes. A-alumoxane alumina was used for alumina production and UF membranes, with 4.5–5.0-layer pairs of polys (styrene sulfonate)/poly(allylamine hydrochloride) deposited atop porous alumina. These NF membranes demonstrated significant water flow and divalent cation retention (Ca(II) and Mg(II)) retention [76].

7.6 GROUNDWATER

Groundwater can be contaminated by a variety of pollutants, including petroleum hydrocarbons, chlorinated solvents, and heavy metals:

1. **Groundwater Nano-Remediation:** The use of NPs in water purification dates back to the 1990s, making it one of the more recent methods. The notion of employing NPs for the decontamination of water was first proposed by Gilliam and Hanne Sin [57, 77]. They help in the remediation of halogenated group compounds using nZVI. Nonetheless, Wang and Zhang [58, 78] pioneered the use of NPs to remove organo-chlorines from contaminated groundwater. Using bimetallic NPs, they were able to completely remove many aromatic chlorinated compounds.

2. **Groundwater Remediation Using MNPs:** Because of their ease of separation with a magnet and unusual metal-ion adsorption, groundwater clean-up with MNPs has recently gotten a lot of interest. The effectiveness of MNPs for groundwater restoration has lately been examined in a number of studies. Gong et al. [79] investigated the performance of FeS-coated iron (Fe/FeS) magnetic NPs (MNPs) for the remediation of Cr(VI)-contaminated groundwater [79].

3. **Nano-Remediation of Heavy Metals Contaminated Soils in Agriculture:** Heavy metal (HM) contamination of agricultural

soils is a serious problem for food safety and human health. They have grown in popularity in recent years as a result of enormous industrialization and urbanization [80, 81]. Soil erosion, floods, volcanic activity, sediment re-suspension, metal corrosion, geological weathering, metal evaporation from soil, and wastewater are important natural sources of HMs contamination in agricultural soils. Contamination of terrestrial habitats, particularly agricultural lands, with HMs, has become a serious concern for the developing globe. Foundries, mining, and smelting are examples of human activity. Pesticides, fertilizers, and other items used to increase crop yields are secondary sources of HMs in agricultural regions [82, 83]. Several in-situ and ex-situ rehabilitation strategies for HMs-contaminated soils have been deployed, but they have a number of limitations, including capital expenditure, toxicity, and environmental health threats [79–81]. Because of their potential uses in the environment and agriculture, nanoparticles have received a lot of interest in recent years. Nanoremediation employs nanoparticles to effectively lower heavy metal levels in the soil-plant system.

7.7 LIMITATIONS

Nanoremediation is a new approach that has demonstrated effective techniques of restoration for groundwater, surface water, and soil. However, there are still certain limitations and hazards involved with the usage of nanoparticles [81, 84]. Hence, a lack of proper knowledge of nanoparticles in the environment can lead to possible ecological implications. High concentrations of nano-zero valent species can lead to the formation of agglomerates. Risk to human health and ecology is still under process. Further, the smaller size of these particles possesses a higher rate of dispersal, causing Eco-toxicity in the environment. Nanoparticles also have the risk of bioaccumulation in living organisms. Studies have shown certain bacteria cultures which have sulfate-reducing ability leads to oxidizing nano-zero valent ions. However,However, oxidation of nano-zero valent ions generates the formation of reactive oxygen species (ROS) [85]. Many reports suggest that higher concentration of ZVIs leads to oxidative stress, reduced transpiration, stunted growth, and plant cell membrane damage, which further cause the death of plant after a prolonged period of exposure. In humans, reports have been seen where exposure to nanoparticles caused genotoxicity, inflammation, lipid peroxidation, and pulmonary disease and may lead to death [81, 84, 85].

7.8 CONCLUSION

The contribution of nanotechnology is playing a significant role in research and development. Different nano-based systems are being employed to reinforce the aquatic ecosystems. In spite of that, there is growing concern about the toxicity of nanoparticles, which is still under consideration. In this regard, eco-friendly and safe applications are applied for the restoration of the ecosystems. The current chapter explored the use of zero-valent ions in the restoration of water bodies. The authors have discussed the applications directed toward determining the mechanisms of nanoparticles to cure the contaminants present in aquatic ecosystems. Moreover, use of natural bioactive nano-embed particles could be used. These nanoparticle-based adsorbents and films can be used for the purification of water bodies.

ACKNOWLEDGMENT

The author would like to acknowledge the resources and support provided by the Rashtriya Uchchatar Shiksha Abhiyan (RUSA) 2.0 program.

KEYWORDS

- **contaminants**
- **desalination**
- **environmental pollution**
- **nanomaterials**
- **pond water sterilization**
- **purification**
- **remediation**
- **wastewater treatment**

REFERENCES

1. *Aquatic Ecosystem – Types of Aquatic Ecosystem and its Adaptations* (byjus.com).
2. Bourbonnais, A., Ho, S. L., Kinnard, C., Lenaerts, J., Sugiyama, S., & Altabet, M., (2021). Global change on the blue planet. *Communications Earth & Environment, 2*(1), 1–4.
3. Zhao, H., Li, Q., Wang, Z., Wu, T., & Zhang, M. (2020). Synthesis of MIL-101 (Cr) and its water adsorption performance. *Microporous and Mesoporous Materials, 297*, 110044.
4. Understanding Rivers | National Geographic Society.
5. Wetland | National Geographic Society.
6. Lake | National Geographic Society.
7. Ramachandra, T. V., Ahalya, N., & Murthy, R., (2005). Aquatic ecosystems: Conservation, restoration, and management. *Aquatic Ecosystems-Conservation, Restoration, and Management*, 26–50.
8. Weerasekara, P., (2017). The United Nations World Water Development Report 2017 wastewater. *Future of Food: Journal on Food, Agriculture, and Society, 5*(2), 80, 81.
9. Ingle, A. P., Seabra, A. B., Duran, N., & Rai, M., (2014). Nanoremediation: A new and emerging technology for the removal of toxic contaminants from the environment. In: *Microbial Biodegradation and Bioremediation* (pp. 233–250). Elsevier.
10. Grieger, K. D., Hjorth, R., Rice, J., Kumar, N., & Bang, J., (2015). *Nano-Remediation: Tiny Particles Cleaning up Big Environmental Problems*. Blog entry for IUCN. 2015 Jul 9.
11. Das, P. K., (2018). Phytoremediation and nanoremediation: Emerging techniques for treatment of acid mine drainage water. *Def. Life Sci J., 3*(2), 190–196.
12. Bardos, P., Bone, B., Černík, M., Elliott, D. W., Jones, S., & Merly, C., (2015) Nanoremediation and international environmental restoration markets. *Remediation Journal, 25*(2), 83–94.
13. Li, L., Hu, J., Shi, X., Fan, M., Luo, J., & Wei, X., (2016). Nanoscale zero-valent metals: A review of synthesis, characterization, and applications to environmental remediation. *Environmental Science and Pollution Research, 23*(18), 17880–1900.
14. Corsi, I., Winther-Nielsen, M., Sethi, R., Punta, C., Della, T. C., Libralato, G., Lofrano, G., et al., (2018). Ecofriendly nanotechnologies and nanomaterials for environmental applications: Key issue and consensus recommendations for sustainable and ecosafe nanoremediation. *Ecotoxicology and Environmental Safety, 154*, 237-44.
15. Rane, A. V., Kanny, K., Abitha, V. K., & Thomas, S., (2018). Methods for synthesis of nanoparticles and fabrication of nanocomposites. In: *Synthesis of Inorganic Nanomaterials* (pp. 121–139). Woodhead publishing.
16. Kumar, N., & Kumbhat, S., (2016b). Carbon-based nanomaterials. *Essentials in Nanoscience and Nanotechnology*, 189–236.
17. Kumar, N., & Kumbhat, S., (2016a). *Essentials in Nanoscience and Nanotechnology*. John Wiley & Sons.
18. Jeevanandam, J., Barhoum, A., Chan, Y. S., Dufresne, A., & Danquah, M. K. (2018). Review on nanoparticles and nanostructured materials: History, sources, toxicity, and regulations. *Beilstein Journal of Nanotechnology, 9*(1), 1050–1074.

19. Shen, Y., Tang, J., Nie, Z., Wang, Y., Ren, Y., & Zuo, L., (2009). Preparation and application of magnetic Fe_3O_4 nanoparticles for wastewater purification. *Sep. Purif. Technol., 68*, 312–319.
20. Xu, P., Zeng, G. M., Huang, D. L., Feng, C. L., Hu, S., Zhao, M. H., Lai, C., et al., (2012). Use of iron oxide nanomaterials in wastewater treatment: A review. *Sci. Total Environ., 424*, 1–10.
21. Yadav, M., Gupta, R., & Sharma, R. K. (2019). Green and sustainable pathways for wastewater purification. In: *Advances in Water Purification Techniques* (pp. 355–383). Elsevier.
22. Gupta, V. K., Kumar, R., Nayak, A., Saleh, T. A., & Barakat, M., (2013). Adsorptive removal of dyes from aqueous solution onto carbon nanotubes: A review. *Adv. Colloid Interf. Sci., 193*, 24–34.
23. Sharma, R., Puri, A., Monga, Y., & Adholeya, A., (2014a). Newly modified silica-based magnetically driven nanoadsorbent: A sustainable and versatile platform for efficient and selective recovery of cadmium from water and fly-ash ameliorated soil. *Sep. Purif. Technol., 127*, 121–130.
24. Sharma, R., Puri, A., Monga, Y., & Adholeya, A., (2014b). Acetoacetanilide-functionalized Fe_3O_4 nanoparticles for selective and cyclic removal of Pb^{2+} ions from different charged wastewaters. *J. Mater. Chem. A, 2*, 12888–12898.
25. Sharma, R., Kumar, H., & Kumar, A., (2015). A highly efficient and magnetically retrievable functionalized nanoadsorbent for ultrasonication assisted rapid and selective extraction of Pd^{2+} ions from water samples. *RSC Adv., 5*, 43371–43380.
26. Li, T., Farrell, J., Kason, M., et al., (2000). Investigation of the long-term performance of zerovalent iron for reductive dechlorination of trichloroethylene. *Environmental Science and Technology, 34*, 514–521.
27. Henn, K. W., & Waddill, D. W., (2006). Utilization of nanoscale zerovalent iron for source remediation case study. *Remediation, 16*(2), 57–77.
28. Thome, A., Reddy, K. R., Reginatto, C., et al., (2015). Review of nanotechnology for soil and groundwater remediation: Brazilian perspectives. *Water Air Soil Pollut.*
29. Sharma, J., (2019). Nanoremediation. *International Journal of Life Sciences and Technology, 12*(1), 1–6.
30. Köber, R., Hollert, H., Hornbruch, G., Jekel, M., Kamptner, A., Klaas, N., & Braun, J., (2014). Nanoscale zero-valent iron flakes for groundwater treatment. *Environmental Earth Sciences, 72*(9), 3339–3352.
31. Gheju, M., (2011). Hexavalent chromium reduction with zero-valent iron (ZVI) in aquatic systems. *Water, Air, & Soil Pollution, 222*(1), 103–148.
32. Nie, X., Liu, J., Zeng, X., & Yue, D., (2013). Rapid degradation of hexachlorobenzene by micron Ag/Fe bimetal particles. *Journal of Environmental Sciences, 25*(3), 473–478.
33. Jiao, C., Cheng, Y., Fan, W., & Li, J., (2015). Synthesis of agar-stabilized nanoscale zero-valent iron particles and removal study of hexavalent chromium. *International Journal of Environmental Science and Technology, 12*(5), 1603–1612.
34. Thatai, S., Khurana, P., Boken, J., Prasad, S., & Kumar, D., (2014). Nanoparticles and core–shell nanocomposite based new generation water remediation materials and analytical techniques: A review. *Microchemical Journal, 116*, 62–76.
35. Chen, X., Yao, X., Yu, C., Su, X., Shen, C., Chen, C., & Xu, X., (2014). Hydrodechlorination of polychlorinated biphenyls in contaminated soil from an e-waste

recycling area, using nanoscale zerovalent iron and Pd/Fe bimetallic nanoparticles. *Environmental Science and Pollution Research*, *21*(7), 5201–5210.

36. Fu, F., Dionysiou, D. D., & Liu, H., (2014). The use of zero-valent iron for groundwater remediation and wastewater treatment: A review. *Journal of Hazardous Materials*, *267*, 194–205.
37. Xu, J., & Bhattacharyya, D., (2005). Membrane-based bimetallic nanoparticles for environmental remediation: Synthesis and reactive properties. *Environmental Progress*, *24*(4), 358-366.
38. Avila, A. F., Viviane, C. M., Aline, M. De. O., Mayara, C. G. S., Glenda, R. B. S. L., & Camila, P., (2014). Gonçalves. "Nano-based systems for oil spills control and cleanup." *Journal of Hazardous Materials, 272*, 20-27.
39. Wang, N., Zhou, L., Guo, J., Ye, Q., Lin, J. M., & Yuan, J., (2014). Adsorption of environmental pollutants using magnetic hybrid nanoparticles modified with β-cyclodextrin. *Applied Surface Science*, *305*, 267–273.
40. Travlou, N. A., Kyzas, G. Z., Lazaridis, N. K., & Deliyanni, E. A., (2013). Functionalization of graphite oxide with magnetic chitosan for the preparation of a nanocomposite dye adsorbent. *Langmuir, 29*(5), 1657–1668.
41. Bhattacharya, S., Saha, I., Mukhopadhyay, A., Chattopadhyay, D., Ghosh, U., & Chatterjee, D., (2013). Role of nanotechnology in water treatment and purification: Potential applications and implications. *Int. J. Chem. Sci. Technol.*, *3*, 59–64.
42. Qutub, N., Pirzada, B. M., Umar, K., & Sabir, S., (2016). Synthesis of CdS nanoparticles using different sulfide ion precursors: Formation mechanism and photocatalytic degradation of acid blue-29. *Journal of Environmental Chemical Engineering, 4*(1), 808-817.
43. Xu, J., Cao, Z., Zhou, H., Lou, Z., Wang, Y., Xu, X., & Lowry, G. V. (2019). Sulfur dose and sulfidation time affect the reactivity and selectivity of post-sulfidized nanoscale zerovalent iron. *Environmental Science & Technology*, *53*(22), 13344–13352.
44. Mugunthan, E., Saidutta, M. B., & Jagadeeshbabu, P. E. (2018). Visible light assisted photocatalytic degradation of diclofenac using TiO_2-WO_3 mixed oxide catalysts. *Environmental Nanotechnology, Monitoring & Management*, *10*, 322–330.
45. Chang, D., Chen, T., Liu, H., Xi, Y., Qing, C., Xie, Q., & Frost, R. L., (2014). A new approach to prepare ZVI and its application in the removal of Cr (VI) from aqueous solution. *Chemical Engineering Journal*, *244*, 264–272.
46. Rani, K., Gomathi, T., Vijayalakshmi, K., Saranya, M., & Sudha, P. N., (2019). Banana fiber cellulose nanocrystals grafted with butyl acrylate for heavy metal lead(II) removal. *International Journal of Biological Macromolecules*, *131*, 461–472.
47. Zhang, J., Han, J., Wang, M., & Guo, R., (2017). Fe_3O_4/PANI/MnO_2 core–shell hybrids as advanced adsorbents for heavy metal ions. *Journal of Materials Chemistry A*, *5*(8), 4058-4066.
48. Weng, X., Guo, M., Luo, F., & Chen, Z. (2017). One-step green synthesis of bimetallic Fe/Ni nanoparticles by eucalyptus leaf extract: Biomolecules identification, characterization, and catalytic activity. *Chemical Engineering Journal*, *308*, 904–911.
49. Wan, H., Islam, M. S., Qian, D., Ormsbee, L., & Bhattacharyya, D. (2020). Reductive degradation of CCl_4 by sulfidized Fe and Pd-Fe nanoparticles: Kinetics, longevity, and morphology aspects. *Chemical Engineering Journal*, *394*, 125013.

50. Marcon, L., Oliveras, J., & Puntes, V. F., (2021). In situ nanoremediation of soils and groundwaters from the nanoparticle's standpoint: A review. *Science of the Total Environment*, 148324.
51. Zhang, W. X., & Elliott, D. W., (2006). Applications of iron nanoparticles for groundwater remediation. *Remediation Journal: The Journal of Environmental Cleanup Costs, Technologies & Techniques*, *16*(2), 7–21.
52. Agarwal, A., & Joshi, H., (2010). Application of nanotechnology in the remediation of contaminated groundwater: A short review. *Recent Research in Science and Technology*, *2*(6).
53. Maksoud, M. A., Elgarahy, A. M., Farrell, C., Ala'a, H., Rooney, D. W., & Osman, A. I., (2020). Insight on water remediation application using magnetic nanomaterials and biosorbents. *Coordination Chemistry Reviews*, *403*, 213096.
54. Xia, S., Nan, J., Liu, R., & Li, G., (2004). Study of drinking water treatment by ultrafiltration of surface water and its application to China. *Desalination*, *170*(1), 41–47.
55. Kimura, K., Hane, Y., Watanabe, Y., Amy, G., & Ohkuma, N., (2004). Irreversible membrane fouling during ultrafiltration of surface water. *Water Research*, *38*(14, 15), 3431–3441.
56. Mozia, S., & Tomaszewska, M., (2004). Treatment of surface water using hybrid processes—Adsorption on PAC and ultrafiltration. *Desalination*, *162*, 23–31.
57. Ganguly, R. K., Chakraborty, S. K., Midya, S., & Ravindran, B. (2022). Impact of nanomaterials on waste management: An insight into the modern concept of waste abatement. In: *Emerging Nanomaterials for Advanced Technologies* (pp. 621–641). Springer, Cham.
58. Agustina, T. E., Ang, H. M., & Vareek, V. K., (2005). A review of synergistic effect of photocatalysis and ozonation on wastewater treatment. *Journal of Photochemistry and Photobiology C: Photochemistry Reviews*, *6*(4), 264–273.
59. Robertson, P. K., Robertson, J. M., & Baermann, D. W., (2012). Removal of microorganisms and their chemical metabolites from water using semiconductor photocatalysis. *Journal of Hazardous Materials*, *211*, 161–171.
60. Faust, S. D., & Aly, O. M., (1983). Removal of inorganic contaminants. *Chemistry of Water Treatment*. Butterworth Publishers, Stoncham, Mass.
61. Qu, X., Alvarez, P. J., & Li, Q., (2013). Applications of nanotechnology in water and wastewater treatment. *Water Research*, *47*(12), 3931–3946.
62. Ali, I., (2012). New generation adsorbents for water treatment. *Chemical Reviews*, *112*(10), 5073–5091.
63. Yang, K., & Xing, B., (2010). Adsorption of organic compounds by carbon nanomaterials in aqueous phase: Polanyi theory and its application. *Chemical Reviews*, *110*(10), 5989–6008.
64. Pan, B., & Xing, B., (2008). Adsorption mechanisms of organic chemicals on carbon nanotubes. *Environmental Science & Technology*, *42*(24), 9005–9013.
65. Humplik, T., Lee, J., O'hern, S. C., Fellman, B. A., Baig, M. A., Hassan, S. F., & Wang, E. N., (2011). Nanostructured materials for water desalination. *Nanotechnology*, *22*(29), 292001.
66. Manke, A., Wang, L., & Rojanasakul, Y., (2013). Mechanisms of nanoparticle-induced oxidative stress and toxicity. *BioMed Research International*, *2013*.

67. Saha, I., Bhattacharya, S., Mukhopadhyay, A., Chattopadhyay, D., Ghosh, U., & Chatterjee, D., (2013). Role of nanotechnology in water treatment and purification: Potential applications and implications. *Int. J. Chem. Sci. Technol.*, *3*(3), 59–64.
68. Karn, P., (1990). MACA-a new channel access method for packet radio. In: *ARRL/CRRL Amateur Radio 9th Computer Networking Conference* (Vol. 140, pp. 134–140).
69. Saha, I., Bhattacharya, S., Mukhopadhyay, A., Chattopadhyay, D., Ghosh, U., & Chatterjee, D., (2013). Role of nanotechnology in water treatment and purification: Potential applications and implications. *Int. J. Chem. Sci. Technol.*, *3*(3), 59–64.
70. Patil, S. S., Shedbalkar, U. U., Truskewycz, A., Chopade, B. A., & Ball, A. S., (2016). Nanoparticles for environmental clean-up: A review of potential risks and emerging solutions. *Environmental Technology & Innovation*, *5*, 10–21.
71. Sarkar, B., Mahanty, A., Gupta, S. K., Choudhury, A. R., Daware, A., & Bhattacharjee, S., (2022). Nanotechnology: A next-generation tool for sustainable aquaculture. *Aquaculture*, *546*, 737330.
72. Guerra, F. D., Attia, M. F., Whitehead, D. C., & Alexis, F. (2018). Nanotechnology for environmental remediation: Materials and applications. *Molecules*, *23*(7), 1760.
73. Khan, N. A., Khan, S. U., Ahmed, S., Farooqi, I. H., Dhingra, A., Hussain, A., & Changani, F., (2019). Applications of nanotechnology in water and wastewater treatment: A review. *Asian Journal of Water, Environment and Pollution*, *16*(4), 81–86.
74. National Center for Biotechnology Information (nih.gov).
75. Theron, J., Walker, J. A., & Cloete, T. E., (2008). Nanotechnology and water treatment: Applications and emerging opportunities. *Critical Reviews in Microbiology*, *34*(1), 43–69.
76. DeFriend, K. A., Wiesner, M. R., & Barron, A. R., (2003). Alumina and aluminate ultrafiltration membranes derived from alumina nanoparticles. *Journal of Membrane Science*, *224*(1, 2), 11–28.
77. Gillham, R. W., O'Hannesin, S. F., (1994). Enhanced degradation of halogenated aliphatics by zero-valent iron. *Groundwater*, *32*, 958–967.
78. Wang, C. B., & Zhang, W. X., (1997). Synthesizing nanoscale iron particles for rapid and complete dechlorination of TCE and PCBs. *Environ. Sci. Technol.*, *31*, 2154–2156.
79. Gong, C., Li, L., Li, Z., Ji, H., Stern, A., Xia, Y., & Zhang, X., (2017). Discovery of intrinsic ferromagnetism in two-dimensional van der Waals crystals. *Nature*, *546*(7657), 265–269.
80. Uchimiya, M., Bannon, D., Nakanishi, H., McBride, M. B., Williams, M. A., & Yoshihara, T. (2020). Chemical speciation, plant uptake, and toxicity of heavy metals in agricultural soils. *Journal of Agricultural and Food Chemistry*, *68*(46), 12856–12869.
81. Nasr-Eldahan, S., Nabil-Adam, A., Shreadah, M. A., Maher, A. M., & El-Sayed, A. T., (2021). A review article on nanotechnology in aquaculture sustainability as a novel tool in fish disease control. *Aquaculture International*, *29*(4), 1459–1480.
82. Jeevanandam, J., Barhoum, A., Chan, Y. S., Dufresne, A., & Danquah, M. K., (2018). Review on nanoparticles and nanostructured materials: History, sources, toxicity, and regulations. *Beilstein Journal of Nanotechnology*, *9*(1), 1050–1074.
83. Ramachandra, T. V., Ahalya, N., & Murthy, R., (2005). Aquatic ecosystems: Conservation, restoration, and management. *Aquatic Ecosystems-Conservation, Restoration, and Management*, 26–50.

84. Choudhary, B. K., Majumdar, K., & Deb, S., (2016). An overview of application of nanotechnology in environmental restoration. Status of Research in Physics in North-East India. First edition. *Sungraphics.*
85. Manke, A., Wang, L., & Rojanasakul, Y., (2013). Mechanisms of nanoparticle-induced oxidative stress and toxicity. *BioMed Research International, 2013.*

CHAPTER 8

Remediation of Heavy Metals from Contaminated Soils Using Nanomaterials and Hyperaccumulator Plants

ABIDA PARVEEN, KHALID SULTAN, SHAGUFTA PERVEEN, SARA ZAFAR, and NAEEM IQBAL

Department of Botany, Government College University, Faisalabad, Pakistan

ABSTRACT

Effluents from the industries are very harmful. These contain a large number of heavy metals which cause damage in fields. In biology, the phrase "heavy" comprises metalloids and metals that, at extremely low levels, can be harmful to both plants and animals. Several contaminants like selenium lead, mercury, and cadmium may not be necessary for plants because no physiological role is recognized. Others are necessary for plant development and metabolism, including iron, cobalt, molybdenum, manganese, copper, zinc, and nickel. Heavy metals (Pb, Cd, Tl, Cu, Mn, Zn, Co, As, Sb, Se, Ni,) are discovered in roughly 450 angiosperm species, accounting for fewer than 0.2% of all species of the angiosperms. Nanotechnology is emerging in this era. With the help of nanotechnology, small-sized particles are being produced and are helpful for the remediation of polluted soil. When plants are subjected to nanoparticles, they can remove the metals. Many species of plants are like metals and will grow in the metals or accumulate the metals from the soil. Hyperaccumulators are plants that accumulate heavy metals like cadmium, zinc, iron, mercury, and lead from the water as well as from the soil, for the checking of the capacity of plant species for phytoremediation in plant soil samples, shoots, and roots, the bioconcentration factor

Nano-Bioremediation for Water and Soil Treatment: An Eco-Friendly Approach.
Vishnu D. Rajput, Arpna Kumari, and Tatiana M. Minkina (Eds.)

(BCF), biological accumulation coefficient (BAC), and biological transfer coefficient (BCF) were established.

8.1 INTRODUCTION

The development of the industrial areas causes severe hazardous effects on the different population areas. This development causes environmental problems for the people. Through the gas emission, pesticides, ore mining, and wastes of the urban areas humans polluted the soil in large quantities [1]. These pollutants are added to the water and soil and become part of the food chain. When these pollutants become part of the food chain then they have hazardous effects on animals, humans, as plants. In humans, these effects on the immune system, impact the endocrine system and cause cancer disease as well [2]. Methods of chemicals, excavation, precipitation, electro-remediation, and treatment through heat are very costly [3].

In this era nanotechnology for the removal of heavy metals is used globally to restore contaminated soil in an effective way [4, 5]. Nanotechnology is used as a cleaner for the environment and detoxification of the heavy metals and metalloids from the soil [6, 7]. Detoxification of the heavy metals from the soil nanotechnology has the potential to remove hazardous toxicities. On a large scale, the remediation of the soil from heavy metals is one of the main domains. With the help of redox reactions, adsorption, coprecipitation, and precipitations are some ways to remove the metals from the soil [6, 7]. Nanoparticles application on the plants such as hyperaccumulators is a new approach for the removal of heavy metals. In this approach, soil microbes are being used to detoxify the metals from the soil. These microbes enhance the process of biodegradation and also improve the remediation of the metals from the soil. These technologies are called nano-remediation and bioremediation [8]. These nanomaterials are small in size and are beneficial to the soil and to the environment. The foliar application of the nanoparticles reduces the adverse effects of the pollutants which also increases the growth of the plants. Due to these properties, plants accumulate a large number of metals from the soil [9–11].

With the help of phytoremediation which is a green method and has no effects on the soil using the hyperaccumulator to get rid of the hazardous metals from the fields. Microorganisms present in the rhizospheric area of the plant are used to detoxify the metals [12]. Using this strategy is very cheap, eco-friendly, and can be used in efficient ways [13, 14]. On the base of the condition of the soil and plant species and heavy metals in soil, there are five types of phytoremediation: phytoextraction, phytodegradation, Phyto stabilization, Phyto filtration and with the help of phytovolatilization. When plants do not show any signs of the effects of the heavy metals they are divided into tolerance, extraction of the heavy metals and a large amount of accumulation of the heavy metals [15]. Due to these properties, this green strategy is used for the removal of heavy metals from contaminated soil. Using these technologies on large scale not efficient because these have some limitation to use. By the transgenic application, these limitations can be overcome to improve the accumulation of the heavy metals from the soil [1].

8.2 NANO-FERTILIZERS

Fertilizers are a necessary component of agriculture since they help plants grow and develop. Nano-fertilizers, on the other hand, has lately been shown to be more effective than traditional fertilizers. Nanoparticles (NPs) have a smaller surface area, increasing fertilizer availability and allowing for higher absorption by plants. They reduce the leaching of the fertilizers losses, and assimilation of the microbes in the soil [16, 17]. Furthermore, small-size fertilizers are applied in low quantity and increase the fertility of the soil. These fertilizers aid the soil to reduce the adverse effects of the chemical fertilizers [18].

8.3 APPLICATION OF THE NANOMATERIAL ON HYPERACCUMULATORS

Restoration of the heavy metal-contaminated soil is a pressing issue from an ecological point of view [19]. In contaminated areas of the world, there is a need to get rid of hazardous pollutants from the land with the help of a hyperaccumulator plant by the application of nanomaterials. This relies on the soil quality and the contamination of the soil pollutants. These parameters keep in mind to utilize the heavy metal through the hyperaccumulators [20]. Metal bioavailability in the rhizosphere is dependent on the pH of the soil, concentration of the gradients, moving of the microbes, potential of the oxidation and reduction reaction, CO_2:O_2 ratio, and other factors [21].

The root zone of the plants totally depends upon the root exudates according to the plant species and as well as the shape of the root [20]. Some plant species are survived at the optimum level of heavy metals. These species such as *Alternanthera pungens, Pedioplanis burchelli* and *Amaranthus spinosus* has made different mechanism to adapt in this environment [22]. Plant survives in the optimum level of heavy metals, but this concentration of the metals has adverse effects on humans, plants' growth as well as other organisms 23. Beyond the adverse effects, plants absorb a large quantity of the heavy metals from the soil and store it in their parts to clean up the soil from the metal contamination. During hyperaccumulation, these plants have no toxic effects [24, 25].

8.4 ELIMINATION OF THE POLLUTANTS BY NANOBIOREMEDIATION

Plant removes metals from the soil in different ways such as phytoextraction, rhizodegradation, and phytostabilization [26]. Through the activities of the microbes, metal pollutants are deposited in the areas of the root zone or rhizosphere. This process of the deposition of the metal is rhidegradation. When these pollutants are deposited in the rhizosphere they are metabolized by the bacteria to get nutrition and energy (Figure 8.1). When these toxic pollutants are break down by the bacteria, they become nontoxic and nonhazardous [27]. In the rhizosphere, plants release some components which contain carbon like sugar, acids, and alcohol and are helpful for the microorganisms to get more nutrients to improve the rhizodegradation process [28].

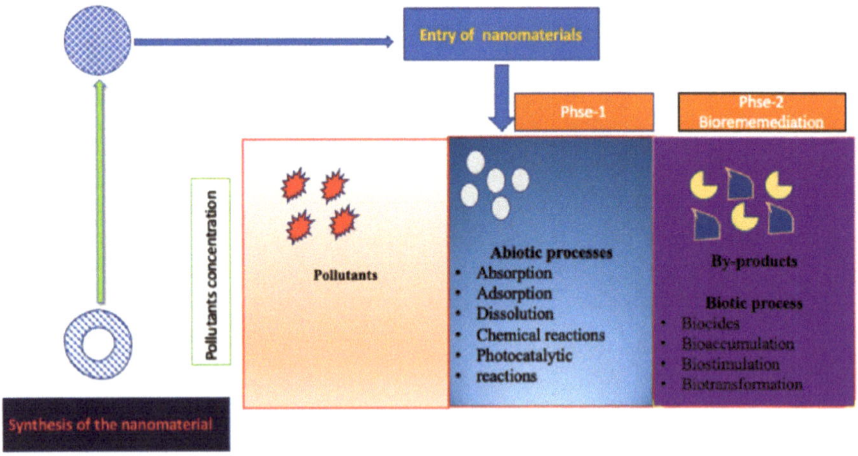

FIGURE 8.1 Schematic diagram of nanobioremediation of heavy metals from the soil [29].

Source: Adapted from Ref. [29]

Nanoparticles of the FeS helpful for the removal of the chromium metal 99.65% from the soil at the pH of 6.0 while the removal of the metals from the soil is decreased at the pH of 10 [30]. The fluctuation in the temperature affects the adsorption of the nanomaterials. The potential of the FeS nanoparticles for the removal of the Hg^{2+} is high when the contact time is 30 min [31]. By the application of the nZVI on the plant species was effective in the removal of hazardous metals from the field areas. These elements contain the shells of the iron like Fe^{3+}. These two elements have the potential for the removal of metals from the soil. The application of the nZVI removes the different amounts of the different metals from the soil. In the affected areas of the soil nano-valent zero iron particles remove the metals at the highest rate [32].

In the application of the nano-valent zero iron particles in the pot experiment, it was shown that the moving of the chromium efficiency was increased after two weeks of the application [23]. It was reported that the removal of the zinc, cadmium, and lead from the polluted soil increase the uptake of metals with the help of the OA-nano-valent zero iron nanoparticles at the rate of 47.01%, and 46.66%, 48.88% respectively [33]. If the concentration of the nZVI is low but it improves the growth of the plant, a seedling of the plant, leaf areas and root length in white willow plant *Salix alba* L. In this plant an increase in the concentration of the bio-concentration factor of the cadmium. If the concentration of the nZVI is high, then it affects the plant growth and may increase the BCF for the lead and the cadmium [34].

Removal of the Cd, Zn, and Pb with the combination of the nZVI and biochar immobilizes the cadmium, zinc, and lead [35]. Application of the nono-valent zero iron particles on the two species of the plant-like edible rapeseed *Brassica napus* and *Brassica rapa* shown that these species increase the immobilization of the chromium metal [36].

Application of the nano-hydroxyapatite to the plant of the ryegrass for 1–12 months for the elimination of the pollutants of the field. After one month of the application, it was shown that the lead was removed 30% from the soil and a three-month application showed that the metal from the soil 44.39% [37]. On the other hand, the application of the nZVI particles to the ryegrass it was observed that after 45 days of the treatment, the accumulation of the lead was increased [38]. A tree named as *Isatis capapdocia* applied the nanoparticles of the salicylic acid to remove the adverse effects of the heavy metals from the soil. It improves plant growth as well as removes arsenic from contaminated soil [39].

It was shown that the application of the TiO_2 nanoparticles increases the photolytic and has a high reactivity which helps to absorb the contaminants. Due to this adsorption these nanomaterials lessen the toxic effects of the heavy metals and maintain the mobility of the contaminated soil. TiO_2 particles absorb an amount of 70.67% of the cadmium and 88.01% of the copper metal and it is more effective at the pH of 7.0 [40]. Azolla plants are water-living plants. This plant belongs to the fern family and is to be known for the accumulation of heavy metals from the water. These plants have unique metallothioneins and phytochelatins. These are two metal-binding ligands that hyperaccumulating the heavy metals from the soil. These species accumulate a large number of heavy metals like cadmium from the soil as shown in Table 8.1 [41]. The species of the *Thalaspi* are more tolerant and hyperaccumulators. These plants are nearly related to the Arabidopsis species' properties. This plant hyper accumulates the zinc metal through the roots system (Table 8.1) [42]. *Alyssum lesbiacum* plants have histidine in roots which hyper accumulates the nickel-metal from the soil plant [43]. Some phosphate transporters are found in the *Pteris vitata* plant species. These transporters remove the arsenic from the soil. *Solanum photeinocarpum* has the potential to eliminate Cd from land pollution. While the rate of photosynthesis and the chlorophyll contents were not affected by the cadmium metal as shown in Table 8.1.

TABLE 8.1 Hyperaccumulation of Several Metals by Different Plant Parts

Plants Name	Name of the Metals	Metal Quantity in mg/kg	Parts of the Plants Metal Accumulation	References
Azolla pinnata	Cadmium	740	Roots	[44]
Thalaspi caerulescens	Zinc	19,410	Shoots	[45]
Alyssum bertolonii	Nickel	10,900	Shoots	[46]
Schina superba	Manganese	62,412.3	Leaves	[47]
Euphorbia cheriadenia	Lead	1,138	Shoots	[48]
Pteris vittata	Arsenic	2,200–3,030	Frond and root	[49]
Alyssum morale	Nickel	4,730–20,100	Roots	[50]
Solanum photeinocarpum	Cadmium	158	Root	[51]
Alyssum corsicum	Nickel	18,100	Shoot	[46]
Corrigiola telephiifola	Arsenic	2,110	All parts above the soil	[52]
Eleocharis acicularis	Copper	20,200	Shoots	[53]

8.5 EFFECT OF HEAVY METAL

Plants cannot escape or hide from bad climatic circumstances including drought, salinity, waterlogging, extreme temperature, UV radiation, and so on because they are sessile organisms. Plants have a system of enzymatic antioxidants, such as ascorbate peroxidase, superoxide dismutase, glutathione reductase, catalase, and non-enzymatic antioxidants, such as glutathione and ascorbate, to conflict with oxidative stress-caused by the metals. For survival in osmotic stress accumulate some compounds such as amino acids (glycine, proline, and taurine) as well as some osmolytes like polyols (sorbitol, glycerols, inositols) and trehalose. These compounds maintain the plant water and keep the cells of the plants hydrated.

Plants are deprived of adequate oxygen in hypoxic conditions, resulting in energy depletion and low energy status. To maintain energy levels, plants change their metabolism and transition from glucose to fermentation metabolism [54]. To alleviate mental stress, plants produce metal chelates, organic acids, and polyphosphates, which limit and sequester harmful metals in apoplastic and symplasm, respectively.

The phytotoxicity that comes from the toxic metals may occur which can cause the membrane integrity at the molecular level. Due to these metals can cause the blocking of metabolites or change the role of the nutrients which may phytotoxic. Increased ROS produces oxidative stress in cells, which may cause peroxidation of the lipids, degradation of the large molecules, breaking of the membranes, leakage of the ions, and breakage of the DNA strands [55–57].

8.6 IMPROVEMENT OF PLANT GROWTH BY NANOMATERIALS

In the roots, these metals are combined with other organic substances, or the amino acids are detoxified by the plant's 58. Reducing heavy metal transfer to above-ground organs reduces heavy metal harm to the photosynthetic tissues or in the tissues of the leaves [55, 59]. Plants can grow in heavy metal-polluted environments caused by anthropogenic activities. The diversity of the plants acts to move out the heavy metals from their body parts. The plants which are kick out the heavy metals or the toxic metals from their parts are excluded. The majority of heavy metals are kept and detoxified in root tissues, with only minor transfer to leaves, which are nevertheless susceptible to phytotoxic effects 58.

When it comes to heavy metal uptake and dispersion in plants, however, many hyper-tolerant species, also known as "hyperaccumulators," have the opposite behavior. They don't have any phytotoxicity [60, 61]. The tolerance of the hyperaccumulators differentiates them from non-accumulator plants [62].

Heavy metals (Mn, Se, Ni, Co, As, Pb, Ti, Cu, Zn, Cd) have been discovered in roughly 450 angiosperm species. New reports of this type of plant continue to appear implying that there are likely to be many more hyperaccumulators in nature that have yet to be discovered. Cu and Co hyperaccumulation in various cuprophytes was discovered to be owing to leaf surface contamination by field sample prompting a full re-evaluation of the Cu/Co hyperaccumulators [63, 64].

Nano-TiO_2 enhances spinach growth by improving the function of the antioxidant activities of the plant [65]. Nano-TiO_2 can help nitrate reductase convert inorganic nitrogen (NO_3eN and NH_4eN) into an organic form [66]. Rice biomass in Cd-contaminated soil was boosted using TiO_2 NPs [67, 68] has reported the successful immobilization of several heavy metals in soil utilizing Fe-based additives such as zerovalent iron grit (Fe°). The use of 10% iron nanoparticles (nZVI) boosted Pb, As, and Cr immobilization by over 82%, but did not affect Cd mobilization in soil (about 13–42%) [69].

Phosphate-based compounds have been reported to have a significant impact on the immobilization of Pb in contaminated soil and have lately become a popular approach because of their cost-effectiveness and low disruption [70]. Hydroxyapatite (HAP) is widely employed in Pb-contaminated soil and is considered the most efficient supplement among P-based materials. HAP is successful in immobilizing additional toxic metals such as cadmium lead and copper of the zinc remediation [71, 72].

The hyperaccumulation of the heavy metals in one or more species differs greatly from each other. This hyperaccumulation differs greatly between the species or the population of a group of species [73, 74]. Hyperaccumulation, on the other hand, is distinguished from comparable non-hyperaccumulator species by three essential characteristics. The leaves of these plants have a greater ability to detoxify the metals from the soil.

These transporters are responsible for the roots of hyperaccumulators plants such as for the uptake of zinc and in the non-hyperaccumulator plants these transporters do not work like these plants [75].

The single physiological involvement of this heavy metal has previously been recognized because it was discovered as a rare Cd-containing carbonic anhydrase in a marine diatom's active metal-binding site *Thalassiosira weissglogii* is a species of Thalassiosira [76, 77]. According to a lot of data, arsenate can reach plant roots via phosphate transporters [78].

Phosphate/arsenate transporters were found in higher density in plasma membranes of As hyperaccumulator *Pteris vittata* than in plasma membranes of non-hyperaccumulator *Pteris tremula*, which could be due to constitutive gene overexpression [79]. Additionally, the hyperaccumulating fern's increased As uptake is dependent on phosphate/arsenate transport systems have a greater affinity for the arsenate [80]. A large amount of the organic carbon present in the root exudates of the plants helps them to more availability of Arsenate bioavailability in root zones by decreasing of pH levels 81. As the pH decreases, the amount of water-soluble As that the roots may absorb rises [81, 82].

8.7 TOLERANCE MECHANISM BY PLANT GROWING IN HEAVY METAL SOIL

The hyperaccumulators, on the other hand, transfer these elements to the shoot swiftly and efficiently through the xylem. According to root cell tonoplast-specific features, this needs a high degree of poor sequestration into and fast outflow out of the vacuoles resulting in low metal availability for xylem loading [83]. In cell root vacuoles, the amount of Zn absorbed is two to three times lower in hyperaccumulators *T. caerulescens* [83] and *S. alfredii* [84] Zn efflux from vacuoles is near twice as quick in hyperaccumulating ancestors. Lower sequestration into root vacuoles explains greater.

Hyperaccumulator roots contain substantial concentrations of small chemical compounds which function as bindings of the metals. On the other hand, the role of different chelators in hyperaccumulation approaches is still unknown [85]. Some amino acids like histidine and nicotinamide help the hyperaccumulators for the uptake of heavy metals. These amino acids make strong and stable complexes with cations of divalent bonds [86]. In the hyperaccumulation of nickel from the contaminated soil, histidine plays a major role [86].

Under non-toxic conditions, root-to-shoot Ni transfer rates are similar in both the Nikel detoxifying Thlaspi species or in the non-hyperaccumulating *T. arvense* species [87]. The scientists infer that *T. goes* intense's hyperaccumulation ability is due to highly functional rather than increased heavy metal transport, Ni decontamination and/or storage methods were used. In hyperaccumulator *P. vittata*, arsenite contributes for roughly 90% of the overall of the As in the xylem sap, compared to non-hyperaccumulator ferns [88].

Alyssum lesbiacum plant which is a hyperaccumulator has a high concentration of histidine in roots as compared to non-hyperaccumulator plants. ATP-phosphoribosyl is a transferase enzyme that has more histidine in the roots of the hyperaccumulators plants as compared to no hyperaccumulators which act as loading of Ni and form Ni-His complex [43].

Because significant His concentrations have been found in the roots of other Ni hyperaccumulating Thlaspi species, the amino acid may have a similar impact in other hyperaccumulators [42]. *T. caerulescens* and *A. halleri*, Zn/Ni hyperaccumulators with a 3-fold greater nicotinamide content than non-hyperaccumulating congeners, have overexpressed genes encoding nicotinamide biosynthesis pathway enzymes in their roots [89–91].

Increased nicotinamide production and nicotinamide-metal chelation has been associated with Ni hyperaccumulation in *T. caerulescens*, but they have been linked to Zn hyperaccumulation in *A. halleri*, suggesting that cytosol nicotinamide–Zn complexes may play a role in keeping metals [92].

Overexpression of the HMA4 protein (which belongs to the Zn/Co/Cd/Pb HMA subclass and is found at the plasma membranes of xylem parenchyma) suggests that the HMA4 protein is engaged in the outflow of zinc and cadmium comes from the root cytoplasm into xylem vessels, which also is required for a hyperaccumulation of Cd and Zn in the shoot. According to QTL analysis, the HMA4 gene is co-located with a substantial QTL for Zn and Cd tolerance in *A. halleri* [93–95].

Phosphate transporters can load the residual arsenate into the xylem, but the efflux of the predominant arsenite toward the vascular tissues requires alternative transport mechanisms that have yet to be found. The plasma membrane has been shown to play a role in arsenite trafficking in mammals [96] as well as in plants [97, 98].

8.8 CONCLUSIONS AND FUTURE PERSPECTIVES

In the developing era of technology, a large number of hazardous chemicals and pollutants are being released from industries that are harmful to the environment as well as humans. Hence for the removal of these toxic pollutants different techniques are being used without their negative impacts on the environment. Some techniques are not efficient to remove the pollution and cause effects. Beyond the different techniques, nanotechnology is one of them for the removal of metals from the soil. From different studies, this method was proven to get desirable results. This technology for the elimination of hazardous metals from the field is helpful in future perspectives. This technology eliminates the metals from the land without the wait for the experiment on it. Through different methods, this land is clean from metals. But in this era of the developing world, this nano-remediation of the soil is only limited to lab experiments. This nano-remediation of the soil is not being used on a large scale. In elimination of the metals from the soil different techniques and approaches should be applied. Different disciplinary programs should be started to eliminate the pollution from the soil. But keeping in mind the use of these nanomaterials effects or should be used carefully to make this technique in a wide range. For the removal of hazardous pollutants from the soil different strategies should be applied. These strategies select the matching hyperaccumulator plants on which the nanomaterials will be applied.

KEYWORDS

- **heavy metals**
- **hyperaccumulators**
- **nano-fertilizers**
- **nanoremediation**
- **persistence**
- **pollutants**
- **tolerance mechanism**
- **toxicity**

REFERENCES

1. Shah, V., & Daverey, A., (2020). Phytoremediation: A multidisciplinary approach to clean up heavy metal contaminated soil. *Environ. Technol. Innov.*, *18*, 100774.
2. Nedjimi, B., (2009). Can calcium protect *Atriplex halimus* subsp. schweinfurthii against cadmium toxicity? *Acta Bot. Gall*, *156*(3), 391–397.
3. Medina-Pérez, G., Fernández-Luqueño, F., Vazquez-Nuñez, E., López-Valdez, F., Prieto-Mendez, J., Madariaga-Navarrete, A., & Miranda-Arámbula, M., (2019). Remediating polluted soils using nanotechnologies: Environmental benefits and risks. *Polish Journal of Environmental Studies*, *28*(3).

4. Nedjimi, B., & Daoud, Y., (2009). Ameliorative effect of $CaCl_2$ on growth, membrane permeability and nutrient uptake in *Atriplex halimus* subsp. schweinfurthii grown at high (NaCl) salinity. *Desalination, 249*(1), 163–166.
5. Farooqi, A. A., & Sreeramu, B., (2004). *Cultivation of Medicinal and Aromatic Crops.* Universities Press.
6. Kumari, A., Kumari, P., Rajput, V. D., Sushkova, S. N., & Minkina, T., (2022). Metal (loid) nanosorbents in restoration of polluted soils: Geochemical, ecotoxicological, and remediation perspectives. *Environ. Geochem. Health, 44*(1), 235–246.
7. Raffa, C. M., Chiampo, F., & Shanthakumar, S., (2021). Remediation of metal/metalloid-polluted soils: A short review. *Appl. Sci., 11*(9), 4134.
8. Alazaiza, M. Y., Albahnasawi, A., Ali, G. A., Bashir, M. J., Copty, N. K., Amr, S. S. A., Abushammala, M. F., & Al Maskari, T., (2021). Recent advances of nanoremediation technologies for soil and groundwater remediation: A review. *Water, 13*(16), 2186.
9. Ghani, M. I., Saleem, S., Rather, S. A., Rehmani, M. S., Alamri, S., Rajput, V. D., Kalaji, H. M., et al., (2022). Foliar application of zinc oxide nanoparticles: An effective strategy to mitigate drought stress in cucumber seedling by modulating antioxidant defense system and osmolytes accumulation. *Chemosphere, 289*, 133202.
10. Faizan, M., Sehar, S., Rajput, V. D., Faraz, A., Afzal, S., Minkina, T., Sushkova, S., et al., (2021). Modulation of cellular redox status and antioxidant defense system after synergistic application of zinc oxide nanoparticles and salicylic acid in rice (*Oryza sativa*) plant under arsenic stress. *Plants, 10*(11), 2254.
11. Faizan, M., Rajput, V. D., Al-Khuraif, A. A., Arshad, M., Minkina, T., Sushkova, S., & Yu, F., (2021). Effect of foliar fertigation of chitosan nanoparticles on cadmium accumulation and toxicity in *Solanum lycopersicum. Biology, 10*(7), 666.
12. Liu, S., Yang, B., Liang, Y., Xiao, Y., & Fang, J., (2020). Prospect of phytoremediation combined with other approaches for remediation of heavy metal-polluted soils. *Environ. Sci. Pollut. Res., 27*(14), 16069–16085.
13. Ashraf, M., Ozturk, M., & Ahmad, M. S. A., (2010). Toxins and their phytoremediation. In: *Plant Adaptation and Phytoremediation* (pp. 1–32). Springer.
14. Nedjimi, B., (2020). Germination characteristics of *Peganum harmala* L. (*Nitrariaceae*) subjected to heavy metals: Implications for the use in polluted dryland restoration. *Int J. Environ. Sci. Technol., 17*(4), 2113–2122.
15. Krämer, U., (2010). Metal hyperaccumulation in plants. *Annu. Rev. Plant Biol., 61*, 517–534.
16. DeRosa, M. C., Monreal, C., Schnitzer, M., Walsh, R., & Sultan, Y., (2010). Nanotechnology in fertilizers. *Nat. Nanotechnol., 5*(2), 91–91.
17. Liu, F., Wen, L. X., Li, Z. Z., Yu, W., Sun, H. Y., & Chen, J. F., (2006). Porous hollow silica nanoparticles as controlled delivery system for water-soluble pesticide. *Mater. Res. Bull., 41*(12), 2268–2275.
18. Prasad, R., Jain, V., & Varma, A., (2010). Role of nanomaterials in symbiotic fungus growth enhancement. *Curr. Sci., 99*(9), 1189–1191.
19. Palansooriya, K. N., Shaheen, S. M., Chen, S. S., Tsang, D. C., Hashimoto, Y., Hou, D., Bolan, N. S., et al., (2020). Soil amendments for immobilization of potentially toxic elements in contaminated soils: A critical review. *Environ. Int., 134*, 105046.
20. Ahkami, A. H., White, III. R. A., Handakumbura, P. P., & Jansson, C., (2017). Rhizosphere engineering: Enhancing sustainable plant ecosystem productivity. *Rhizosphere, 3*, 233–243.

21. Upadhyay, S. K., Ahmad, M., Srivastava, A. K., Abhilash, P. C., & Sharma, B., (2021). Optimization of eco-friendly novel amendments for sustainable utilization of fly ash based on growth performance, hormones, antioxidant, and heavy metal translocation in chickpea (*Cicer arietinum* L.) plant. *Chemosphere*, *267*, 129216.
22. Rizwan, M., Ali, S., Ur Rehman, M. Z., Adrees, M., Arshad, M., Qayyum, M. F., Ali, L., et al., (2019). Alleviation of cadmium accumulation in maize (*Zea mays* L.) by foliar spray of zinc oxide nanoparticles and biochar to contaminated soil. *Environ. Pollut.*, *248*, 358–367.
23. Sun, T. Y., Bornhöft, N. A., Hungerbühler, K., & Nowack, B., (2016). Dynamic probabilistic modeling of environmental emissions of engineered nanomaterials. *Environ. Sci. Technol.*, *50*(9), 4701–4711.
24. Jabeen, R., Ahmad, A., & Iqbal, M., (2009). Phytoremediation of heavy metals: Physiological and molecular mechanisms. *Bot. Rev.*, *75*(4), 339–364.
25. Rascio, N., & Navari-Izzo, F., (2011). Heavy metal hyperaccumulating plants: How and why do they do it? And what makes them so interesting? *Plant Sci.*, *180*(2), 169–181.
26. Yan, A., Wang, Y., Tan, S. N., Mohd, Y. M. L., Ghosh, S., & Chen, Z., (2020). Phytoremediation: A promising approach for revegetation of heavy metal-polluted land. *Front. Plant Sci.*, *11*, 359.
27. Ojuederie, O. B., & Babalola, O. O., (2017). Microbial and plant-assisted bioremediation of heavy metal polluted environments: A review. *Int. J. Environ. Res. Public Health*, *14*(12), 1504.
28. Rai, G. K., Bhat, B. A., Mushtaq, M., Tariq, L., Rai, P. K., Basu, U., Dar, A. A., et al., (2021). Insights into decontamination of soils by phytoremediation: A detailed account on heavy metal toxicity and mitigation strategies. *Physiol. Plant.*, *173*(1), 287–304.
29. Rajput, V. D., Minkina, T., Upadhyay, S. K., Kumari, A., Ranjan, A., Mandzhieva, S., Sushkova, S., et al., (2022). Nanotechnology in the restoration of polluted soil. *Nanomaterials*, *12*(5), 769.
30. Wu, X., Hu, J., Wu, F., Zhang, X., Wang, B., Yang, Y., Shen, G., et al., (2021). Application of TiO_2 nanoparticles to reduce bioaccumulation of arsenic in rice seedlings (*Oryza sativa* L.): A mechanistic study. *J. Hazard. Mater.*, *405*, 124047.
31. Gong, Y., Liu, Y., Xiong, Z., & Zhao, D., (2014). Immobilization of mercury by carboxymethyl cellulose stabilized iron sulfide nanoparticles: Reaction mechanisms and effects of stabilizer and water chemistry. *Environ. Sci. Technol.*, *48*(7), 3986–3994.
32. Klimkova, S., Cernik, M., Lacinova, L., Filip, J., Jancik, D., & Zboril, R., (2011). Zero-valent iron nanoparticles in treatment of acid mine water from in situ uranium leaching. *Chemosphere*, *82*(8), 1178–1184.
33. Cao, Y., Zhang, S., Zhong, Q., Wang, G., Xu, X., Li, T., Wang, L., Jia, Y., & Li, Y., (2018). Feasibility of nanoscale zero-valent iron to enhance the removal efficiencies of heavy metals from polluted soils by organic acids. *Ecotoxicol. Environ. Saf.*, *162*, 464–473.
34. Mokarram-Kashtiban, S., Hosseini, S. M., Tabari, K. M., & Younesi, H., (2019). The impact of nanoparticles zero-valent iron (nZVI) and rhizosphere microorganisms on the phytoremediation ability of white willow and its response. *Environ. Sci. Pollut. Res.*, *26*(11), 10776–10789.
35. Mitzia, A., Vítková, M., & Komárek, M., (2020). Assessment of biochar and/or nano zero-valent iron for the stabilization of Zn, Pb and Cd: A temporal study of solid phase geochemistry under changing soil conditions. *Chemosphere*, *242*, 125248.

36. Wang, Y., Fang, Z., Kang, Y., & Tsang, E. P., (2014). Immobilization and phytotoxicity of chromium in contaminated soil remediated by CMC-stabilized nZVI. *J. Hazard. Mater.*, *275*, 230–237.
37. Liang, S. X., Jin, Y., Liu, W., Li, X., Shen, S. G., & Ding, L., (2017). Feasibility of Pb phytoextraction using nano-materials assisted ryegrass: Results of a one-year field-scale experiment. *J. Environ. Manage.*, *190*, 170–175.
38. Huang, D., Qin, X., Peng, Z., Liu, Y., Gong, X., Zeng, G., Huang, C., et al., (2018). Nanoscale zero-valent iron assisted phytoremediation of Pb in sediment: Impacts on metal accumulation and antioxidative system of *Lolium perenne*. *Ecotoxicol. Environ. Saf.*, *153*, 229–237.
39. Souri, Z., Karimi, N., Sarmadi, M., & Rostami, E., (2017). Salicylic acid nanoparticles (SANPs) improve growth and phytoremediation efficiency of *Isatis cappadocica* Desv., under as stress. *IET Nanobiotechnol.*, *11*(6), 650–655.
40. Mahmoud, M. E., Abou, A. S. A., & Elweshahy, S. M., (2018). Microwave functionalization of titanium oxide nanoparticles with chitosan nanolayer for instantaneous microwave sorption of Cu(II) and Cd(II) from water. *Int. J. Biol. Macromol.*, *111*, 393–399.
41. Talebi, M., Tabatabaei, B. E. S., & Akbarzadeh, H., (2019). Hyperaccumulation of Cu, Zn, Ni, and Cd in azolla species inducing expression of methallothionein and phytochelatin synthase genes. *Chemosphere*, *230*, 488–497.
42. Assunção, A. G., Schat, H., & Aarts, M. G., (2003). *Thlaspi caerulescens*, an attractive model species to study heavy metal hyperaccumulation in plants. *New Phytol.*, *159*(2), 351–360.
43. Kerkeb, L., & Kramer, U., (2003). The role of free histidine in xylem loading of nickel in *Alyssum lesbiacum* and *Brassica juncea*. *Plant Physiol.*, *131*(2), 716–724.
44. Rai, P. K., (2008). Phytoremediation of Hg and Cd from industrial effluents using an aquatic free-floating macrophyte *Azolla pinnata*. *Int. J. Phytorem.*, *10*(5), 430–439.
45. Lombi, E., Zhao, F., McGrath, S., Young, S., & Sacchi, G., (2001). Physiological evidence for a high-affinity cadmium transporter highly expressed in a *Thlaspi caerulescens* ecotype. *New Phytol.*, *149*(1), 53–60.
46. Li, Y. M., Chaney, R., Brewer, E., Roseberg, R., Angle, J. S., Baker, A., Reeves, R., & Nelkin, J., (2003). Development of a technology for commercial phytoextraction of nickel: Economic and technical considerations. *Plant Soil*, *249*(1), 107–115.
47. Yang, S., Deng, H., & Li, M., (2008). Manganese uptake and accumulation in a woody hyperaccumulator, *Schima superba*. *Plant Soil Environ.*, *54*(10), 441–446.
48. Chehregani, A., & Malayeri, B. E., (2007). Removal of heavy metals by native accumulator plants. *Int. J. Agric. Biol. (Pakistan)*.
49. Kalve, S., Sarangi, B. K., Pandey, R. A., & Chakrabarti, T., (2011). Arsenic and chromium hyperaccumulation by an ecotype of *Pteris vittata*–prospective for phytoextraction from contaminated water and soil. *Curr. Sci.*, 888–894.
50. Bani, A., Pavlova, D., Echevarria, G., Mullaj, A., Reeves, R. D., Morel, J. L., & Sulçe, S., (2010). Nickel hyperaccumulation by the species of *Alyssum* and *Thlaspi* (*Brassicaceae*) from the ultramafic soils of the Balkans. *Bot. Serb.*, *34*(1), 3–14.
51. Zhang, X., Xia, H., Zhuang, P., & Gao, B., (2011). Identification of a new potential Cd-hyperaccumulator *Solanum photeinocarpum* by soil seed bank-metal concentration gradient method. *J. Hazard. Mater.*, *189*(1, 2), 414–419.

52. García-Salgado, S., García-Casillas, D., Quijano-Nieto, M., & Bonilla-Simón, M., (2012). Arsenic and heavy metal uptake and accumulation in native plant species from soils polluted by mining activities. *Water Air Soil Pollu.*, *223*(2), 559–572.
53. Sakakibara, M., Sugawara, M., Sano, S., & Sera, K., (2013). Phytoremediation of heavy metal-contaminated river water by aquatic macrophyte *Eleocharis acicularis* in a mine site, southwestern Japan. *NMCC Annual Report*, *20*, 226–33.
54. Banti, V., Giuntoli, B., Gonzali, S., Loreti, E., Magneschi, L., Novi, G., Paparelli, E., et al., (2013). Low oxygen response mechanisms in green organisms. *Int. J. Mol. Sci.*, *14*(3), 4734–4761.
55. Navari-Izzo, F., Pinzino, C., Quartacci, M. F., & Sgherri, C. L., (1999). Superoxide and hydroxyl radical generation, and superoxide dismutase in PII membrane fragments from wheat. *Free Radical Res.*, *31*(sup1), 3–9.
56. Navari-Izzo, F., Quartacci, M. F., Pinzino, C., Vecchia, F. D., & Sgherri, C. L., (1998). Thylakoid-bound and stromal antioxidative enzymes in wheat treated with excess copper. *Physiol. Plant.*, *104*(4), 630–638.
57. Quartacci, M. F., Cosi, E., & Navari-Izzo, F., (2001). Lipids and NADPH-dependent superoxide production in plasma membrane vesicles from roots of wheat grown under copper deficiency or excess. *J. Exp. Bot.*, *52*(354), 77–84.
58. Hall, J. Á., (2002). Cellular mechanisms for heavy metal detoxification and tolerance. *J. Exp. Bot.*, *53*(366), 1–11.
59. Sgherri, C., Cosi, E., & Navari-Izzo, F., (2003). Phenols and antioxidative status of *Raphanus sativus* grown in copper excess. *Physiol. Plant*, *118*(1), 21–28.
60. Rascio, N., (1977). Metal accumulation by some plants growing on zinc-mine deposits. *Oikos*, 250–253.
61. Reeves, R., (2006). Hyperaccumulation of trace elements by plants. In: *Phytoremediation of Metal-Contaminated Soils* (pp 25–52). Springer.
62. Chaney, R. L., Malik, M., Li, Y. M., Brown, S. L., Brewer, E. P., Angle, J. S., & Baker, A. J., (1997). Phytoremediation of soil metals. *Curr. Opin. Biotechnol.*, *8*(3), 279–284.
63. Faucon, M. P., Shutcha, M. N., & Meerts, P., (2007). Revisiting copper and cobalt concentrations in supposed hyperaccumulators from SC Africa: Influence of washing and metal concentrations in soil. *Plant Soil*, *301*(1), 29–36.
64. Faucon, M. P., Colinet, G., Mahy, G., Luhembwe, M. N., Verbruggen, N., & Meerts, P., (2009). Soil influence on Cu and Co uptake and plant size in the cuprophytes *Crepidorhopalon perennis* and *C. tenuis* (*Scrophulariaceae*) in SC Africa. *Plant Soil*, *317*(1), 201–212.
65. Gao, J., Xu, G., Qian, H., Liu, P., Zhao, P., & Hu, Y., (2013). Effects of nano-TiO_2 on photosynthetic characteristics of *Ulmus elongata* seedlings. *Environ. Pollut.*, *176*, 63–70.
66. Yang, X., Li, T., Yang, J., He, Z., Lu, L., & Meng, F., (2006). Zinc compartmentation in root, transport into xylem, and absorption into leaf cells in the hyper accumulating species of *Sedum alfredii* Hance. *Planta*, *224*(1), 185–195.
67. Kumpiene, J., Ore, S., Renella, G., Mench, M., Lagerkvist, A., & Maurice, C., (2006). Assessment of zerovalent iron for stabilization of chromium, copper, and arsenic in soil. *Environ. Pollut.*, *144*(1), 62–69.
68. Rizwan, M., Ali, S., Malik, S., Adrees, M., Qayyum, M. F., Alamri, S. A., Alyemeni, M. N., & Ahmad, P., (2019). Effect of foliar applications of silicon and titanium dioxide

nanoparticles on growth, oxidative stress, and cadmium accumulation by rice (*Oryza sativa*). *Acta Physiolo. Plantarum*, *41*(3), 1–12.
69. Gil-Díaz, M., Pinilla, P., Alonso, J., & Lobo, M., (2017). Viability of a nanoremediation process in single or multi-metal (loid) contaminated soils. *J. Hazard. Mater.*, *321*, 812–819.
70. Cao, R. X., Ma, L. Q., Chen, M., Singh, S. P., & Harris, W. G., (2003). Phosphate-induced metal immobilization in a contaminated site. *Environ. Pollu.*, *122*(1), 19–28.
71. Chen, S., Xu, M., Ma, Y., & Yang, J., (2007). Evaluation of different phosphate amendments on availability of metals in contaminated soil. *Ecotoxicol. Environ. Saf.*, *67*(2), 278–285.
72. Misra, V., & Chaturvedi, P. K., (2007). Plant uptake/bioavailability of heavy metals from the contaminated soil after treatment with humus soil and hydroxyapatite. *Environ. Monit. Assess*, *133*(1), 169–176.
73. Deng, D., Shu, W., Zhang, J., Zou, H., Lin, Z., Ye, Z., & Wong, M. H., (2007). Zinc and cadmium accumulation and tolerance in populations of *Sedum alfredii*. *Environ. Pollut.*, *147*(2), 381–386.
74. Roosens, N., Verbruggen, N., Meerts, P., Ximenez-Embun, P., & Smith, J., (2003). Natural variation in cadmium tolerance and its relationship to metal hyperaccumulation for seven populations of *Thlaspi caerulescens* from western Europe. *Plant Cell Environ.*, *26*(10), 1657–1672.
75. Assunção, A., Martins, P. D. C., De Folter, S., Vooijs, R., Schat, H., & Aarts, M., (2001). Elevated expression of metal transporter genes in three accessions of the metal hyperaccumulator *Thlaspi caerulescens*. *Plant Cell Environ.*, *24*(2), 217–226.
76. Lane, T. W., & Morel, F. M., (2000). A biological function for cadmium in marine diatoms. *Proc. Natl. Acad. Sci.*, *97*(9), 4627–4631.
77. Lane, T. W., Saito, M. A., George, G. N., Pickering, I. J., Prince, R. C., & Morel, F. M., (2005). A cadmium enzyme from a marine diatom. *Nature*, *435*(7038), 42–42.
78. Meharg, A. A., & Hartley-Whitaker, J., (2002). Arsenic uptake and metabolism in arsenic resistant and nonresistant plant species. *New Phytol.*, *154*(1), 29–43.
79. Caille, N., Zhao, F., & McGrath, S., (2005). Comparison of root absorption, translocation, and tolerance of arsenic in the hyperaccumulator *Pteris vittata* and the nonhyperaccumulator *Pteris tremula*. *New Phytol.*, *165*(3), 755–761.
80. Poynton, C. Y., Huang, J. W., Blaylock, M. J., Kochian, L. V., & Elless, M. P., (2004). Mechanisms of arsenic hyperaccumulation in *Pteris* species: Root as influx and translocation. *Planta*, *219*(6), 1080–1088.
81. Gonzaga, M. I. S., Ma, L. Q., Santos, J. A. G., & Matias, M. I. S., (2009). Rhizosphere characteristics of two arsenic hyperaccumulating *Pteris* ferns. *Sci. Total Environ.*, *407*(16), 4711–4716.
82. Fitz, W. J., & Wenzel, W. W., (2002). Arsenic transformations in the soil–rhizosphere–plant system: Fundamentals and potential application to phytoremediation. *J. Biotechnol.*, *99*(3), 259–278.
83. Lasat, M. M., Pence, N. S., Garvin, D. F., Ebbs, S. D., & Kochian, L. V., (2000). Molecular physiology of zinc transport in the Zn hyperaccumulator *Thlaspi caerulescens*. *J. Exp. Bot.*, *51*(342), 71–79.
84. Yang, X., Long, X., Ye, H., He, Z., Calvert, D., & Stoffella, P., (2004). Cadmium tolerance and hyperaccumulation in a new Zn-hyperaccumulating plant species (*Sedum alfredii* Hance). *Plant Soil*, *259*(1), 181–189.

85. Haydon, M. J., & Cobbett, C. S., (2007). Transporters of ligands for essential metal ions in plants. *New Phytol.*, *174*(3), 499–506.
86. Callahan, D. L., Baker, A. J., Kolev, S. D., & Wedd, A. G., (2006). Metal ion ligands in hyperaccumulating plants. *JBIC J. Biol. Inorg. Chem.*, *11*(1), 2–12.
87. Kramer, U., Smith, R. D., Wenzel, W. W., Raskin, I., & Salt, D. E., (1997). The role of metal transport and tolerance in nickel hyperaccumulation by *Thlaspi goesingense* Halacsy. *Plant Physiol.*, *115*(4), 1641–1650.
88. Su, Y., McGrath, S., Zhu, Y., & Zhao, F., (2008). Highly efficient xylem transport of arsenite in the arsenic hyperaccumulator *Pteris vittata*. *New Phytol.*, *180*(2), 434–441.
89. Mari, S., Gendre, D., Pianelli, K., Ouerdane, L., Lobinski, R., Briat, J. F., Lebrun, M., & Czernic, P., (2006). Root-to-shoot long-distance circulation of nicotianamine and nicotianamine–nickel chelates in the metal hyperaccumulator *Thlaspi caerulescens*. *J. Exp. Bot.*, *57*(15), 4111–4122.
90. Talke, I. N., Hanikenne, M., & Krämer, U., (2006). Zinc-dependent global transcriptional control, transcriptional deregulation, and higher gene copy number for genes in metal homeostasis of the hyperaccumulator *Arabidopsis halleri*. *Plant Physiol.*, *142*(1), 148–167.
91. Weber, M., Harada, E., Vess, C., Roepenack-Lahaye, E. V., & Clemens, S., (2004). Comparative microarray analysis of *Arabidopsis thaliana* and *Arabidopsis halleri* roots identifies nicotianamine synthase, a ZIP transporter, and other genes as potential metal hyperaccumulation factors. *The Plant J.*, *37*(2), 269–281.
92. Vacchina, V., Mari, S., Czernic, P., Marquès, L., Pianelli, K., Schaumlöffel, D., Lebrun, M., & Łobiński, R., (2003). Speciation of nickel in a hyperaccumulating plant by high-performance liquid chromatography− inductively coupled plasma mass spectrometry and electrospray MS/MS assisted by cloning using yeast complementation. *Anal. Chem.*, *75*(11), 2740–2745.
93. Courbot, M., Willems, G., Motte, P., Arvidsson, S., Roosens, N., Saumitou-Laprade, P., & Verbruggen, N., (2007). A major quantitative trait locus for cadmium tolerance in *Arabidopsis halleri* colocalizes with HMA4, a gene encoding a heavy metal ATPase. *Plant Physiol.*, *144*(2), 1052–1065.
94. Willems, G., Dräger, D. R. B., Courbot, M., Godé, C., Verbruggen, N., & Saumitou-Laprade, P., (2007). The genetic basis of zinc tolerance in the metallophyte *Arabidopsis halleri* ssp. Halleri (*Brassicaceae*): An analysis of quantitative trait loci. *Genetics*, *176*(1), 659–674.
95. Willems, G., Frérot, H., Gennen, J., Salis, P., Saumitou-Laprade, P., & Verbruggen, N., (2010). Quantitative trait loci analysis of mineral element concentrations in an *Arabidopsis helleri* × *Arabidopsis lyrata* petraea F2 progeny grown on cadmium-contaminated soil. *New Phytol.*, *187*(2), 368–379.
96. Liu, Z., Shen, J., Carbrey, J. M., Mukhopadhyay, R., Agre, P., & Rosen, B. P., (2002). Arsenite transport by mammalian aquaglyceroporins AQP7 and AQP9. *Proc. Natl. Acad. Sci.*, *99*(9), 6053–6058.
97. Kamiya, T., Tanaka, M., Mitani, N., Ma, J. F., Maeshima, M., & Fujiwara, T., (2009). NIP1; 1, an aquaporin homolog, determines the arsenite sensitivity of *Arabidopsis thaliana*. *J. Biol. Chem.*, *284*(4), 2114–2120.
98. Ma, J. F., Yamaji, N., Mitani, N., Xu, X. Y., Su, Y. H., McGrath, S. P., & Zhao, F. J., (2008). Transporters of arsenite in rice and their role in arsenic accumulation in rice grain. *Proc. Natl. Acad. Sci.*, *105*(29), 9931–9935.

CHAPTER 9

New Dimensions into the Removal of Pesticides Using an Innovative Ecofriendly Technique: Nanoremediation

ARPNA KUMARI,[1] SNEH RAJPUT,[2] SHIV VENDRA SINGH,[3] GAURI SHARMA,[2] ANTON ZHUMBEI,[1] VISHNU D. RAJPUT,[1] SAGLARA S. MANDZHIEVA,[1] TATIANA M. MINKINA,[1] NEHA SAHU,[4] ANUJ RANJAN,[1] SVETLANA N. SUSHKOVA,[1] and RAJINDER KAUR[2]

[1]*Academy of Biology and Biotechnology, Southern Federal University, Rostov-on-Don, Russia*

[2]*Department of Botanical and Environmental Sciences, Guru Nanak Dev University, Amritsar, Punjab, India*

[3]*School of Agriculture, Graphic Era Hill University, Dehradun, Uttarakhand, India*

[4]*Department of Botany, University of Lucknow, Uttar Pradesh, India*

ABSTRACT

Pesticides are being used more often in agriculture to protect crops against pests, weeds, and pathogens, but a significant proportion of applied pesticides strike non-target vegetation and remain as pesticide residue in the agroecosystems, recklessly harming the plants. A significant portion, often as high as 80% of the sprayed pesticides could be detected, along with 50% of their residues from the agricultural soils. However, to ensure food security and safety, healthy soil systems are the need of the hour to fulfill the increasing demands of the ever-increasing population around the world. Soils play a crucial role as ultimate sinks for contaminants, making it reasonable to

Nano-Bioremediation for Water and Soil Treatment: An Eco-Friendly Approach.
Vishnu D. Rajput, Arpna Kumari, and Tatiana M. Minkina (Eds.)

expect their impacts on vegetation and soil properties. Therefore, researchers worldwide are exploring a variety of current and cutting-edge strategies for the reclamation of contaminants, including pesticides. In this context, in this context, the peculiar properties of nanomaterials are attracting considerable interest in decontaminating polluted sites. Based on their physicochemical characteristics, nanomaterials are commonly selected for the detection, degradation, and removal of pesticides. Thus, this chapter addresses the present state of pesticide pollution, its consequences on agroecosystems, and the remediation approaches using various nanomaterials. For pesticide elimination, several nanoparticles, including metals, bimetallic, metal oxide, nanotubes, etc., have been used. However, to achieve a comprehensive application of nanoremediation to alleviate the health of pesticide-contaminated soils, more research and measures are required.

9.1 INTRODUCTION

As the human population is increasing rapidly a steep increase in anthropogenic activities leads to the addition of several types of pollutants to the environment. Such an increase in population has resulted in the extensive use of pesticides in agricultural, commercial, residential, and industrial practices, which has adversely affected the major components of the environment such as soil and water [1]. Pesticides are the synthetic organic compounds sprayed on crops for preventing or deterring pest infestation. Issues like loss of soil biodiversity, contamination of groundwater, pollution of soil water and soil air, and desertification are directly linked to pesticide contamination [2, 3]. Moreover, the increase in population is also resulting in industrialization and urbanization. A pesticide is a toxic chemical or mixture of toxic chemicals or biological agents that are intentionally released into the environment to control and/or kill and destroy populations of insects, weeds, rodents, fungi, or other harmful pests. As we know, pesticides degrade biocoenosis of soils which effect on biodegradation of organic matter present in the soil by microorganisms [4, 5]. Over the last few decades, the use of pesticides is considered one of the most effective methods for crop protection, but these chemicals eventually enter the natural ecosystems, including the agroecosystem, and thus hinder their structural and functional aspects that are also associated with the well-being of humans [5]. Further, a growing body of literature suggests that exposure to toxic pesticides for an extended period increases the risk of human multi-organ dysfunction, which can result in a variety of chronic diseases, including diabetes, cancer, Parkinson's, Alzheimer's, etc. [6–8].

Globally used pesticides include organochlorines, organophosphates, and pyrethroid insecticides, all of which have recently been shown to be mutagenic, carcinogenic, cytotoxic, genotoxic, or immunotoxic [7, 9]. Besides, pesticides are documented to induce several morpho-physiological implications in plants due to the accelerated generation of reactive oxygen species, which consequently led to a decline in plants' productivity [8, 10]. Also, the cue for the stunned growth and development in plants is reported owing to the accumulation of pesticides in plants. For example, applications of pesticides are undertaken in 168 countries, and a spatially distributed environmental model is utilized to evaluate the risk of environmental contamination due to 92 active pesticide ingredients over the world. According to the model's observations, 31% of the world's agricultural land is extremely sensitive to pesticide contamination by more than one active ingredient, and 64% is at the urge of contamination [11].

Several approaches have been explored for the reclamation of sites contaminated with pesticides. Recently, nanotechnology is described as a cutting-edge technology that is advancing speedily and spreading its tender areas throughout all scales, including the atomic, molecular, and supramolecular ones, with a minimum one-dimensional size of 1 to 100 nm [12]. Besides, its application in the area of remediation, it is also relevant to a broad range of other fields, including agriculture for delivery of genetic material in plants, nano-fertilizer, nanopesticides, sensing, and monitoring applications and in diagnostics and therapeutics applications, pharmaceuticals, drug delivery, the treatment of cancer, etc. [13–16]. The cumulative use and applications of nanomaterials, techniques, or tools for remediation likely arise from the fundamental requirement of technology that is realistic, eco-friendly, unquestionably attainable, cost-effective, and at the same time, faster in conveying results without additional problems to the removal process in the form of deposits and ecological insistence [17]. In addition to these, it has successfully completed the most anticipated environmental remediation bids, especially for the decontamination of water and soil resources [18]. In recent studies, the NPs were found to effectively remediate the pesticides by 80% from the aquatic environment [19]. Thus, with this background in mind, the current chapter has been designed to mainly envisage the current status of pesticide consumption around the world, its effect on soil and plants' health, and roles of nanoremediation in the removal of pesticides along with research gaps and future prospects.

9.2 GLOBAL OVERVIEW OF PESTICIDE CONTAMINATION

Pesticide usage can be divided into three eras. In the first era (before the 1870s), natural pesticides were used for pest control. In the second era (1870–1945), inorganic pesticides were synthesized and used. In the third era (after 1940), organic pesticides like DDT and 2,4-D were synthesized [20]. The worldwide consumption of pesticides is about 2 MT/Year and Europe accounts for about 45% of this consumption alone [21]. These organic chemicals have inevitable effects on both biotic and abiotic factors as they not only cause mutilation in the targeted organisms but also affect the non-targeted ones [22]. Each year about billion kilograms of pesticide is used out of which only 1% reaches the targeted organisms [23]. Remaining pesticide residues enter into different environmental matrices through sorption, leaching, spray drift, volatilization, surface runoff, etc., and cause negative impacts on human and environmental health [24]. In 1962, Rachel Louise Carson in her book "Silent Spring" described the harmful effects of pesticides on the ecosystem. The report was critically analyzed, and researchers analyzed the contamination of pesticide residues in various environmental matrices [19]. Moreover, pesticides are also reported to migrate through the grasshopper effects and get distributed across the globe and as far as Antarctica [25, 26]. Table 9.1 provides an insight into pesticide contamination around the world based on literature studies.

9.3 IMPACT ON AGROECOSYSTEM

A considerable amount of agrochemicals is used in agroecosystems to manage or deter pests and weeds to improve the yield for meeting the ever-growing food demands and food security [44]. Undoubtedly, this approach is somehow useful in the short term but in the long-term, it has been a potential threat to the agroecosystem and environment due to the serious impacts on non-targeted organisms, soil microbial community, soil fertility, and food security concerns due to their accumulation in edible products [45]. Thus, the extensive application of pesticides causes great public health concern about the negative impacts on the environment and human health. Therefore, this section will provide an overview of pesticide-induced effects on soil microbial communities and subsequent consequences on plants.

TABLE 9.1 Summary of Pesticide Contamination and Detection around the World based on Recent Literature

Study Area	Commodity	Method Used	Detected Pesticides	Range	References
Greece	Bee pollen	LC and GC-MS/MS	Coumaphos propargite, azoxystrobin, dimethoate, and cypermethrin.	0.8–35 μg/kg	[27]
Chile and Mexico	Vegetables	GC–MS/MS	Lambda-cyhalothrin, dimethoate, chlorpyrifos, and methamidophos.	–	[28]
China	Litchi	LC-MS/MS GC-MS	Pyraclostrobin	1,412 μg/kg to 2,030 μg/kg	[29]
Serbia	Fruits juice	LC-MS/MS	Carbendazim, acetamiprid, and pyrimethanil.	0.001 to 0.629 mg/kg	[30]
Portugal	Vine-cane	GC-MS	Aldrin, p,p′-DDE, α-HCH, PCB101 and PCB28)	5.85 ± 0.32 to 5.99 ± 0.25 μg/kg and aldrin 2.44 ± 0.15 μg/kg.	[31]
South America (Uruguay and Negro Rivers)	Muscle tissue of wild fish	Mass spectrometry	Trifloxystrobin, metolachlor, and pyraclostrobin.	<1 to 194 μg kg^{-1}	[32]
Brazil	Drinking water	LC-MS	Atrazine	5 to 49 ng L^{-1}	[33]
India	Pond water	RP-HPLC	Carbofuran, Parathion methyl, atrazine, carbendazim.	2.50–182.78 mg/L	[34]
Malaysia	River water	UHPLC-MS/MS	Imidacloprid, propiconazole, and buprofezin, tebuconazole.	57.7–4493.1 ng/L	[35]
China	Lake	HPLC	Carbendazim, Imidacloprid	438–508 ng/l	[36]
Argentina	Surface water, suspended particulate matter, and sediments.	GC-MS	Glyphosate and (aminomethyl) phosphonic acid	Maximum in water, 20.04 and 4.86 $\mu g \cdot L^{-1}$; in sediment, 1146.5 and 4032.7 $\mu g \cdot kg_{dw}^{-1}$, respectively.	[37]

TABLE 9.1 *(Continued)*

Study Area	Commodity	Method Used	Detected Pesticides	Range	References
Mediterranean Region of Turkey	Tomato	LC-MS/MS	Chlorpyrifos methyl, cyfluthrin, deltamethrin, and acetamiprid.	0.002 to 0.009 mg kg^{-1}	[38]
India	Fish	GC-MS	Endosulfan, aldrin, and endrin.	0.02–26.05 μg/kg	[39]
Italy	Sediments	HPLC-ESI-MS/MS	DDT, heptachlor, aldrin, dieldrin, Endosulfan.	0.54 to 99.6 ng/g	[40]
South Africa	Fish tissue	GC/GC-TOF-M	Heptachlor epoxide and dieldrin.	2,953–8,740 ng g^{-1} lw	[41]
Tunisia	Adipose tissue	GC-MS	DDT	74.49 to 1834.76 ng g^{-1}	[42]
Lebanon	Loquat	GC-MS/MS and LC-MS/MS	Cypermethrin, dimethoat, methomyl, carbendazim, and thiametoxam.	0.0162 to 0.0963 mg/kg	[43]

9.3.1 IMPACT ON SOIL HEALTH

The presence of xenobiotics including pesticides affects the structural and functional aspects of soil microbial structure. There are two main types of pesticides: synthetic and natural. However, natural pesticides are not as harmful as synthetic, but both have significant effects on microbial activity and pathways of chemical elements when they are metabolized or degraded by soil microorganisms. Thus, both types of pesticides have been reported to exert harmful effects on soil microbiota with the main difference, i.e., the extent of toxicity. According to previous reports, it has been established that the long-term applications of pesticides adversely affected the nitrification rate by interfering with the activities of nitrifying microorganisms [46, 47]. Besides, it is reported that microbial communities are prone to change their extracellular DNA after exposure to persistent organic pesticides (i.e., organochlorinated pesticides) due to the enhanced activity of deoxyribonuclease I [48]. Another major concern related to the pesticides' abundance and persistence in the agro-environment is the increasing multidrug resistance among soil microorganisms which in long run may emerge as an attractive and agronomically feasible alternative for the augmentation of edible crops in contaminated environment [49]. However, such pesticide-resistant and antibiotic-resistant microbial flora can find a route to enter the human body through agricultural products or animals, thus can induce serious health issues [50].

The contamination of soil with synthetic pesticides is often related to the increased risk of heavy metals contamination because of their interaction mediated considerable effects on soil properties. In this context, a study showed that the co-occurrence of synthetic pesticides and heavy metals slightly modulated the distribution of metals in soil fractions that were found to closely be associated with soil characteristics [51]. Furthermore, the application of soil fumigation has been found to decline bacterial diversity and abundance after 7 days of application. However, such effects on soil microbiota were recorded to be declined or even disappeared following this period. Plus, soil fumigation was also reported to affect the normal metabolism of non-targeted organisms which further might impact crop production [52] (Figure 9.1).

9.3.2 PESTICIDES AND PLANT HEALTH

The rising use of pesticides has a variety of effects on plants either terrestrial or aquatic. Pesticides are reported to penetrate plants and modify their physiology, such as by decreasing the net photosynthetic rate, growth, and

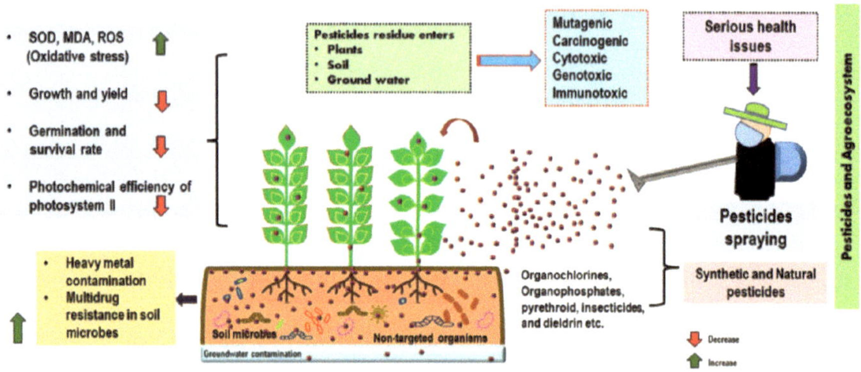

FIGURE 9.1 Schematic representation of the consequences induced be pesticides ubiquity on agroecosystem.

biomass production, which, in turn, decreases productivity and yield [53]. As well, pesticides are documented to interfere with the defense system of the plants *via* modulation of the activities of antioxidative enzymes [54, 55]. A detailed summary of the adverse effects of pesticides on the plants based on the recent literature is presented in Table 9.2.

9.4 NANOREMEDIATION INSIGHTS

In the past decades numerous physical, chemical, and biological methods have been developed for the removal of pesticides from environmental resources. The physical methods primarily rely on membranous separation and filtration of pesticides having larger particle sizes [62]. However, if the employed membranes have a pore size larger than the pesticide, the affinity of the membrane rejecting the contaminants gets adversely affected. Though there are other challenges as well to be as fouling of membranes, particularly, pollutants like pesticides, interfaces among the membranes and solute may be altered due to foulants. Chemical remediation typically uses radicals or reagents to convert the noxious pesticides into less harmful forms. The oxidation processes involve the use of oxidizing agents, i.e., hydroxy radicals capable of disintegrating aimed contaminants. Such oxidizing agents are highly efficient in the degradation of pesticides and organic contaminants removal [63]. The biological remediation approaches primarily involve microbial decomposition and phytoremediation. Nevertheless, conventional remediation methods face challenges in commercialization and efficient

TABLE 9.2 Effect of Different Pesticides on Plants' Morphological, Biochemical, and Physiological Responses along with the Accumulation

Pesticide(s) and Dose; Plant; Plants Consequences	References
Pesticide: Chlorpyrifos (CP); **Doses:** 100 and 200 mg/kg; **Plant:** Maize (*Zea mays* L.).	[54]
Plant Responses: Declined plants biomass and enzymes associated with plants' growth and development, i.e., 4-hydroxyl phenyl pyruvate dioxygenase.	
Enhanced the activities of SOD and POX of plants as compared to the control.	
There was a significant enhancement in the accumulation of CP in shoots.	
Pesticides: Chlorpyrifos, dimethoate, and dieldrin; **Doses:** 100 kg N/ha; **Plant:** *Spinacia oleracea* L.	[56]
Plant Responses: Caused the reduction in all morphometric parametric in comparison to the control and maximal inhibitor effect in yield was recorded in the treatment with dieldrin.	
Showed inhibitory effects photosynthetic pigment under the exposure of all pesticides when compared to the control.	
All induced oxidative stress via increasing superoxide radical, electrolyte leakage and the content of MDA and H_2O_2 over the control.	
The activities of all antioxidative enzymes were found to be significantly altered with the treatment all pesticides.	
Pesticide: Imidacloprid (IMI); **Doses:** 0, 250, 300, and 350 mg IMI/kg soil; **Plant:** *Brassica juncea* L.	[55]
Plant Responses: Dry weight of shoots and pigments content were recorded to be reduced by the increasing doses of the IMI in the plants of *B. juncea*.	
The contents of H_2O_2 and MDA have enhanced with the treatment of IMI in *B. juncea* leaves.	
Besides, the activities of all the antioxidative enzymes (SOD, CAT, APOX, GPOX, and DHAR) showed significant augmentation.	
Also, the organic acids were found to be increased under the exposure of IMI stress in *B. juncea* plants.	
Thus, the research outcomes revealed that IMI imposed the oxidative strs in brassica plants which confirmed the elevation in the contents of oxidative stress bioindicators, organic acids, and increment in the activities of antioxidative enzymes due to the accelerated ROS production.	
Pesticides: Mancozeb (MZ) and CP; **Doses:** 0, 10, 30, 50, 70, 90, 110, 130, and 150 ppm for MZ and 0, 5, 10, 15, 20, 25, and 30 ppm for CP; **Plant:** *Allium cepa*	[57]
Plant Responses: The exposure to pesticides led to a decline in the germination rate and survival rate.	
A significant decrement was recorded with the increase in the doses of pesticides.	
There was a dose-dependent increase was recorded in the case of activities of SOD, POD, and CAT in comparison to the control.	

TABLE 9.2 *(Continued)*

Pesticide(s) and Dose; Plant; Plants Consequences	References
Pesticides: Propaquizafop, quizalofop-p-ethyl, clopyralid, and picloram; **Doses:** At rate 70 g a.i. (active ingredient)* ha^{-1} (propaquizafop); 50 g a.i.* ha^{-1} (quizalofop-p-ethyl), 90 g a.i. ha^{-1} (clopyralid), and 24 g a.i.* ha^{-1} (picloram); **Plant:** *Camelina sativa* **Plant Responses:** In this study, the exposure of quizalofop-p-ethyl and propaquizafop was found to exert minimal toxic effects on all camelina varieties and genotypes. The photosynthetic parameters were also found to be adversely affected, e.g., the maximal decline in maximum photochemical efficiency of photosystem II including other parameters reported with the exposure of picloram for 21 days.	[58]
Pesticides (15): Consisted of 13 insecticides, 1 acaricide, and 1 botanical [acephate, alathion, profenophos, methomyl, chlorphyriphos, cypermethrin, alphamethrin, fipronil, carbofuron, phorate, imidacloprid, chlorantraniliprole, *E. benzoate*, Spiromesifen, NSKE] were evaluated for toxicity analyzes. **Doses:** 3 doses [recommended (RCD), 2-times of RCD, and 4-times of RCD]. **Plant:** *Sorghum bicolor* (L.) Moench. **Plant Responses:** The maximal phytotoxicity index (%age) was recorded with the treatment of profenophos at seedling, vegetative, and reproductive stages of crop growth. Further, the yield was contrarywise correlated with the phytotoxicity index.	[59]
Pesticides: 4-chlorophenol (4-CP), 2,4-dichlorophenol (2,4-DCP); **Doses:** 0, 0.06, 0.1, 1.0, 1.5 g/L (4-chlorophenol) and 0, 0.06, 0.1, 1.0 g/L (2,4-dichlorophenol); **Plants:** *Zea mays* and *Phaseolus vulgaris* **Plant Responses:** This study revealed that 2,4-DCP showed more toxicity for both plants that 4-CP. The germination index was dependent on doses for both plant models tested. The results indicate that 4-CP and 2,4-DCP affected the germination rate but not influenced other indices associated with the germination of both plants. Maize was found to be more tolerant to both chlorophenols at tested doses.	[60]
Pesticides: Isoxaben, chlorpyrifos, and oxyfluorfen; **Doses:** 0, 0.15, 0.35, 0.7, and 1.4 mg/L (isoxaben), 0, 0.05, 0.1, 0.2, and 0.4 mg/L (chlorpyrifos), and 0, 0.005, 0.01, 0.015, and 0.02 mg/L (oxyfluorfen). **Plants:** *Hydrangea paniculata* 'Limelight,' *Cornus obliqua* 'Powell Garden,' and *Hosta* 'Gold standard.' **Plant Responses:** There was a considerable extent of detrimental effects recorded based on the observation of morphometric parameters. The treatment of pesticides also affected the parameters related to photosynthetic physiology. Thus, this study concluded that the toxicity of the pesticides was pesticide-specific and taxa-specific.	[61]

residual pesticides bioremediation due to the concurrent presence of numerous pollutants. Besides, due to the highly hydrophilic and refractory nature of pesticides, the efficiency of conventional remediation methods is low [64]. In the last few years, the advanced approach of nanotechnology-based efficient and sustainable water treatment has attracted a lot of attention and particularly among researchers and industries. Several physicochemical properties make nanomaterial functional with various chemical groups to increase their efficiency for the removal of targeted compounds [65]. Nanotechnological approaches are primarily using adsorption, filtration, and degradation mechanisms for the removal of pesticides from the ecosystem.

9.4.1 ADSORPTION

Among the number of remediation techniques developed in recent years including biodegradation, photodegradation, chemical oxidation and precipitation, flocculation, neutralization, adsorption, membranous separation and purification, ultrafiltration, etc., the adsorption technique is the most attractive, simple, and effective tool being applied for water purification. Several studies have used metal-organic frameworks (MOFs), polymers, carbon tubes, activated carbon composites, graphene oxide (GO), biochar, and nanochar, and carbon nanotubes (CNTs) as efficient adsorbents to distinguish pesticides and toxic metals polluted soil and water ecosystem [66]. Adsorption is a surface phenomenon that dependent on surface charges, porosity, functionality, and specific surface area of the adsorbent as well as resident time and interactions between adsorbent, adsorbate, and microclimatic conditions.

The materials like carbon nanotubes (CNTs), reduced graphene oxides (RGO), nanocrystalline metal oxides (NMOs), etc., are a new class of nanocomposites proving their potential in pesticides removal. Further, Shi et al. [67] investigated the adsorption capacity of modified CNTs, SWNTs, and MWNTs for atrazine and reported a significant reduction due to the hydrophilic fraction of the surfactant micelles in the water and becoming more hydrophilic [67]. The activated carbon is also being explored for the adsorption of hydrophobic compounds [68]. The granulated activated carbon (GAC) and pulverized activated carbon (PAC) are also one of the most used sorbent materials in industrial effluents treatment due to high surface area and porosity [69]. Pimenta et al. [70] compeered the atrazine, simazine, and diuron adsorption efficiency of PAC, CNT, and GAC. They reported the efficacy of GAT for atrazine, simazine, and diuron by 91.1%, 98.8%, and 99.6%, respectively while removal efficiency of PAC and CNT for atrazine, simazine, and diuron was 76 and 93%, 63 and 94%, and 96 and 45%, respectively [70]. Bahrami et al. [71] as well reported the adsorption of 2,4-D onto the BRH, GAC, and MWCNTs was examined was >90% [71]. One of such adsorbents is chitosan/carbon nanotube that successfully removed 82.5% of diazinon under a near neutral pH [72]. Wanjeri et al. [73] reported a significant increment in organophosphate chlorpyrifos, parathion, and malathion adsorption from aqueous solution by 2-phenylethylamine functionalized graphene oxide-based silica-coated magnetic nanoparticles (Fe_3O_4@SiO_2@GO-PEA) [73]. Other nanomaterials being used for the removal/degradation of pesticides and their removal efficiency are listed in Table 9.3.

9.4.2 NANOMEMBRANE FILTRATION

Nanofiltration (NF) is one of the recently developed techniques of micropollutants (pesticides, dyes, heavy metals, microplastics) purification using a pressure-driven membrane with the properties of reverse osmosis and ultrafiltration. These nanomembranes offer a high degree of filtration by ultra-small pore size than contaminant practical. Particularly, polymer-based nanomembranes are being widely employed to eradicate contaminants from the aqueous phase by detouring elements by the chemical interface between the pollutants and membranes [74]. The stability and efficacy of membranes can be improved by integrating synthetic or natural polymers like polyamide, cellulose, and chitosan into the membrane matrices in combination with components such as triethanolamine, metal oxide, and CNTs [75]. In this context, Mukherjee et al. [76] examined the removal effectiveness of pesticides from an aqueous solution by thin-film composite (TFC) polyamide membrane prepared by polymerization with 1,3-phenylenediamine and 1,3,5-trimesoyl chloride [76]. The efficiency of the membrane was evaluated for 43 pesticides and out of these, 33 were removed by >80% (some have been listed in Table 9.3). Plattner et al. [77] also worked out the removal of pesticides, (Phorate, Parathion-methyl, Atrazine, Dichlorvos, and Clofibric acid) from the brackish groundwater direct membrane distillation (DCMD) system [77]. They reported lower rejection capacity (10–60%) of Phorate and Dichlorvos due to their highly hydrophobic and volatile nature whereas, the filtration efficacy was >80% for Parathion-methyl, Atrazine, and Clofibric acid. The inclusion of CNTs-based nanomembrane filtration offers strong antimicrobial activity, tuneable pore size and higher water flux compared to other porous materials [78]. Besides, an electrochemical CNT filter where the CNTs combine with an anodic membrane offers the electrochemical oxidation degradation of the pollutants and is recommended for desalination with increased efficiency [79]. With enhanced hydrophilicity along with mechanical properties, functionalized CNTs-based smart membranes provide a greater pesticide purification potential.

9.4.3 DEGRADATION

Many nanomaterials are semiconductive and they are known to easily degrade pesticides photocatalytically, photo-Fenton reactions, and advanced oxidative processes [80]. The ZnO being used as photocatalysts to degrade organophosphate when incorporated with photocatalyst and adsorbent, its efficiency has been reported to enhance significantly. This graphene oxide embedded in zinc oxide and NPs has shown increased organophosphorus pesticide (quinalphos) degradation (98%) compared to neat ZnO (80%) [81]. The enhancement was observed with GO combination with silver nanoparticles led to dimethoate degradation [82].

Recently, zero-valent iron (ZVI) also have been widely applied for the treatment of contamination because of its easy accessibility, effective degradation of pollutants, generation of very little waste, and secondary pollutants [83]. The nZVI act as an electron donor and is frequently used in pilot trials as they allow the subtraction of organic solvents, phenyls, and pesticides through the oxidative reduction process [84]. These nZVI have been reported for effective bioremediation of trichloroethene, hexavalent chromium, and DDT with great removal efficacy [85]. However, the hydrophobic nature of pesticides appears to limit efficient electron transmission which in turn reduces the pesticides removal capacity of nZVI. Such pollutants include pesticides containing nitrogen heteroatoms *viz.*, atrazine, picloram, chlorpyrifos, diazinon, diuron, etc. [72]. The Fenton process is also one of the emerging organic pollutants degradation processes where hydroxyl radicals get generated through the oxidation of ferrous Fe_2O_4 to Fe_3O_4 using H_2O_2 as an oxidizing agent. These radicals attack the organophosphorus pesticides to degrade them into less or non-toxic compounds. The efficiency of the Fenton process suffered from limitations of the generation of an enormous quantity of trivalent iron sludge. Estorun with Midik et al. [83] evaluated the degradation of 2,4-Dichlorophenoxyacetic through nZVI, Fenton, and Photo-Fenton process and reported the decomposition of 98, 67, and 76%, respectively [83]. An overview of different nanomaterials techniques used for the bioremediation of various noxious pesticides has been listed in Table 9.3.

TABLE 9.3 An Overview of Process and Nanomaterials Application for Pesticides Removal from Contaminated Aqueous Solution

Process	Nanomaterials	Target Pesticides (Contaminant)	Removal Efficiency (%)	References
Adsorption	Graphene oxide	Malathion	89%	[86]
		Chlorpyrifos	88%	
	MWCNTs	Malathion	>99%	[87]
	BRH, GAC, and MWCNTs	2,4-D	~90%	[71]
	Powdered activated carbon (PAC)	Malathion	94%	[88]
	GAT	Atrazine	91.1%	[89]
		Simazine	98.8%	
		Diuron	99.6%	
	MCNTs	Fenuron	90%	[90]
	Magnetic Fe_2O_3/GO	Triazine	74%	[91]
	Fe_3O_4@SiO_2@GO-PEA	Chlorpyrifos	~90%	[73]
		Parathion		
		Malathion		
	Nano-bentonite	Pendimethalin	87.8%	[92]
		Atrazine	71.1%	
Nanofiltration	Thin film composite (Polyamide and polysulfone interaction)	Aldrin	89.6%	[76]
		α-Endosulfan	100%	
		Chlorpyrifos	89.9%	
		Monocrotophos	38%	
		Imidacloprid	89.2%	
		DDT	95–96%	
		Carbofuran	90%	
		Carbendazim	64.2%	
		Metalaxyl	85.6%	
		Isoproturon	84%	
		Dichlorodiphenyltrichloroethane	95%	

TABLE 9.3 *(Continued)*

Process	Nanomaterials	Target Pesticides (Contaminant)	Removal Efficiency (%)	References
	DCMD	Phorate	10–50%	[77]
		Parathion-methyl	80%	
		Atrazine	97%	
		Dichlorvos	10–60%	
		Clofibric acid	99%	
	Membrane distillation	Atrazine	>95%	[93]
	Nanofiltration membranes (Polyamide thin-film composite)	Fenobucarb	96%	[94]
		Isoprothiolane	99%	
		Pretilachlor	99%	
Degradation	Zero-valent Fe	DDT	92% (water)	[95]
			22.4% (soil)	
	Bimetallic Ni/Fe	DDT	~100%	[96]
	Fe/Ni NPs	Profenofos	94.5%	[97]
	Ag/Cu NPs	Chlorpyrifos	100%	[98]
	GO-AgNPs (Catalytic dehalogenation)	Chlorpyrifos	95%	[82]
		Endosulfan		
		DDE		
	Fe_3O_4 dopped halloysite nanotubes	Pentachloro-phenol	100%	[99]
	nZVI	2,4-Dichlorophenoxyacetic	89%	[83]
	Fenton		67%	
	Photo-Fenton		76%	
	AgI/ Ag_3PO_4/g-C_3N_4 (Photocatalyst)	Nitenpyram	95%	[100]
	GO-ZnO	Quinalphos	98%	[81]

9.4.4 NANOPHYTOREMEDIATION

Nano-phytoremediation is an integrated approach that combines nano-technology with phytotechnology for the remediation and revitalization of hazardous contaminants from polluted sites [101]. Besides, nano-phytoremediation is referred to as an eco-friendly method for cleaning and managing soil environments that does not deplete the resource. Recently, in a study, the application of nano-phytoremediation (using iron and silver, green-NPs *Plantago major*) has remediated contaminated soil and water with chlorfenapyr by more than 90% [102]. Despite the numerous advantages of this technological combination, relatively few studies have been conducted on phytoremediation employing NPs for pesticides polluted sites or in some studies they were limited to controlled environments [103]. Therefore, future studies need to use more realistic studies and a better understanding of field practicals is needed.

9.5 CONCLUSION AND FUTURE PERSPECTIVES

In recent times, the occurrence and persistence of pollutants in the environment have become a major challenging issue for the health of humans, plants, and animals. Rapid development, industrialization, and agricultural contamination have led to an increase in the demand for heavy metals, pesticides, PCBs, PAHs, and microplastics in various applications, including automobile fuels, explosives, batteries, pigmentation, steel, and coating industries, toys, cosmetics, etc. These commercial products eventually release environmental contaminants which are harmful for living beings even at small doses. Over the past few years, most studies on nanoparticles have revealed that these particles show significantly reduced contamination through *in vitro* and *in vivo* applications. Engineered nanomaterials are widely used around the globe, which implies a great concern regarding their adverse impact on ecosystems. There have been reports that NPs can accumulate in plants and damage their cells. NPs are known to enter a variety of biochemical reactions that pose a serious threat to human health. On the basis of their physicochemical characteristics, nanomaterials are commonly selected for the detection, degradation, and removal of pesticides. For pesticide elimination, several nanoparticles, including metal, bimetallic, metal oxide, nanotubes, etc., have been used. However, to achieve a comprehensive application of nanoremediation to cleanse pesticide-contaminated soils, more research and measures are required.

As for future perspectives, NPs must be synthesized or developed from biodegradable and biocompatible compounds. There is a need to focus research on the smart design of NPs which can mitigate stress and ensure sustainable agriculture practices. The researchers should also support the development of eco-safe and sustainable nanotechnology for this remediation. Currently, nanoremediation techniques are restricted to laboratory settings, and their commercialization is highly concerning. Standardized methods should be evaluated for applications of nanoremediation, and awareness and benefits of this technology should be promoted.

ACKNOWLEDGMENT

The research was financially supported by the Ministry of Science and Higher Education of the Russian Federation (no. FENW-2023-0008) and the Strategic Academic Leadership Program of the Southern Federal University ("Priority 2030")

KEYWORDS

- **adsorption**
- **agrochemicals**
- **agroecosystem**
- **crops**
- **nanomembrane filtration**
- **nanoremediation**
- **pesticides**
- **pests**
- **sustainable approach**

REFERENCES

1. Nicolopoulou-Stamati, P., Maipas, S., Kotampasi, C., Stamatis, P., & Hens, L., (2016). Chemical pesticides and human health: The urgent need for a new concept in agriculture. *Frontiers in Public Health, 4*, 148. doi: 10.3389/FPUBH.2016.00148/BIBTEX.
2. Brevik, E. C., Slaughter, L., Singh, B. R., et al., (2020). *Soil and Human Health: Current Status and Future Needs.* https://doi.org/101177/1178622120934441.;13. doi: 10.1177/1178622120934441.

3. Pérez-Lucas, G., Vela, N., Aatik, A. El., & Navarro, S., (2018). Environmental risk of groundwater pollution by pesticide leaching through the soil profile. *Pesticides – Use and Misuse and Their Impact in the Environment*. doi: 10.5772/INTECHOPEN.82418.
4. Aktar, W., Sengupta, D., & Chowdhury, A., (2009). Impact of pesticides use in agriculture: Their benefits and hazards. *Interdisciplinary Toxicology, 2*(1), 1. doi: 10.2478/V10102-009-0001-7.
5. Mahmood, I., Imadi, S. R., Shazadi, K., Gul, A., & Hakeem, K. R., (2016). Effects of pesticides on environment. *Plant, Soil and Microbes: Implications in Crop Science* (Vol. 1, pp. 253–269). doi: 10.1007/978-3-319-27455-3_13/FIGURES/2.
6. Mostafalou, S., & Abdollahi, M., (2013). Pesticides, and human chronic diseases: Evidences, mechanisms, and perspectives. *Toxicology and Applied Pharmacology, 268*(2), 157–177. doi: 10.1016/J.TAAP.2013.01.025.
7. Shah, R., (2020). Pesticides and human health. *Emerging Contaminants*. doi: 10.5772/INTECHOPEN.93806.
8. Ranjan, A., & Jindal, T., (2022). Toxicology of organophosphate poisoning. *Toxicology of Organophosphate Poisoning*. Published online. doi: 10.1007/978-3-030-79128-5.
9. Rajput, S., Sharma, R., Kumari, A., et al., (2022). Pesticide residues in various environmental and biological matrices: Distribution, extraction, and analytical procedures. *Environment, Development and Sustainability, 24*(5), 6032–6052. doi: 10.1007/S10668-021-01722-4.
10. Hatamleh, A. A., Danish, M., Al-Dosary, M. A., El-Zaidy, M., & Ali, S., (2022). Physiological and oxidative stress responses of *Solanum lycopersicum* (L.) (tomato) when exposed to different chemical pesticides. *RSC Advances, 12*(12), 7237–7252. doi: 10.1039/D1RA09440H.
11. Tang, F. H. M., Lenzen, M., McBratney, A., & Maggi, F., (2021). Risk of pesticide pollution at the global scale. *Nature Geoscience, 14*(4), 206–210. doi: 10.1038/s41561-021-00712-5.
12. Kumari, A., Kumari, P., Rajput, V. D., Sushkova, S. N., & Minkina, T., (2021). Metal(loid) nanosorbents in restoration of polluted soils: Geochemical, ecotoxicological, and remediation perspectives. *Environmental Geochemistry and Health, 44*(1), 235–246. doi: 10.1007/S10653-021-00996-X.
13. Lowry, G. V., Avellan, A., & Gilbertson, L. M., (2019). Opportunities and challenges for nanotechnology in the agri-tech revolution. *Nature Nanotechnology, 14*(6), 517–522. doi: 10.1038/s41565-019-0461-7.
14. Kirtane, A. R., Verma, M., Karandikar, P., Furin, J., Langer, R., & Traverso, G., (2021). Nanotechnology approaches for global infectious diseases. *Nature Nanotechnology, 16*(4), 369–384. doi: 10.1038/s41565-021-00866-8.
15. Alvarez, P. J. J., Chan, C. K., Elimelech, M., Halas, N. J., & Villagrán, D., (2018). Emerging opportunities for nanotechnology to enhance water security. *Nature Nanotechnology, 13*(8), 634–641. doi: 10.1038/s41565-018-0203-2.
16. Zhang, Y., Li, M., Gao, X., Chen, Y., & Liu, T., (2019). Nanotechnology in cancer diagnosis: Progress, challenges, and opportunities. *Journal of Hematology and Oncology, 12*(1), 1–13. doi: 10.1186/S13045-019-0833-3/FIGURES/2.
17. Ganie, A. S., Bano, S., Khan, N., et al., (2021). Nanoremediation technologies for sustainable remediation of contaminated environments: Recent advances and challenges. *Chemosphere, 275*, 130065. doi: 10.1016/J.CHEMOSPHERE.2021.130065.

18. Hussain, A., Rehman, F., Rafeeq, H., et al., (2022). In-situ, ex-situ, and nano-remediation strategies to treat polluted soil, water, and air: A review. *Chemosphere, 289*, 133252. doi: 10.1016/J.CHEMOSPHERE.2021.133252.
19. Zandsalimi, Y., Maleki, A., Shahmoradi, B., Dehestani, S., Rezaee, R., & McKay, G., (2020). Photocatalytic removal of 2,4-dichlorophenoxyacetic acid from aqueous solution using tungsten oxide doped zinc oxide nanoparticles immobilized on glass beads. *Environmental Technology (United Kingdom)* (pp. 1–36). doi: 10.1080/09593330.2020.1797901/SUPPL_FILE/TENT_A_1797901_SM1382.DOCX.
20. Zhang, W., Jiang, F., & Ou, J., (2011). Global pesticide consumption and pollution: With China as a focus. *Proceedings of the International Academy of Ecology and Environmental Sciences, 1*(2), 125–144. www.iaees.org (accessed on 23 August 2023).
21. De, A., Bose, R., Kumar, A., & Mozumdar, S., (2014). *Targeted Delivery of Pesticides Using Biodegradable Polymeric Nanoparticles.* doi: 10.1007/978-81-322-1689-6.
22. Poudel, S., Poudel, B., Acharya, B., & Poudel, P., (2020). Pesticide use and its impacts on human health and environment. *Environment & Ecosystem Science, 4*(1), 47–51. doi: 10.26480/ees.01.2020.47.51.
23. Bernardes, M. F. F., Pazin, M., Pereira, L. C., & Dorta, D. J., (2015). Impact of pesticides on environmental and human health. *Toxicology Studies – Cells, Drugs and Environment.* doi: 10.5772/59710.
24. Tudi, M., Ruan, H. D., Wang, L., et al., (2021). Agriculture development, pesticide application and its impact on the environment. *Int J Environ Res Public Health, 18*(3), 1–24. doi: 10.3390/IJERPH18031112.
25. Bhardwaj, L., Chauhan, A., Ranjan, A., & Jindal, T., (2018). Persistent organic pollutants in biotic and abiotic components of Antarctic pristine environment. *Earth Systems and Environment, 2*(1), 35–54. doi: 10.1007/S41748-017-0032-8/FIGURES/1.
26. Bhardwaj, L., Sharma, S., Ranjan, A., & Jindal, T., (2019). Persistent organic pollutants in lakes of Broknes peninsula at Larsemann Hills area, East Antarctica. *Ecotoxicology, 28*(5), 589–596. doi: 10.1007/S10646-019-02045-X/FIGURES/2.
27. Zafeiraki, E., Kasiotis, K. M., Nisianakis, P., Manea-Karga, E., & Machera, K., (2022). Occurrence and human health risk assessment of mineral elements and pesticides residues in bee pollen. *Food Chem Toxicol, 161*. doi: 10.1016/J.FCT.2022.112826.
28. Calderon, R., García-Hernández, J., Palma, P., et al., (2022). Assessment of pesticide residues in vegetables commonly consumed in Chile and Mexico: Potential impacts for public health. *Journal of Food Composition and Analysis, 108*, 104420. doi: 10.1016/J.JFCA.2022.104420.
29. Wang, S., Zeng, X., Wang, X., Chang, H., Sun, H., & Liu, Y., (2022). A survey of multiple pesticide residues on litchi: A special fruit. *Microchemical Journal, 175*, 107175. doi: 10.1016/J.MICROC.2022.107175.
30. Torović, L., Vuković, G., & Dimitrov, N., (2021). Pesticide residues in fruit juice in Serbia: Occurrence and health risk estimates. *Journal of Food Composition and Analysis, 99*, 103889. doi: 10.1016/J.JFCA.2021.103889.
31. Dorosh, O., Fernandes, V. C., Moreira, M. M., & Delerue-Matos, C., (2021). Occurrence of pesticides and environmental contaminants in vineyards: Case study of Portuguese grapevine canes. *Science of The Total Environment, 791*, 148395. doi: 10.1016/J.SCITOTENV.2021.148395.

32. Ernst, F., Alonso, B., Colazzo, M., et al., (2018). Occurrence of pesticide residues in fish from south American rainfed agroecosystems. *Science of The Total Environment, 631, 632*, 169–179. doi: 10.1016/J.SCITOTENV.2018.02.320.
33. Caldas, S. S., Arias, J. L. O., Rombaldi, C., et al., (2019). Occurrence of pesticides and PPCPs in surface and drinking water in Southern Brazil: Data on 4-year monitoring. *J. Braz Chem. Soc., 30*(1), 71–80. doi: 10.21577/0103-5053.20180154.
34. Rajput, S., Kumari, A., Arora, S., & Kaur, R., (2018). Multi-residue pesticides analysis in water samples using reverse phase high performance liquid chromatography (RP-HPLC). *MethodsX, 5*, 744–751. doi: 10.1016/J.MEX.2018.07.005.
35. Elfikrie, N., Ho, Y. B., Zaidon, S. Z., Juahir, H., & Tan, E. S. S., (2020). Occurrence of pesticides in surface water, pesticides removal efficiency in drinking water treatment plant and potential health risk to consumers in Tengi River Basin, Malaysia. *Sci Total Environ, 712*. doi: 10.1016/J.SCITOTENV.2020.136540.
36. Zhou, Y., Wu, J., Wang, B., et al., (2020). Occurrence, source and ecotoxicological risk assessment of pesticides in surface water of Wujin District (Northwest of Taihu Lake), China. *Environ Pollut., 265*(Pt A). doi: 10.1016/J.ENVPOL.2020.114953.
37. Mac, L. T. M., Peluso, M. L., & Marino, D. J. G., (2022). Multiple pesticides occurrence, fate, and environmental risk assessment in a small horticultural stream of Argentina. *Sci. Total Environ., 802*. doi: 10.1016/J.SCITOTENV.2021.149893.
38. Hepsağ, F., & Kizildeniz, T., (2021). Pesticide residues and health risk appraisal of tomato cultivated in greenhouse from the Mediterranean region of Turkey. *Environ. Sci. Pollut. Res. Int., 28*(18), 22551–22562. doi: 10.1007/S11356-020-12232-7.
39. Arisekar, U., Jeya, S. R., Shalini, R., & Jeyasekaran, G., (2021). Pesticides contamination in the Thamirabarani, a perennial river in peninsular India: The first report on ecotoxicological and human health risk assessment. *Chemosphere, 267*. doi: 10.1016/J.CHEMOSPHERE.2020.129251.
40. Qu, C., Sun, Y., Albanese, S., et al., (2018). Organochlorine pesticides in sediments from Gulfs of Naples and Salerno, Southern Italy. *Journal of Geochemical Exploration, 195*, 87–96. doi: 10.1016/J.GEXPLO.2017.12.010.
41. Buah-Kwofie, A., Humphries, M. S., & Pillay, L., (2018). Bioaccumulation, and risk assessment of organochlorine pesticides in fish from a global biodiversity hotspot: ISimangaliso Wetland Park, South Africa. *Sci. Total Environ., 621*, 273–281. doi: 10.1016/J.SCITOTENV.2017.11.212.
42. Achour, A., Derouiche, A., Barhoumi, B., et al., (2017). Organochlorine pesticides and polychlorinated biphenyls in human adipose tissue from northern Tunisia: Current extent of contamination and contributions of socio-demographic characteristics and dietary habits. *Environ. Res., 156*, 635–643. doi: 10.1016/J.ENVRES.2017.04.021.
43. Abou, Z. M. I., Awad, M. K., Melki, K. C., Jawdah, Y. A., & Jammoul, A. M., (2021). Pesticides residues on loquat: A minor crop in lebanon. *Food Control, 130*, 108297. doi: 10.1016/J.FOODCONT.2021.108297.
44. Imfeld, G., & Vuilleumier, S., (2012). Measuring the effects of pesticides on bacterial communities in soil: A critical review. *European Journal of Soil Biology, 49*, 22–30. doi: 10.1016/J.EJSOBI.2011.11.010.
45. Aktar, W., Sengupta, D., & Chowdhury, A., (2009). Impact of pesticides use in agriculture: Their benefits and hazards. *Interdisciplinary Toxicology, 2*(1), 1. doi: 10.2478/V10102-009-0001-7.

46. Pahalvi, H. N., Rafiya, L., Rashid, S., Nisar, B., & Kamili, A. N., (2021). Chemical fertilizers and their impact on soil health. *Microbiota and Biofertilizers, 2*, 1–20. doi: 10.1007/978-3-030-61010-4_1.
47. Akter, R., Mukhles, M. B., Rahman, M. M., et al., (2022). Effect of pesticides on nitrification activity and its interaction with chemical fertilizer and manure in long-term paddy soils. *Chemosphere, 304*, 135379. doi: 10.1016/J.CHEMOSPHERE.2022.135379.
48. Qin, C., Yang, B., Zhang, W., et al. (2019). *Organochlorinated Pesticides Expedite the Enzymatic Degradation of DNA*. doi: 10.1038/s42003-019-0326-5.
49. Shahid, M., & Khan, M. S., (2022). Tolerance of pesticides and antibiotics among beneficial soil microbes recovered from contaminated rhizosphere of edible crops. *Current Research in Microbial Sciences, 3*, 100091. doi: 10.1016/J.CRMICR.2021.100091.
50. Rangasamy, K., Murugan, A., Devarajan, N., et al., (2018). Pesticide degrading natural multidrug resistance bacterial flora. *Microbial Pathogenesis, 114*, 304–310. doi: 10.1016/J.MICPATH.2017.12.013.
51. Meite, F., Granet, M., & Imfeld, G., (2022). Ageing of copper, zinc, and synthetic pesticides in particle-size and chemical fractions of agricultural soils. *Science of The Total Environment, 824*, 153860. doi: 10.1016/J.SCITOTENV.2022.153860.
52. Castellano-Hinojosa, A., Boyd, N. S., & Strauss, S. L., (2022). Impact of fumigants on non-target soil microorganisms: A review. *Journal of Hazardous Materials, 427*, 128149. doi: 10.1016/J.JHAZMAT.2021.128149.
53. Ankit, Saha, L., Kishor, V., & Bauddh, K., (2020). Impacts of synthetic pesticides on soil health and non-targeted flora and fauna. *Ecological and Practical Applications for Sustainable Agriculture* (pp. 65–88). doi: 10.1007/978-981-15-3372-3_4/TABLES/3.
54. Aziz, H., Murtaza, G., Saleem, M. H., et al., (2021). Alleviation of chlorpyrifos toxicity in maize (*Zea mays* L.) by reducing its uptake and oxidative stress in response to soil-applied compost and biochar amendments. *Plants, 10*(10), 2170. doi: 10.3390/PLANTS10102170.
55. Sharma, A., Kumar, V., Kanwar, M. K., Thukral, A. K., & Bhardwaj, R., (2017). Ameliorating imidacloprid induced oxidative stress by 24-epibrassinolide in *Brassica juncea* L. *Russian Journal of Plant Physiology, 64*(4), 509–517. doi: 10.1134/S1021443717040124.
56. Singh, P., & Prasad, S. M., (2018). Antioxidant enzyme responses to the oxidative stress due to chlorpyrifos, dimethoate and dieldrin stress in palak (*Spinacia oleracea* L.) and their toxicity alleviation by soil amendments in tropical croplands. *Science of The Total Environment, 630*, 839–848. doi: 10.1016/J.SCITOTENV.2018.02.203.
57. Fatma, F., Verma, S., Kamal, A., & Srivastava, A., (2018). Phytotoxicity of pesticides mancozeb and chlorpyrifos: Correlation with the antioxidative defense system in Allium cepa. *Physiology and Molecular Biology of Plants, 24*(1), 115–123. doi: 10.1007/S12298-017-0490-3/TABLES/4.
58. Sobiech, Ł., Grzanka, M., Kurasiak-Popowska, D., & Radzikowska, D., (2020). Phytotoxic effect of herbicides on various camelina (*Camelina sativa* (L.) Crantz) genotypes and plant chlorophyll fluorescence. *Agriculture, 10*(5), 185. doi: 10.3390/AGRICULTURE10050185.
59. *Effect of Phytotoxicity of Pesticides on Grain Yield of Rabi Sorghum (Sorghum bicolor (L.) Moench)*. Retrieved from:. https://www.phytojournal.com/archives/2017.v6.i5.1914/effect-of-phytotoxicity-of-pesticides-on-grain-yield-of-rabi-sorghum-sorghum-bicolor-l-moench (accessed on 23 August 2023).

60. Pérez-Portuondo, I., Serrat-Díaz, M., Pérez-Silva, R. M., & Ábalos-Rodríguez, A., (2021). Phytotoxic effects of 4-chlorophenol and 2,4-dichlorophenol in the germination of seeds of *Phaseolus vulgaris* and *Zea mayz*. *American Journal of Plant Sciences, 12*(04), 614–623. doi: 10.4236/AJPS.2021.124041.
61. Poudyal, S., Thomas, F. R., Owen, J., & Cregg, B., (2019). Dose-dependent phytotoxicity of pesticides in simulated nursery runoff on landscape nursery plants. *Water, 11*(11), 2354. doi: 10.3390/W11112354.
62. Ponnuchamy, M., Kapoor, A., Senthil, K. P., et al., (2021). Sustainable adsorbents for the removal of pesticides from water: A review. *Environmental Chemistry Letters, 19*(3), 2425–2463. doi: 10.1007/s10311-021-01183-1.
63. Yang, C., Zhang, C., & Liu, L., (2018). Excellent degradation performance of 3D hierarchical nanoporous structures of copper towards organic pollutants. *Journal of Materials Chemistry A, 6*(42), 20992–21002. doi: 10.1039/C8TA07973K.
64. Rambabu, N., Guzman, C. A., Soltan, J., & Himabindu, V., (2012). Adsorption characteristics of atrazine on granulated activated carbon and carbon nanotubes. *Chemical Engineering & Technology, 35*(2), 272–280. doi:https://doi.org/10.1002/ceat.201100376.
65. Malik, S., Kishore, S., Shah, M. P., & Kumar, S. A., (2022). A comprehensive review on nanobiotechnology for bioremediation of heavy metals from wastewater. *Journal of Basic Microbiology, 62*(3, 4), 361–375. doi:https://doi.org/10.1002/jobm.202100555.
66. Garba, Z. N., Abdullahi, A. K., Haruna, A., & Gana, S. A., (2021). Risk assessment and the adsorptive removal of some pesticides from synthetic wastewater: A review. *Beni-Suef University Journal of Basic and Applied Sciences, 10*(1), 1–18. doi: 10.1186/S43088-021-00109-8.
67. Shi, B., Zhuang, X., Yan, X., Lu, J., & Tang, H., (2010). Adsorption of atrazine by natural organic matter and surfactant dispersed carbon nanotubes. *Journal of Environmental Sciences, 22*(8), 1195–1202. doi: 10.1016/S1001-0742(09)60238-2.
68. Piai, L., Dykstra, J. E., Adishakti, M. G., Blokland, M., Langenhoff, A. A. M., & Van, D. W. A., (2019). Diffusion of hydrophilic organic micropollutants in granular activated carbon with different pore sizes. *Water Research, 162*, 518–527. doi:https://doi.org/10.1016/j.watres.2019.06.012.
69. Jaria, G., Calisto, V., Silva, C. P., Gil, M. V., Otero, M., & Esteves, V. I., (2019). Fixed-bed performance of a waste-derived granular activated carbon for the removal of micropollutants from municipal wastewater. *Science of the Total Environment, 683*, 699–708. doi:https://doi.org/10.1016/j.scitotenv.2019.05.198.
70. Alves Pimenta, J. A., Francisco Fukumoto, A. A., Madeira, T. B., Alvarez Mendez, M. O., Nixdorf, S. L., Cava, C. E., & Kuroda, E. K., (2022). Adsorbent selection for pesticides removal from drinking water. *Environmental Technology, 43*(11), 1672–1683.
71. Bahrami, M., Amiri, M. J., & Beigzadeh, B., (2018). Adsorption of 2,4-dichlorophenoxyacetic acid using rice husk biochar, granular activated carbon, and multi-walled carbon nanotubes in a fixed bed column system. *Water Science and Technology, 78*(8), 1812–1821. doi: 10.2166/wst.2018.467.
72. Taghizade, F. T., Mehrdadi, N., Baghdadi, M., & Nabi, B. G. R., (2018). Application of nanotechnology in pesticides removal from aqueous solutions: A review. *International Journal of Nanoscience and Nanotechnology, 14*(1), 43–56. http://www.ijnnonline.net/article_30199.html.

73. Wanjeri, V. W. O., Sheppard, C. J., Prinsloo, A. R. E., Ngila, J. C., & Ndungu, P. G., (2018). Isotherm and kinetic investigations on the adsorption of organophosphorus pesticides on graphene oxide based silica coated magnetic nanoparticles functionalized with 2-phenylethylamine. *Journal of Environmental Chemical Engineering, 6*(1), 1333–1346. doi:https://doi.org/10.1016/j.jece.2018.01.064.
74. Long, Q., Zhang, Z., Qi, G., Wang, Z., Chen, Y., & Liu, Z. Q., (2020). Fabrication of chitosan nanofiltration membranes by the film casting strategy for effective removal of dyes/salts in textile wastewater. *ACS Sustainable Chemistry & Engineering, 8*(6), 2512–2522. doi: 10.1021/acssuschemeng.9b07026.
75. Lakhotia, S. R., Mukhopadhyay, M., & Kumari, P., (2018). Cerium oxide nanoparticles embedded thin-film nanocomposite nanofiltration membrane for water treatment. *Scientific Reports, 8*(1), 4976. doi: 10.1038/s41598-018-23188-7.
76. Mukherjee, A., Mehta, R., Saha, S., Bhattacharya, A., Biswas, P. K., & Kole, R. K., (2020). Removal of multiple pesticide residues from water by low-pressure thin-film composite membrane. *Applied Water Science, 10*(12), 244. doi: 10.1007/s13201-020-01315-y.
77. Plattner, J., Kazner, C., Naidu, G., Wintgens, T., & Vigneswaran, S., (2018). Removal of selected pesticides from groundwater by membrane distillation. *Environmental Science and Pollution Research, 25*(21), 20336–20347. doi: 10.1007/s11356-017-8929-1.
78. Chung, J. H., Hasyimah, N., & Hussein, N., (2021). Application of carbon nanotubes (CNTs) for remediation of emerging pollutants: A review. *Tropical Aquatic and Soil Pollution, 2*(1), 13–26. doi: 10.53623/tasp.v2i1.27.
79. Das, R., Abd, H. S. B., Ali, Md. E., Ismail, A. F., Annuar, M. S. M., & Ramakrishna, S., (2014). Multifunctional carbon nanotubes in water treatment: The present, past and future. *Desalination, 354*, 160–179. doi:https://doi.org/10.1016/j.desal.2014.09.032.
80. Rani, M., & Shanker, U., (2017). Degradation of traditional and new emerging pesticides in water by nanomaterials: Recent trends and future recommendations. *International Journal of Environmental Science and Technology, 15*(6), 1347–1380. doi: 10.1007/S13762-017-1512-Y.
81. Garg, R., Gupta, R., Singh, N., & Bansal, A., (2022). Eliminating pesticide quinalphos from surface waters using synthesized GO-ZnO nanoflowers: Characterization, degradation pathways and kinetic study. *Chemosphere, 286*, 131837. doi:https://doi.org/10.1016/j.chemosphere.2021.131837.
82. Koushik, D., Sen, G. S., Maliyekkal, S. M., & Pradeep, T., (2016). Rapid dehalogenation of pesticides and organics at the interface of reduced graphene oxide–silver nanocomposite. *Journal of Hazardous Materials, 308*, 192–198. doi:https://doi.org/10.1016/j.jhazmat.2016.01.004.
83. Midik, E. F., Cellat, K., Eren, O., Gül, Ş., Kuşvuran, E., & Şen, F., (2019). Comparison of nanoscale zero-valent iron, fenton, and photo-fenton processes for degradation of pesticide 2,4-dichlorophenoxyacetic acid in aqueous solution. *SN Applied Sciences, 1*(11), 1491. doi: 10.1007/s42452-019-1554-5.
84. Stefaniuk, M., Oleszczuk, P., & Ok, Y. S., (2016). Review on nano zerovalent iron (nZVI): From synthesis to environmental applications. *Chemical Engineering Journal, 287*, 618–632. doi:https://doi.org/10.1016/j.cej.2015.11.046.
85. Guerra, F. D., Attia, M. F., Whitehead, D. C., & Alexis, F., *(2018)*. Nanotechnology for environmental remediation: Materials and applications. *Molecules, 23*(7), 1760. doi: 10.3390/MOLECULES23071760.

86. Yadav, S., Goel, N., Kumar, V., & Singhal, S., (2019). *Graphene Oxide as Proficient Adsorbent for the Removal of Harmful Pesticides: Comprehensive Experimental Cum DFT Investigations, 9*(3), 291–310. https://doi.org/101080/22297928201916 29999 doi: 10.1080/22297928.2019.1629999.
87. Dehghani, M. H., Niasar, Z. S., Mehrnia, M. R., et al., (2017). Optimizing the removal of organophosphorus pesticide malathion from water using multi-walled carbon nanotubes. *Chemical Engineering Journal, 310*, 22–32. doi:https://doi.org/10.1016/j.cej.2016.10.057.
88. Vukčević, M. M., Kalijadis, A. M., Vasiljević, T. M., Babić, B. M., Laušević, Z. V., & Laušević, M. D., (2015). Production of activated carbon derived from waste hemp (*Cannabis sativa*) fibers and its performance in pesticide adsorption. *Microporous and Mesoporous Materials, 214*, 156–165. doi: 10.1016/J.MICROMESO.2015.05.012.
89. Alves, P. J. A., Francisco, F. A. A., Madeira, T. B., et al., (2020). *Adsorbent Selection for Pesticides Removal from Drinking Water, 43*(11), 1672–1683. https://doi.org/101080/09593330202018 47203. doi: 10.1080/09593330.2020.1847203.
90. Ali, I., Alharbi, O. M. L., ALOthman, Z. A., Al-Mohaimeed, A. M., & Alwarthan, A., (2019). Modeling of fenuron pesticide adsorption on CNTs for mechanistic insight and removal in water. *Environmental Research, 170*, 389–397. doi:https://doi.org/10.1016/j.envres.2018.12.066.
91. Boruah, P. K., Sharma, B., Hussain, N., & Das, M. R., (2017). Magnetically recoverable Fe_3O_4/graphene nanocomposite towards efficient removal of triazine pesticides from aqueous solution: Investigation of the adsorption phenomenon and specific ion effect. *Chemosphere, 168*, 1058–1067. doi:https://doi.org/10.1016/j.chemosphere.2016.10.103.
92. Shabeer, T. P. A., Saha, A., Gajbhiye, V. T., Gupta, S., Manjaiah, K. M., & Varghese, E., (2015). Exploitation of nano-bentonite, nano-halloysite and organically modified nano-montmorillonite as an adsorbent and coagulation aid for the removal of multi-pesticides from water: A sorption modeling approach. *Water, Air, and Soil Pollution, 226*(3), 1–14. doi: 10.1007/S11270-015-2331-8/TABLES/4.
93. Wijekoon, K. C., Hai, F. I., Kang, J., Price, W. E., Cath, T. Y., & Nghiem, L. D., (2014). Rejection and fate of trace organic compounds (TrOCs) during membrane distillation. *Journal of Membrane Science, 453*, 636–642. doi:https://doi.org/10.1016/j.memsci.2013.12.002.
94. Nguyen, L. D., Gassara, S., Bui, M. Q., Zaviska, F., Sistat, P., & Deratani, A., (2019). Desalination and removal of pesticides from surface water in Mekong delta by coupling electrodialysis and nanofiltration. *Environmental Science and Pollution Research, 26*(32), 32687–32697. doi: 10.1007/s11356-018-3918-6.
95. El-Temsah, Y. S., Sevcu, A., Bobcikova, K., Cernik, M., & Joner, E. J., (2016). DDT degradation efficiency and ecotoxicological effects of two types of nano-sized zero-valent iron (nZVI) in water and soil. *Chemosphere, 144*, 2221–2228. doi:https://doi.org/10.1016/j.chemosphere.2015.10.122.
96. Tian, H., Li, J., Mu, Z., Li, L., & Hao, Z., (2009). Effect of pH on DDT degradation in aqueous solution using bimetallic Ni/Fe nanoparticles. *Separation and Purification Technology, 66*(1), 84–89. doi: 10.1016/J.SEPPUR.2008.11.018.
97. Mansouriieh, N., Sohrabi, M. R., & Khosravi, M., (2019). Optimization of profenofos organophosphorus pesticide degradation by zero-valent bimetallic nanoparticles using response surface methodology. *Arabian Journal of Chemistry, 12*(8), 2524–2532. doi:https://doi.org/10.1016/j.arabjc.2015.04.009.

98. Rosbero, T. M. S., & Camacho, D. H., (2017). Green preparation and characterization of tentacle-like silver/copper nanoparticles for catalytic degradation of toxic chlorpyrifos in water. *Journal of Environmental Chemical Engineering, 5*(3), 2524–2532. doi:https://doi.org/10.1016/j.jece.2017.05.009.
99. Tsoufis, T., Katsaros, F., Kooi, B. J., et al., (2017). Halloysite nanotube-magnetic iron oxide nanoparticle hybrids for the rapid catalytic decomposition of pentachlorophenol. *Chemical Engineering Journal, 313*, 466–474. doi:https://doi.org/10.1016/j.cej.2016.12.056.
100. Tang, M., Ao, Y., Wang, C., & Wang, P., (2020). Facile synthesis of dual Z-scheme g-C_3N_4/Ag_3PO_4/AgI composite photocatalysts with enhanced performance for the degradation of a typical neonicotinoid pesticide. *Applied Catalysis B: Environmental, 268*, 118395. doi:https://doi.org/10.1016/j.apcatb.2019.118395.
101. Gul, M. Z., Rupula, K., & Beedu, S. R., (2022). Nano-phytoremediation for soil contamination: An emerging approach for revitalizing the tarnished resource. *Phytoremediation* (pp. 115–138). doi: 10.1016/B978-0-323-89874-4.00014-5.
102. Romeh, A. A., & Ibrahim, S. R. A., (2020). Green nano-phytoremediation and solubility improving agents for the remediation of chlorfenapyr contaminated soil and water. *Journal of Environmental Management, 260*, 110104. doi: 10.1016/J.JENVMAN.2020.110104.
103. Asante-Badu, B., Kgorutla, L. E., Li, S. S., Danso, P. O., Xue, Z., & Qiang, G. (2020). Phytoremediation of organic and inorganic compounds in a natural and an agricultural environment: a review. *Applied Ecology & Environmental Research, 18*(5), 6875–6904.

CHAPTER 10

Nanoremediation: A Promising Reclamation Method for the Removal of Organic Pollutants From Different Environmental Sites

PRANGYA RATH,[1] ANUJ RANJAN,[2] ARPNA KUMARI,[2] VISHNU D. RAJPUT,[2] EVGENYA V. PRAZDNOVA,[2] SAGLARA S. MANDZHIEVA,[2] SVETLANA N. SUSHKOVA,[2] TATIANA M. MINKINA,[2] JAYATI ARORA,[1] ABHISHEK CHAUHAN,[3] and TANU JINDAL[3]

[1]*Amity Institute of Environmental Sciences, Amity University, Noida, Uttar Pradesh, India*

[2]*Academy of Biology and Biotechnology, Southern Federal University, Stachki, Rostov-on-Don, Russia*

[3]*Amity Institute of Environmental Toxicology Safety and Management, Amity University, Noida, Uttar Pradesh, India*

ABSTRACT

Eliminating organic pollutants from the environment has become a serious concern as they have become a global hazard. Such pollutants have been observed to risk the health of several flora, fauna, humans, and the overall environment. Nanoremediation for dealing with organic pollutants is a potentially novel strategy that is significantly effective and safer than conventional remedial approaches. Varieties of nanomaterials are employed to clean up contaminants from soil and water due to their versatility. Nanomaterials such as metal/metal oxides, carbon tubes, and polymers have been largely explored for bioremediation purposes. This chapter deals with various types

Nano-Bioremediation for Water and Soil Treatment: An Eco-Friendly Approach.
Vishnu D. Rajput, Arpna Kumari, and Tatiana M. Minkina (Eds.)

of organic pollutants, their environmental fate, and hazardous effects on biotic and abiotic components, and a strategy for viable long-term approaches for mitigation of these pollutants from the environment using nanoremediation.

10.1 INTRODUCTION

In the past few decades, the rapid increase in population, the consumption of fuel, industrial chemicals, fertilizers, pesticides, and pharmaceutical by-products have increased multifolds, resulting in the release of significant amounts of organic pollutants into the environment [1, 2]. Organic pollutants such as volatile organic compounds (VOCs), poly-aromatic hydrocarbons (PAHs), polycarbonated biphenyl (PCBs), several dyes, petroleum hydrocarbons, pesticides, etc., have become major threats to the environment [3]. They are significantly toxic and pose a greater risk to many life forms on the earth. According to the reports of the World Health Organization (WHO), one-third of the diseases afflicting humanity are caused by extended exposure to such persistent organic pollutants (POPs) exceeding the allotted permissible limits [4]. Their excess presence has been observed to cause bioaccumulation and biomagnification, thereby causing diseases in humans as well as wildlife and disturbing biodiversity.

Because of the various point and non-point sources of emission, the POPs are widely dispersed and reported in many biological and environmental samples across the globe [5]. Industrial effluents, vehicular and industrial emissions, airflow drift, intentional dumping of pollutants, and direct spray are the few most common sources of emission of POPs. After they are released into the environment, they are prone to enter soil, waterbodies, and air, and then they find a way to reach non-target organisms [6]. They mostly reach waterbodies after the run-off from the soil of nearby fields and agricultural farms. Leaching of the POPs through the soil column also helps these compounds to reach groundwater [7–9].

From the environment, they are then taken up by microorganisms, plankton, several lower organisms, plants, etc., and get transported along the food web [10, 11]. Organic pollutants bioaccumulate and biomagnify at every trophic level and are also responsible for causing health hazards to human and animal tissues [12].

Their excess presence has been reported to cause severe disorders such as allergies, gastric problems, immune system dysfunction, cancers, defects in the reproduction system, birth defects, loss and gain of body weight, diabetes, and impaired nervous and endocrine systems [12–14]. The fate and persistence of these pollutants depend upon their physicochemical properties, such as low water solubility and volatility, high lipid solubility and molecular mass, and their interaction with other environmental components [2, 15].

Nanotechnology (NT) has a widespread application in various fields, including environmental sciences, and it has gained significant attention and application in bioremediation processes which ,is commonly termed nanoremediation. It is an effective, efficient, and rapid strategy that utilizes engineered nanomaterials (NMs) to eliminate environmental pollutants such as persistent hazardous compounds, pesticides, pharmaceutics, halogenated chemicals, chlorinated solvents, heavy metals, etc. [16]. Such techniques are comparatively economical, efficient, and effective as compared to most conventional methods [17–19]. Nanoparticles (NPs) have also been developed as adsorbents to remove harmful contaminants from various effluents and soil [20, 21]. Apart from engineered NPs, naturally prepared NPs are also gaining popularity for their application in remediating environmental contaminants. Various microorganisms (fungi, bacteria, actinomycetes, and viruses), algae, plants, and their extracts are used for eco-friendly biosynthesis of NPs that are used for remediation of toxicants and organic wastes [17, 22]. Similarly, adsorbents made from metal/metal oxide-based NPs play an important role in removing toxicants from wastewater [17, 23]. Carbon NPs are great adsorbents for hazardous chemicals released from industrial or pharmaceutical wastewater [24].

NT acts as a boon for various sectors, and scientific research is still being done to make and control their beneficial usage. Therefore, it is very important to recognize the comprehensive chemistry of organic pollutants in ecosystem and come up with advanced techniques to mitigate the issues of organic pollutants in the environment. This chapter highlights the different environmental pollutants and stressors, their sources, adverse effects, and minimizing the adverse effects by utilizing nanoremediation techniques.

10.2 ENVIRONMENTAL POLLUTION AND STRESSORS

One of the major classes of hazardous organic compounds is persistent organic pollutants (POPs). It has become a global issue because of its toxic impact on both abiotic and biotic components of the ecosystem. The toxic chemicals present have adversely affected the health of the microbiome [25], humans, flora, and fauna [6], and cumulatively on the environment (Figure

10.1) [26, 27]. They persist in nature for a longer duration and progressively enter the food chain, followed by their bioaccumulation and biomagnification in several food chains and food webs. POPs, or the "dirty dozen," were identified during the Convention of Stockholm in 2001 for their adverse impacts on the environment and living beings [28]. POPs are majorly classified into three categories: (i) pesticides, (ii) polychlorinated biphenyls (PCB), and (iii) unintended industrial by-products, for example, dibenzo-p-dioxins (PCDD), polychlorinated dibenzofurans (PCDF), polychlorinated dibenzofurans (PCDFs), and polycyclic aromatic hydrocarbons (PAHs) [28, 29].

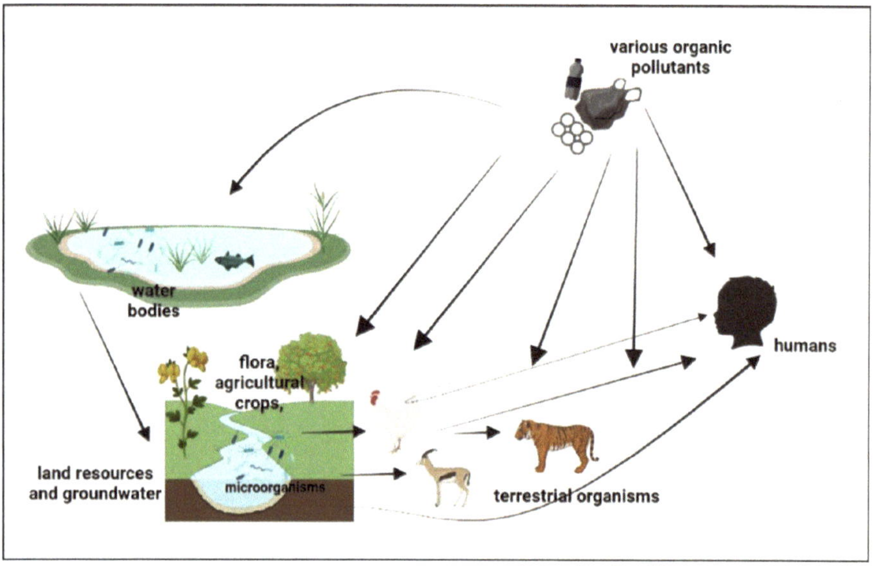

FIGURE 10.1 Fate and bioaccumulation of organic pollutants.

10.2.1 EFFECT OF VARIOUS ORGANIC POLLUTANTS ON WATERBODIES AND LAND

1. **Pesticides:** These are the most used organic chemicals that come in direct contact with the environment, and they are classified as herbicides, insecticides, and fungicides [30]. They have high persistence and penetration in environment, which is attributed to their chemical structure and nature, dosage, and targets [31]. However, excessive usage of pesticides has been associated with several problems [32].

Agricultural activities allow the direct entry of pesticides into the soil, followed by waterbodies through surface runoff, leaching, erosion, etc. [7]. Drift, evaporation, and wind erosion transfer pesticide residues into the atmosphere [31].

Pesticides enter into the body of organisms in different ways and cause susceptibility to toxins. It causes metabolic peculiarities, neurological impairments, and dysfunction of the immune, reproductive, and developmental systems in humans [33]. Dithiocarbamate pesticides were observed to strongly inhibit acetylcholinesterase, affect neuromuscular junctions, and damage brain cholinergic synapses [34]. Exposure to pesticides plays a role in cognitive deficiencies in insects (such as honeybees) [35]. It elevates the steady-state levels of reactive oxygen species (ROS) and stimulates ROS-induced oxidation of proteins and lipids or inactivates certain enzymes, resulting in neurotoxic effects and liver and kidney damage [34]. Such high levels of ROS have been associated with DNA damage and cause genotoxicity due to severe mutations in fishes [36, 37]. Exposure of imidacloprid pesticide exposure reduces calcium signaling in model organisms like Drosophila, honeybees, and human neuronal cell lines [38, 39]. Exposure of sublethal doses of thiamethoxam pesticides to insects during developmental stages resulted in reduced body weight, slow reproductive cycles, fewer viable and mature spermatozoa and fertilized eggs [35]. Organophosphorus pesticides were observed to generate many health issues in higher organisms, including humans [40]. Accumulation of DDT [1,1,1-trichloro-2,2-bis(4-chlorophenyl) ethane] in tissues of organisms leads to troubles associated with absorption, metabolism, and excretion [41]. DDT and its metabolites have been studied to be linked to an increased risk of health issues like obesity, type 2 diabetes mellitus, chronic renal disorders, and autoimmune diseases in adults [42, 43].

2. **Polycyclic Aromatic Hydrocarbons (PAHs):** These are a group of chemicals that are present in crude oil, coal, gasoline, wood, etc. [44]. They generally enter the environment on burning [45] and combine together with smoke/fumes to form fine particles [46]. These are neutral and non-polar organic molecules that are ubiquitous environmental pollutants and known to be carcinogenic and capable of inducing mutations at molecular levels [44]. Due to the presence of large number of π electrons on both sides of the ring

structure of PAHs, they are persistent in nature [45]. Their hydrophobic characteristics allow it to accumulate as aquatic sediments and concentrate in aquatic organisms, leading to bioaccumulation [46]. Benzo(a)pyrene and 7,12-dimethyl-benz(a)anthracene bioaccumulates in aquatic phytoplanktons, zooplanktons, and microalgae such as *Dunaliella tertiolecta*, *Mytilus galloprovincialis*, and fishes like *Dicentrarchus labrax* [46]. Reports have suggested a direct correlation between the bioaccumulation of PAHs in fish and PAH concentration in aquatic plants, thereby entering the food web [47]. Excessive exposure to PAHs is linked to severe health impacts [27]. PAHs as ambient air pollution particles trigger ROS in skin, lung, and cardiovascular cells and cause severe inflammation. It has been linked with asthma, atherosclerosis, premature tissue aging, DNA damage, eye irritation, vomiting, severe diarrhea, and increased cancer risk [48, 49]. They interact with multiple intracellular receptors and affect signaling pathways [50]. They also induce the expression of ROS/early stress, genotoxic markers, altered nuclear receptor signaling, and disturbed endogenous metabolism signaling in HepaRG cell lines [50]. They easily get absorbed into gastrointestinal tracts because of their high lipid solubility [44]. Some PAHs are good photosensitizers and are also mild allergens to human skin, for example, phenanthrene, anthracene, benzo(a)pyrene (BAP) [47, 50]. The majority of PAHs, such as BAP and dibenzo[def,p]chrysene (DBC), alone or in combinations, get absorbed by the skin layers and are reported for disrupting epithelial barrier integrity in primary human bronchial cell lines by downregulating genes responsible for cell adhesion and functional measurements [51].

3. **Polychlorinated Biphenyls (PCBs):** These are industrial products or chemicals that are mostly persistent in the environment. Their production and use have been banned in India since the early 2000s as they are very toxic and persistent [52]. They are found to a larger extent and find their space in air, water, and soil during manufacture and usage [53]. PCBs traces have been detected in living organisms in densely populated areas and also remote areas like the Arctic. These findings of such widespread and persistent contamination contributed to the banning of PCBs by many countries [53]. Once they are released into the environment, they enter various tropic levels and bioaccumulate in the food chain. This is because of their high affinity towards organic materials. They are consumed through

meat, fish, and dairy products and are found in the tissues, breast milk, and blood of humans [54].

Excessive exposure to PCBs Aroclor 1254 in rats during developmental stages leads to hearing loss, optical dysfunctions, motor deficits, retarded mental development, and damaged Golgi bodies [55]. Due to its neurotoxicity, it poses a significant risk to the developing human brain [39]. They are associated with disruption of thyroid hormone (TH), altered signaling of neurotransmitters, modulation of intracellular dynamics of Ca^{2+}, and increase in the production of ROS [56]. PCB153 and PCB180 have been studied to be a factor causing autism disorders by modifying the GSTM1 genotype in Jamaican children [57]. Prenatal exposure to PCB affected cognition (intelligence levels, attention power, verbal memory) and motor performances (fine motor, balance) that had clinically relevant consequences in adolescence. It increased attention-deficit/hyperactivity disorder (ADHD) in children [58]. Toxic chemicals like Bisphenol A, vinclozolin, phthalates, and other PCBs are potent endocrine disrupters [17, 59]. They are capable of affecting hormone receptors and associated downstream signaling pathways. This severely impairs hormone synthesis, metabolism, and mode of action [60]. Lower-chlorinated PCBs have been observed to alter their toxicities in a receptor-specific manner, thereby affecting signaling pathways, altering metabolism, and ATP production, generating ROS, and causing tumor promotion and cancers in human lung epithelial cell lines and in *in-vitro* and *in-vivo* disease models [61, 62].

10.2.2 NANOREMEDIATION OF ORGANIC POLLUTANTS

Nanomaterials (NMs) have gained interest in environmental remediation as they provide comparatively better outcomes than conventional remediation approaches like coagulation, flocculation, adsorption, and advanced oxidation [63]. This is because NMs possess a large surface-area-to-volume ratio, the presence of large number of reactive sites, and high reactivity [23]. Such characteristic features allow higher and enhanced interaction with the pollutants, thereby leading to rapid reduction in pollutant concentration levels [63]. Past studies have highlighted several types of NPs for nanoremediation approaches, and each type has its own limitations with respect to the pollutants and environmental conditions [16, 20]. Some of them are being discussed below (Table 10.1):

TABLE 10.1 Role of Metal-based NPs in Bioremediation of Organic Pollutants like Dyes, PCBs, PAHs, etc.

Type of Meal Oxide NPs	Details of NPs	Pollutants	Remediation Achieved	References
Ag NPs	9.7 ± 3.2 nm Characterized by X-ray diffraction and X-ray photoelectron spectroscopy.	Dyes: Methyl orange, Methyl orange, and Chicago Sky Blue 6B.	• Methyl orange: 96.4% in 30 minutes. • Methyl orange: 96.5% in 18.5 minutes. • Chicago Sky Blue 6B: 99.8% in 21 minutes.	[72]
MgO NPs	Nanoporous MgO with 16 nm pore size	Anthraquinone reactive dye and methyl orange.	1,000 mg/g	[73, 74]
Au NPs	9–32 nm Biosynthesized Au NPs using *B. amyloliquefaciens* SJ14.	Successful catalyst in anionic azo dyes, triphenylmethane, Victoria blue B and R degradation.	Photocatalytic degradation up to 65% and 52% achieved of VBB and VBR in 8 hrs.	[75, 76]
	AuNP on graphene oxide sheets.	PCB 77	100 removals by adsorption.	[77]
Fe NPs	5–10 nm	Total petroleum hydrocarbons (TPHs) from water.	88.34%	[21]
	50 nm (nZVI) 50 nm (nZVI-Pd) and 20–30 nm (nFe_3O_4)	PCB	nFe_3O_4 showed a reversible method for PCB adsorption. nZVI and nZVI-Pd displayed alike rates of PCB degradation in the course of 45 days of treatment.	[68]
	nZVI with 1% surfactant (saponin/Tween 80) couples with electrodialysis 0–1 V/cm	PCBs	76% PCB removal was accomplished with saponin and nZVI when coupled with electrodialysis.	[78]
	Carboxymethyl cellulose (CMC) stabilized ZVI NPs of 11.2 ± 7.9 nm.	Perchlorate	The rate of degradation of perchlorate is enhanced by 53% in saline water or IX brine.	[79]
TiO_2 / SiO_2–carbon nanotubes	0.4 nm	PAH	>90% removed by adsorption.	[80]
Sr–TiO_2	–	Phenanthrene	100% degradation by pseudo-first order photocatalytic degradation.	[81]
TiO_2	30 nm	PAH	Degradation by pseudo-first order photocatalytic degradation.	[82]

1. **Metal/Metal-Oxide-based NPs:** These are widely used for the removal of toxic pollutants such as Ni, Ag, and Fe and have been used for the removal of pollutants from wastewater [16, 19, 64]. This is attributed to their large specific surface areas, affinity towards various chemical groups present in pollutants, high reactivity, photolytic capabilities, and good adsorbent properties [65]. They act as good agents for the removal of organic pollutants like chlorinated organic solvents, PCBs, POPs, etc. [16, 20]. For example, zero valent iron NPs (nZVI) are popularly used for the nanoremediation of soil pollutants [16].

   Similarly, Fe-oxide (Fe_2O_3 and Fe_3O_4), zinc oxide (ZnO), and titanium dioxide (TiO_2) NPs are widely used for the purification processes of water. ZnO NPs are used to remove organic pollutants like chlorobenzenes and dioxin-like PCBs from water samples [66]. A novel method comprising metal/metal oxide NPs having magnetic core-shell successfully removes 3,3′,4,4′,5-Pentachlorobiphenyl (PCB-126) through adsorption [67]. Fe-NPs also recovered a co-contaminated soil with Cr and PCBs [68]. Cu-NPs are efficiently utilized for the removal of azo-dyes from polluted textile water in a minimum duration of time [69]. TiO_2-NPs are used as photocatalysts for micropollutant removal and pharmaceutical drugs present in water [70]. They are effective eco-friendly fertilizers [21]. It is also used to reduce the excessive nitrogen and phosphorus produced by harmful algae and reduce its bloom [71].

2. **Nanosized Carbon NPs (Graphene/Graphene Oxide, Single/Multiwall Carbon Nanotubes (S/MWCNT):** Nanoporous materials made up of carbon products like activated carbon materials, carbon nanotubes (CNTs), single-walled carbon nanotubes (SWCNTs), multi-walled carbon nanotubes (MWCNTs), graphene/graphene oxides, showcase physicochemical characteristics which makes them a suitable approach for treatment of water by removing contaminants like fluorides, heavy metals, textile dyes, pharmaceutical products, etc. (Table 10.2) [59]. Total petroleum hydrocarbons, DDT, crude oil, and hexachlorocyclohexane are removed from the soil, thereby enhancing the microbial and plant growth in soils [11]. CNTs based membranes have a high capability for separating CO_2 from other gases and are capable of large-scale applications in the separation of air pollutants. They are thus efficiently utilized for

separating and purifying gases and various pollutant vapors (such as benzene vapors, vapors of toluene, ethylbenzene, and xylene) and to prevent their discharge into the environment [83–85]. Gases like H_2, He, O_2, N_2, metals such as Ar, water, hexane, ethanol, and kerosene can be effectively filtered using CNT membranes [83]. About 25 mg/g of CNTs effectively removed ibuprofen, carbamazepine, and clofibric acid by 68%, 49%, and 46%, respectively [59]. They are also widely applied in water gasoline removal projects. Similarly, MWCNTs with functionalized polyethersulfone membranes, graphene, and other polymers have been observed to be excellent methods for rejection of heavy metals and dyes rejection from aqueous sources [16]. Modified SWCNTs/MWCNTs are widely utilized for detecting H_2S, SO_2, NH_3, VOCs, COx, and NOx [16, 86]. SWCNTs/MWCNTs have been used to remove diuron pesticides from contaminated water and soil samples [87]. Toxic chemical pesticides such as fenpropathrin, alphacypermethrin, deltamethrin, etc., are removed through adsorption using MWCNTs prepared with polyvinyl pyrrolidone and polysilazane liquid [88].

TABLE 10.2 Nanoparticles Mediated Degradation of Organic Pollutants

Type of NPs	Size of the NPs	Pollutants	Remediation Achieved	References
Activated carbon, CNTs	18.7 nm	Caffeine, ibuprofen, and triclosan.	90%	[89]
	8–15 nm diameter, length 10–50 μm.	Diclofenac, ibuprofen, bisphenol A, clofibric acid, carbamazepine.	75–90%	[59]
Fullerenes C_{60}	–	Naphthalene	100% degradation by Adsorption employing the Freundlich reaction	[90]
Graphitic carbon nitride (g-C_3N_4)	–	Nitro-PAH	100% removal by adsorption.	[91]
TiO_2 nanodots on CNT (TiO_2/Co@ NCT)	1–4 μm	PAH	98.48% degradation by photocatalytic	[92]
Carbon modified TiO_2	–	PCB	93% photocatalytic degradation Langmuir.	[93]
CNT	CNT is impregnated with metallic NPs.	Glyphosate	Adsorption	[94]

10.3 NANOBIOREMEDIATION: HARNESSING THE MICROBE-NANOPARTICLE SYNERGY

A number of microorganisms (bacteria, algae, yeast, fungi) have been widely used for synthesizing NPs and studying the synergistic effects [65]. NMs are used in combination with microbes/microbial enzymes, such methods are referred to as nanobioremediation. The approach includes the synergy of microbes with NPs to improve the efficiency and rate of bioremediation. Nanobioremediation is gaining acceptance in the past few years because of its cost effectiveness and efficiency. The combination of NPs with microbes proved a synergistic agents capable of enhancing the removal capabilities of microbes in wastewater treatment [19]. The targeted pollutants get successfully adsorbed, degraded, and modified by NPs as they possess unique physicochemical properties. The process involves using NPs as catalysts and helps by decreasing the activation energy that is required for breaking down the contaminants [95].

Excessive exposure to POPs pollutes a potent carcinogens. Therefore, its remediation is an important requirement. The nanobioremediation process using carbon-based and metal-based NMs has been therefore explored [96–98]. Polymeric NPs, such as nanocapsules and nanospheres, are widely used in the removal of persistent pesticide compounds and pollutants having long-chain hydrocarbons [99]. A study was conducted on analyzing the synergistic effect produced by the combination of CNTs and enzymes of *Delftia* sp. XYJ6 for the removal of aniline from wastewater. The study showed that the biodegradation rate increased rapidly with the augment of *Delftia* sp. XYJ6 protein concentration. This effect further increased upon the usage of SWCNTs and MWCNTs [100]. The fungal strain *Phanerochaete chrysosporium*, widely studied for possessing amazing bioremediation properties, was combined with Ag-NPs (0.1–1 mg/L) and enhanced the removal capability of fungus to degrade 2,4-dichlorophenol (2,4-DCP) [10, 64].

Applications of NPs with microbes show great capability in soil remediation contaminated with herbicides. Fe_3O_4 NPs have been observed to stimulate the population of soil microbes and enhance their enzymatic activity for reductive dechlorination of herbicide 2,4-Dichlorophenoxyacetic acid [69]. Similarly, magnetic nanoparticles (MNPs) immobilize microbial enzymes or microbes and are efficiently used in the treatment of wastewater. This tendency is attributed to their larger surface area and super-paramagnetism [19, 63].

Sphingomonas NM05 strain, a gram-negative bacterium, upon treatment with Pd/nFe0 bimetallic NPs (CMC-Pd/nFe0), synergistically degrades hexachlorocyclohexane (HCH) pesticide. The degradation efficiency was enhanced by nearly 1.7–2.1 folds as compared with controls containing only *Sphingomonas* sp. [101]. This synergistic degradation process was dependent upon environmental conditions such as temperature, pH, and concentration of HCH, etc. [102].

The effect of NPs of Perovskite ($LaFeO_3$) and biochar prepared from water caltrop shells was studied on marine sediment. The study revealed an enhanced level of PAH degradation up to 90%. The study comprised of using 0.75 g L^{-1} of lignocellulosic fibers reinforced with biodegradable composites (LFBC) at pH 6.0 and activating the 3×10^{-4} M of peroxymonosulfate. This led to the oxidation of PAH oxides present in the sediments. However, individual degradation of PAHs with 2-rings (52%), PAHs with 3-rings (61%), PAHs with 4-rings (66%), PAHs with 5-rings (56%), and PAHs with 6 rings (29%) was observed. This process improved sediment microbial diversity. Initially, high levels of *Proteobacteria* were observed; and later predominance of *Hyphomonas* was observed [103]. Continuous-flow experiment systems used for the degradation of naphthalene in groundwater used 400 mgL^{-1} of synthesized calcium peroxide (CaO_2) NPs that degraded naphthalene at an optimum concentration of 20 mgL^{-1}. Environmental conditions have been observed to affect the stability of NPs. Increasing the pH from 3 to 12 led to a rise in dissolved oxygen levels from 4 to 13.6 mgL^{-1}. It also enhanced the NPs stability for approximately 70 days. Temperature change from 4 to 30°C has been observed to reduce NP stability from 120 to 30 days. Complete remediation of naphthalene contaminant was observed with CaO_2 NPs and *Coccobacilli* sp. microbes in a span of 50 days [104].

Improving microbial community with NP application has been observed to remove the load of toxic pollutants present in the soil. Si-NPs improved biomass and colonization of rhizospheric microbes, which improved the overall health of soil [105, 106]. However, the nutritional and organic matter content gets affected upon prolonged accumulation and exposure of NPs [107].

10.4 CONCLUSION AND FUTURE PROSPECTS

Nanotechnology has grabbed significant attention to deal with bioremediation of an array of toxic compounds. As nanoremediation is emerging as the best remediation strategy for organic pollutants, the serious challenge for the application remains unsolved unless the NMs are studied well for their impact on biotic and abiotic components, flora, and fauna, and also for possible impact on human lives. Nanoparticles are favorably used for water treatment, in that case, NMs can be used multiple times, which would certainly be appreciated; hence, its recovery should be easier. If they are used directly on the soil, in that case, recovery cannot be expected and also would not be possible, hence, their half-lives, impact on lower animal that improves the fertility of the soil, and also on rhizospheric microbes should be preferentially studied. Moreover, the emerging trend of using microbes in synergy with NPs, i.e., nanobioremediation, is getting huge success. In such cases, normal soil microbes (including rhizospheric microbes) would be significantly helpful in alleviating soil health as well as plant health. Certainly, NMs are gaining great attention, but with their success reports and growing popularity, they would be demanded highly. It would also demand large-scale manufacturing. For that instance, we require a clearer roadmap that would allow control of handling and manufacturing, emission, distribution, disposal, etc., for a safe and sustainable world.

ACKNOWLEDGMENT

The study was supported by the Ministry of Science and Higher Education of the Russian Federation, agreement No. 075-15-2023-587 and the Strategic Academic Leadership Program of the Southern Federal University ("Priority 2030")

KEYWORDS

- **environmental pollution**
- **microbe-nanoparticle synergy**
- **nanobioremediation**
- **nanomaterials**
- **nanoremediation**
- **organic pollutants**
- **PAHs**
- **PCBs**
- **POPs**
- **waterbodies**

REFERENCES

1. Meng, F., Huang, Q., Yuan, G., Cai, Y., & Han, F. X., (2021). The beneficial applications of humic substances in agriculture and soil environments. In: *New Trends in Removal of Heavy Metals from Industrial Wastewater* (pp. 131–160). Elsevier, ISBN 9780128229651.
2. Ranjan, A., & Jindal, T., (2022). Overview of organophosphate compounds. In: *Toxicology of Organophosphate Poisoning* (pp. 1–25). Springer.
3. Bhardwaj, L., Sharma, S., Ranjan, A., & Jindal, T., (**2019**). Persistent organic pollutants in lakes of Broknes peninsula at Larsemann Hills area, East Antarctica. *Ecotoxicology, 28*, 589–596.
4. World Health Organization (WHO) Food Safety: Persistent Organic Pollutants (POPs), (2020).
5. United States Environmental Protection Agency (US EPA), *Persistent Organic Pollutants: A Global Issue, A Global Response.* Available online: https://www.epa.gov/international-cooperation/persistent-organic-pollutants-global-issue-global-response (accessed on Jun 4, 2022).
6. Ashraf, M. A., (**2017**). Persistent organic pollutants (POPs): A global issue, a global challenge. *Environ. Sci. Pollut. Res., 24*, 4223–4227. doi: 10.1007/S11356-015-5225-9/FIGURES/3.
7. Kumar, A., Ranjan, A., Gulati, K., Thakur, S., & Jindal, T., (**2016**). Assessment of chemical and microbial contamination in groundwater through leaching of sewage waste in Delhi, India. *Environ. Earth Sci., 75*, 1–11. doi: 10.1007/s12665-015-5016-0.
8. Kumar, A., Nirpen, L., Ranjan, A., Gulati, K., Thakur, S., & Jindal, T., (**2014**). Microbial groundwater contamination and effective monitoring system. *Asian. J. Eviron. Sci, 9*, 37–48.
9. Gulati, K., Thakur, S., Kumar, A., Ranjan, A., Hameed, S. A., & Jindal, T., (2015). A Lysimeter Study to Access the Risk of Chlorpyrifos Contaminated Groundwater in a Sandy Loam Soil With Different pH. *Indian J. Env. Prot. 35*(8), 696–703.
10. Zuo, Y., Chen, G., Zeng, G., Li, Z., Yan, M., Chen, A., Guo, Z., Huang, Z., & Tan, Q., (**2014**). Transport, fate, and stimulating impact of silver nanoparticles on the removal of Cd(II) by *Phanerochaete chrysosporium* in aqueous solutions. *J. Hazard. Mater., 285*, 236–244. doi: 10.1016/J.JHAZMAT.2014.12.003.
11. Shan, J., Ji, R., Yu, Y., Xie, Z., & Yan, X., (**2015**). Biochar, activated carbon, and carbon nanotubes have different effects on the fate of 14C-catechol and microbial community in soil. *Sci. Rep., 5*, 1–11. doi: 10.1038/srep16000.
12. Ranjan, A., & Jindal, T., (2022). Toxicology of organophosphate poisoning in humans. In: *Toxicology of Organophosphate Poisoning* (pp. 27–43). Springer.
13. Boudh, S., Singh, J. S., & Chaturvedi, P., (2019). Microbial resources mediated bioremediation of persistent organic pollutants. In: *New and Future Developments in Microbial Biotechnology and Bioengineering: Microbes in Soil, Crop and Environmental Sustainability* (pp. 283–294). Elsevier. ISBN 9780128182581.
14. Ranjan, A., & Jindal, T., (2022). Enzymatic targets of organophosphates. In: *Toxicology of Organophosphate Poisoning* (pp. 45–66). Springer.
15. Oonnittan, A., & Sillanpää, M., (2020). Application of electrokinetic Fenton process for the remediation of soil contaminated with HCB. In: *Advanced Water Treatment.* ScienceDirect.

16. Del Prado-Audelo, M. L., García, K. I., Escutia-Guadarrama, L., Reyna-González, J. M., Magaña, J. J., & Leyva-Gómez, G., (**2021**). Nanoremediation: Nanomaterials and nanotechnologies for environmental cleanup. *Front. Environ. Sci.*, *9*, 645. doi: 10.3389/FENVS.2021.793765/BIBTEX.
17. Mandeep, & Shukla, P., (**2020**). Microbial nanotechnology for bioremediation of industrial wastewater. *Front. Microbiol.*, *11*. doi: 10.3389/FMICB.2020.590631.
18. Lico, D., Vuono, D., Siciliano, C., Nagy, B. J., & De Luca, P., (**2019**). Removal of unleaded gasoline from water by multi-walled carbon nanotubes. *J. Environ. Manage.*, *237*, 636–643. doi: 10.1016/J.JENVMAN.2019.02.062.
19. Salvadori, M. R., Ando, R. A., Muraca, D., Knobel, M., Oller, N. C. A., & Corrêa, B., (**2016**). Magnetic nanoparticles of Ni/NiO nanostructured in film form synthesized by dead organic matrix of yeast. *RSC Adv.*, *6*, 60683–60692. doi: 10.1039/C6RA07274G.
20. Baragaño, D., Forján, R., Welte, L., & Gallego, J. L. R., (**2020**). Nanoremediation of As and metals polluted soils by means of graphene oxide nanoparticles. *Sci. Rep.*, *10*, 1–10. doi: 10.1038/s41598-020-58852-4.
21. Murgueitio, E., Cumbal, L., Abril, M., Izquierdo, A., Debut, A., & Tinoco, O., (**2018**). Green synthesis of iron nanoparticles: Application on the removal of petroleum oil from contaminated water and soils. *J. Nanotechnol.*, *2018*. doi: 10.1155/2018/4184769.
22. Bolade, O. P., Williams, A. B., & Benson, N. U. (**2020**). Green synthesis of iron-based nanomaterials for environmental remediation: A review. *Environ. Nanotechnology, Monit. Manag.*, *13*, 100279. doi: 10.1016/J.ENMM.2019.100279.
23. Aragaw, T. A., Bogale, F. M., & Aragaw, B. A., (**2021**). Iron-based nanoparticles in wastewater treatment: A review on synthesis methods, applications, and removal mechanisms. *J. Saudi Chem. Soc.*, *25*, 101280. doi: 10.1016/J.JSCS.2021.101280.
24. Kariim, I., Abdulkareem, A. S., & Abubakre, O. K., (**2020**). Development and characterization of MWCNTs from activated carbon as adsorbent for metronidazole and levofloxacin sorption from pharmaceutical wastewater: Kinetics, isotherms, and thermodynamic studies. *Sci. African*, *7*, e00242. doi: 10.1016/J.SCIAF.2019.E00242.
25. Chakraborty, J., & Das, S., (**2016**). Molecular perspectives and recent advances in microbial remediation of persistent organic pollutants. *Environ. Sci. Pollut. Res. Int.*, *23*, 16883–16903. doi: 10.1007/S11356-016-6887-7.
26. Li, Q. Q., Loganath, A., Chong, Y. S., Tan, J., & Obbard, J. P., (**2006**). Persistent organic pollutants and adverse health effects in humans. *J. Toxicol. Environ. Health*, *69*, 1987–2005. doi: 10.1080/15287390600751447.
27. Nadal, M., Schuhmacher, M., & Domingo, J. L., (**2011**). Long-term environmental monitoring of persistent organic pollutants and metals in a chemical/petrochemical area: Human health risks. *Environ. Pollut.*, *159*, 1769–1777. doi: 10.1016/J.ENVPOL.2011.04.007.
28. Convention, S., (**2011**). *Stockholm Convention on Persistent Organic Pollutants (POPs)*.
29. Parrado, C., Mercado-Saenz, S., Perez-Davo, A., Gilaberte, Y., Gonzalez, S., & Juarranz, A., (**2019**). Environmental stressors on skin aging. Mechanistic insights. *Front. Pharmacol.*, *10*, 759. doi: 10.3389/FPHAR.2019.00759/BIBTEX.
30. Aktar, W., Sengupta, D., & Chowdhury, A., (**2009**). Impact of pesticides use in agriculture: Their benefits and hazards. *Interdiscip. Toxicol.*, *2*, 1. doi: 10.2478/V10102-009-0001-7.
31. Lushchak, V. I., Matviishyn, T. M., Husak, V. V., Storey, J. M., & Storey, K. B., (**2018**). Pesticide toxicity: A mechanistic approach. *EXCLI J.*, *17*, 1136. doi: 10.17179/EXCLI2018-1710.

32. Mandal, A., Sarkar, B., Mandal, S., Vithanage, M., Patra, A. K., & Manna, M. C. (**2020**). Impact of agrochemicals on soil health. In: Vara, P. M. N., (ed.), *Agrochemicals Detection, Treatment and Remediation* (pp. 161–187). Butterworth-Heinemann.
33. Khan, M. Z., & Law, F. C., (**2005**). Adverse effects of pesticides and related chemicals on enzyme and hormone systems of fish, amphibians, and reptiles: A review. *Proc. Pakistan Acad. Sci.*, *42*, 315–323.
34. Kaul, L., Süss, R., Zannettino, A., & Richter, K., (**2021**). The revival of dithiocarbamates: From pesticides to innovative medical treatments. *iScience*, *24*, 102092. doi: 10.1016/J.ISCI.2021.102092.
35. Chmiel, J. A., Daisley, B. A., Pitek, A. P., Thompson, G. J., & Reid, G. (**2020**). Understanding the effects of sublethal pesticide exposure on honey bees: A role for probiotics as mediators of environmental stress. *Front. Ecol. Evol.*, *8*, 22. doi: 10.3389/FEVO.2020.00022/BIBTEX.
36. Yang, C., Lim, W., & Song, G., (**2020**). Mediation of oxidative stress toxicity induced by pyrethroid pesticides in fish. *Comp. Biochem. Physiol. Part C Toxicol. Pharmacol.*, *234*, 108758. doi: 10.1016/J.CBPC.2020.108758.
37. Franco, R., Li, S., Rodriguez-Rocha, H., Burns, M., & Panayiotidis, M. I., (**2010**). Molecular mechanisms of pesticide-induced neurotoxicity: Relevance to Parkinson's disease. *Chem. Biol. Interact.*, *188*, 289–300. doi: 10.1016/J.CBI.2010.06.003.
38. Andrione, M., Vallortigara, G., Antolini, R., & Haase, A., (**2016**). Neonicotinoid-induced impairment of odor coding in the honeybee. *Sci. Rep.*, *6*, 1–9. doi: 10.1038/srep38110.
39. Loser, D., Grillberger, K., Hinojosa, M. G., Blum, J., Haufe, Y., Danker, T., Johansson, Y., Möller, C., Nicke, A., Bennekou, S. H., et al., (**2021**). Acute effects of the imidacloprid metabolite desnitro-imidacloprid on human nACh receptors relevant for neuronal signaling. *Arch. Toxicol.*, *95*, 3695–3716. doi: 10.1007/S00204-021-03168-Z.
40. Ranjan, A., Chauhan, A., & Jindal, T., (**2022**). Toxicology of organophosphate and recent trends in prophylactic approaches. In: *New Frontiers in Environmental Toxicology*; Springer, 2022; pp. 103–123.
41. Ruus, A., Ugland, K. I., & Skaare, J. U., (**2002**). Influence of trophic position on organochlorine concentrations and compositional patterns in a marine food web. *Environ. Toxicol. Chem.*, *21*, 2356–2364. doi: 10.1002/ETC.5620211114.
42. Al-Othman, A. A., Abd-Alrahman, S. H., & Al-Daghri, N. M., (2015). DDT and its metabolites are linked to increased risk of type 2 diabetes among Saudi adults: A cross-sectional study. *Environ. Sci. Pollut. Res. Int.*, *22*, 379–386. doi: 10.1007/S11356-014-3371-0.
43. Czajka, M., Matysiak-Kucharek, M., Jodłowska-Jędrych, B., Sawicki, K., Fal, B., Drop, B., Kruszewski, M., & Kapka-Skrzypczak, L., (**2019**). Organophosphorus pesticides can influence the development of obesity and type 2 diabetes with concomitant metabolic changes. *Environ. Res.*, *178*, 108685. doi: 10.1016/J.ENVRES.2019.108685.
44. Gupte, A., Tripathi, A., Patel, H., Rudakiya, D., & Gupte, S., (**2016**). Bioremediation of polycyclic aromatic hydrocarbon (PAHs): A perspective. *Open Biotechnol. J.*, *10*, 363–378. doi: 10.2174/1874070701610010363.
45. Haritash, A. K., & Kaushik, C. P., (**2009**). Biodegradation aspects of polycyclic aromatic hydrocarbons (PAHs): A review. *J. Hazard. Mater.*, *169*, 1–15. doi: 10.1016/J.JHAZMAT.2009.03.137.
46. Behera, B. K., Das, A., Sarkar, D. J., Weerathunge, P., Parida, P. K., Das, B. K., Thavamani, P., et al., (**2018**). Polycyclic aromatic hydrocarbons (PAHs) in inland

aquatic ecosystems: Perils and remedies through biosensors and bioremediation. *Environ. Pollut.*, *241*, 212–233. doi: 10.1016/J.ENVPOL.2018.05.016.
47. Zhang, G., Pan, Z., Wang, X., Mo, X., & Li, X., (**2015**). Distribution and accumulation of polycyclic aromatic hydrocarbons (PAHs) in the food web of Nansi Lake, China. *Environ. Monit. Assess.*, *187*, 1–12. doi: 10.1007/S10661-015-4362-4.
48. Vogel, C. F. A., Van, W. L. S., Esser, C., & Haarmann-Stemmann, T., (**2020**). The aryl hydrocarbon receptor as a target of environmental stressors – implications for pollution mediated stress and inflammatory responses. *Redox Biol.*, *34*, 101530. doi: 10.1016/J.REDOX.2020.101530.
49. Hu, C., Hou, J., Zhou, Y., Sun, H., Yin, W., Zhang, Y., Wang, X., Wang, G., Chen, W., & Yuan, J., (**2018**). Association of polycyclic aromatic hydrocarbons exposure with atherosclerotic cardiovascular disease risk: A role of mean platelet volume or club cell secretory protein. *Environ. Pollut.*, *233*, 45–53. doi: 10.1016/J.ENVPOL.2017.10.042.
50. Šimečková, P., Pěnčíková, K., Kováč, O., Slavík, J., Pařenicová, M., Vondráček, J., & Machala, M., (**2022**). In vitro profiling of toxic effects of environmental polycyclic aromatic hydrocarbons on nuclear receptor signaling, disruption of endogenous metabolism, and induction of cellular stress. *Sci. Total Environ.*, *815*, 151967. doi: 10.1016/J.SCITOTENV.2021.151967.
51. Chang, Y., Siddens, L. K., Heine, L. K., Sampson, D. A., Yu, Z., Fischer, K. A., Löhr, C. V., & Tilton, S. C., (**2019**). Comparative mechanisms of PAH toxicity by benzo[a]pyrene and dibenzo[def,p]chrysene in primary human bronchial epithelial cells cultured at the air-liquid interface. *Toxicol. Appl. Pharmacol.*, *379*, 114644. doi: 10.1016/J.TAAP.2019.114644.
52. Venkatasubramanian, K. V., (2016). India bans manufacture and import of PCBs. *Chem. Eng. News*.
53. National Ocean Service, U.D. of C. What are PCBs? (2021).
54. Van Den, B. M., Birnbaum, L. S., Denison, M., De Vito, M., Farland, W., Feeley, M., Fiedler, H., Hakansson, H., Hanberg, A., Haws, L., et al., (**2006**). The 2005 World Health Organization Reevaluation of human and Mammalian Toxic Equivalency Factors for dioxins and dioxin-like compounds. *Toxicol. Sci.*, *93*, 223–241. doi: 10.1093/TOXSCI/KFL055.
55. Goldey, E. S., & Crofton, K. M., (**1998**). Thyroxine replacement attenuates hypothyroxinemia, hearing loss, and motor deficits following developmental exposure to aroclor 1254 in rats. *Toxicol. Sci.*, *45*, 94–105. doi: 10.1006/TOXS.1998.2495.
56. Panesar, H. K., Wilson, R. J., & Lein, P. J., (**2022**). Cellular and molecular mechanisms of PCB developmental neurotoxicity. *Handb. Neurotox.*, 1–30. doi: 10.1007/978-3-030-71519-9_204-1.
57. Bach, M. A., Samms-Vaughan, M., Hessabi, M., Bressler, J., Lee, M. J., Zhang, J., Shakespeare-Pellington, S., et al., (**2020**). Association of polychlorinated biphenyls and organochlorine pesticides with autism spectrum disorder in Jamaican children. *Res. Autism Spectr. Disord.*, *76*, 101587. doi: 10.1016/J.RASD.2020.101587.
58. Berghuis, S. A., Van Braeckel, K. N. J. A., Sauer, P. J. J., & Bos, A. F., (**2018**). Prenatal exposure to persistent organic pollutants and cognition and motor performance in adolescence. *Environ. Int.*, *121*, 13–22. doi: 10.1016/J.ENVINT.2018.08.030.
59. Ndlovu, T., Ba, S., & Malinga, S. P., (**2020**). Overview of recent advances in immobilization techniques for phenol oxidases in solution. *Catalysts*, *10*, 467. doi: 10.3390/CATAL10050467.

60. Gore, A. C., Krishnan, K., & Reilly, M. P. (**2019**). Endocrine-disrupting chemicals: Effects on neuroendocrine systems and the neurobiology of social behavior. *Horm. Behav.*, *111*, 7–22. doi: 10.1016/J.YHBEH.2018.11.006.
61. Pěnčíková, K., Svržková, L., Strapáčová, S., Neča, J., Bartoňková, I., Dvořák, Z., Hýžďalová, M., Pivnička, J., Pálková, L., Lehmler, H. J., et al., (**2018**). In vitro profiling of toxic effects of prominent environmental lower-chlorinated PCB congeners linked with endocrine disruption and tumor promotion. *Environ. Pollut.*, *237*, 473–486. doi: 10.1016/J.ENVPOL.2018.02.067.
62. McCann, M. S., Fernandez, H. R., Flowers, S. A., & Maguire-Zeiss, K. A. (**2021**). Polychlorinated biphenyls induce oxidative stress and metabolic responses in astrocytes. *Neurotoxicology*, *86*, 59–68. doi: 10.1016/J.NEURO.2021.07.001.
63. Roy, A., Sharma, A., Yadav, S., Jule, L. T., & Krishnaraj, R. (**2021**). Nanomaterials for remediation of environmental pollutants. *Bioinorg. Chem. Appl.*, *2021*. doi: 10.1155/2021/1764647.
64. Huang, Z., Chen, G., Zeng, G., Guo, Z., He, K., Hu, L., Wu, J., Zhang, L., Zhu, Y., & Song, Z., (**2017**). Toxicity mechanisms and synergies of silver nanoparticles in 2,4-dichlorophenol degradation by *Phanerochaete chrysosporium*. *J. Hazard. Mater.*, *321*, 37–46. doi: 10.1016/J.JHAZMAT.2016.08.075.
65. Kapoor, R. T., Salvadori, M. R., Rafatullah, M., Siddiqui, M. R., Khan, M. A., & Alshareef, S. A., (**2021**). Exploration of microbial factories for synthesis of nanoparticles – a sustainable approach for bioremediation of environmental contaminants. *Front. Microbiol.*, *12*, 1404. doi: 10.3389/FMICB.2021.658294/BIBTEX.
66. Zhang, Y., Chen, R., Riviere, J. E., & Comer, J., (**2021**). ZnO nanoparticles. *Process. MDPI.*
67. Gutierrez, A. M., Bhandari, R., Weng, J., Stromberg, A., Dziubla, T. D., & Hilt, J. Z., (**2019**). Novel magnetic core-shell nanoparticles for the removal of polychlorinated biphenyls from contaminated water sources. *Mater. Chem. Phys.*, *223*, 68. doi: 10.1016/J.MATCHEMPHYS.2018.10.045.
68. Gil-Díaz, M., Pérez, R. A., Alonso, J., Miguel, E., Diez-Pascual, S., & Lobo, M. C., (**2022**). Iron nanoparticles to recover a co-contaminated soil with Cr and PCBs. *Sci. Rep.*, *12*, 1–14. doi: 10.1038/s41598-022-07558-w.
69. Patra, S. M., Kumari, P., Mahobiya, D., & Kumar, S. S., (**2021**). Nano-bioremediation of environmental contaminants: Applications, challenges, and future prospects. *Bioremediation Environ. Sustain.*, 83–98. doi: 10.1016/B978-0-12-820318-7.00004-6.
70. Mahmoud, W. M. M., Rastogi, T., & Kümmerer, K., (**2017**). Application of titanium dioxide nanoparticles as a photocatalyst for the removal of micropollutants such as pharmaceuticals from water. *Curr. Opin. Green Sustain. Chem.*, *6*, 1–10. doi: 10.1016/J.COGSC.2017.04.001.
71. El-Sheekh, M. M., El-Kassas, H. Y., Shams El-Din, N. G., Eissa, D. I., & El-Sherbiny, B. A., (**2021**). Green synthesis, characterization applications of iron oxide nanoparticles for antialgal and wastewater bioremediation using three brown algae. *Int. J. Phytoremediation*, *23*, 1538–1552. doi: 10.1080/15226514.2021.1915957.
72. Hillyer, M. B., Jordan, J. H., Nam, S., Easson, M. W., & Condon, B. D., (**2022**). Silver nanoparticle-intercalated cotton fiber for catalytic degradation of aqueous organic dyes for water pollution mitigation. *Nanomaterials*, *12*, 1621.
73. Pourrahim, S., Salem, A., Salem, S., & Tavangar, R., (**2020**). Application of solid waste of ductile cast iron industry for treatment of wastewater contaminated by reactive blue

dye via appropriate nano-porous magnesium oxide. *Environ. Pollut.*, *256*, 113454. doi: 10.1016/J.ENVPOL.2019.113454.

74. Qurrat-Ul-Ain, Khurshid, S., Gul, Z., Khatoon, J., Shah, M. R., Hamid, I., Khan, I. A. T., & Aslam, F., (**2020**). Anionic azo dyes removal from water using amine-functionalized cobalt–iron oxide nanoparticles: A comparative time-dependent study and structural optimization towards the removal mechanism. *RSC Adv.*, *10*, 1021–1041. doi: 10.1039/C9RA07686G.
75. Bastús, N. G., Merkoçi, F., Piella, J., & Puntes, V., (**2014**). Synthesis of highly monodisperse citrate-stabilized silver nanoparticles of up to 200 nm: Kinetic control and catalytic properties. *Chem. Mater.*, *26*, 2836–2846. doi: 10.1021/CM500316K/SUPPL_FILE/CM500316K_SI_001.PDF.
76. Jishma, P., Roshmi, T., Snigdha, S., & Radhakrishnan, E. K. (**2018**). Kinetic study of gold nanoparticle mediated photocatalytic degradation of Victoria blue. *3 Biotech*, *8*, 97. doi: 10.1007/S13205-018-1116-3.
77. Wu, L., Lu, X., Fu, X., Wu, L., & Liu, H., (**2017**). Gold nanoparticles dotted reduction graphene oxide nanocomposite-based electrochemical aptasensor for selective, rapid, sensitive, and congener-specific PCB77 detection. *Sci. Rep.*, *7*, 1–7.
78. Gomez-Eyles, J. L., & Ghosh, U., (**2018**). Enhanced biochars can match activated carbon performance in sediments with high native bioavailability and low final porewater PCB concentrations. *Chemosphere*, *203*, 179–187.
79. Xiong, Z., Zhao, D., & Pan, G., (**2007**). Rapid and complete destruction of perchlorate in water and ion-exchange brine using stabilized zero-valent iron nanoparticles. *Water Res.*, *41*, 3497–3505.
80. Rasheed, T., Adeel, M., Nabeel, F., Bilal, M., & Iqbal, H. M. N., (**2019**). TiO_2/SiO_2 decorated carbon nanostructured materials as a multifunctional platform for emerging pollutants removal. *Sci. Total Environ.*, *688*, 299–311.
81. Liu, S., Yu, J., & Jaroniec, M., (**2011**). Anatase TiO_2 with dominant high-energy {001} facets: Synthesis, properties, and applications. *Chem. Mater.*, *23*, 4085–4093.
82. Soni, H., Kumar, N., Patel, K., & Kumar, R. N., (**2017**). Investigation on the heterogeneous photocatalytic remediation of pyrene and phenanthrene in solutions using nanometer TiO_2 under UV irradiation. *Polycycl. Aromat. Compd.*
83. Zhang, L., Zhao, B., Jiang, C., Yang, J., & Zheng, G., (**2015**). Preparation and transport performances of high-density, aligned carbon nanotube membranes. *Nanoscale Res. Lett.*, *10*, 1–8. doi: 10.1186/S11671-015-0970-8/TABLES/2.
84. Abadi, M. B. H., Shirkhanloo, H., & Rakhtshah, J., (**2020**). Air pollution control: The evaluation of TerphApm@MWCNTs as a novel heterogeneous sorbent for benzene removal from air by solid phase gas extraction. *Arab. J. Chem.*, *13*, 1741–1751. doi: 10.1016/J.ARABJC.2018.01.011.
85. Pourzamani, H., Bina, B., Amin, M. M., & Rashidi, A., (**2012**). Monoaromatic pollutant removal by carbon nanotubes from aqueous solution. *Adv. Mater. Res.*, *488*, 934–939. doi: 10.4028/WWW.SCIENTIFIC.NET/AMR.488-489.934.
86. Park, S., Byoun, Y., Kang, H., Song, Y. J., & Choi, S. W., (**2019**). ZnO nanocluster-functionalized single-walled carbon nanotubes synthesized by microwave irradiation for highly sensitive NO_2 detection at room temperature. *ACS Omega*, *4*, 10677–10686. doi: 10.1021/ACSOMEGA.9B00773/ASSET/IMAGES/LARGE/AO-2019-00773X_0002.JPEG.

87. Al-Shaalan, N. H., Ali, I., ALOthman, Z. A., Al-Wahaibi, L. H., & Alabdulmonem, H., (**2019**). High performance removal and simulation studies of diuron pesticide in water on MWCNTs. *J. Mol. Liq.*, *289*, 111039. doi: 10.1016/J.MOLLIQ.2019.111039.
88. Rao, T. N., Gaikwad, S., Naidu, T. M., & Han, S., (**2020**). Extraction and determination of pesticide residues in water using carbon nanotubes coupled with gas chromatography-mass spectroscopy. *Korean J. Chem. Eng.*, *37*, 1042–1049. doi: 10.1007/S11814-020-0511-8.
89. Kaur, H., Bansiwal, A., Hippargi, G., & Pophali, G. R. (**2017**). Effect of hydrophobicity of pharmaceuticals and personal care products for adsorption on activated carbon: Adsorption isotherms, kinetics and mechanism. *Environ. Sci. Pollut. Res.*, *25*, 20473–20485. doi: 10.1007/S11356-017-0054-7.
90. Cheng, X., Kan, A. T., & Tomson, M. B., (**2004**). Naphthalene adsorption and desorption from aqueous C60 fullerene. *J. Chem. Eng. Data*, *49*, 675–683.
91. Lin, Z., Zheng, J., Lin, G., Tang, Z., Yang, X., & Cai, Z., (**2015**). Negative ion laser desorption/ionization time-of-flight mass spectrometric analysis of small molecules using graphitic carbon nitride nanosheet matrix. *Anal. Chem.*, *87*, 8005–8012.
92. Li, X., Yu, J., Li, G., Liu, H., Wang, A., Yang, L., Zhou, W., Chu, B., & Liu, S., (**2018**). TiO_2 nanodots anchored on nitrogen-doped carbon nanotubes encapsulated cobalt nanoparticles as photocatalysts with photo-enhanced catalytic activity towards pollutant removal. *J. Colloid Interface Sci.*, *526*, 158–166.
93. Shaban, Y. A., El Sayed, M. A., El Maradny, A. A., Al Farawati, R. K., Al Zobidi, M. I., & Khan, S. U. M., (**2016**). Photocatalytic removal of polychlorinated biphenyls (PCBs) using carbon-modified titanium oxide nanoparticles. *Appl. Surf. Sci.*, *365*, 108–113.
94. Diel, J. C., Franco, D. S. P., Nunes, I. D. S., Pereira, H. A., Moreira, K. S., Thiago, A. De. L., Foletto, E. L., & Dotto, G. L., (**2021**). Carbon nanotubes impregnated with metallic nanoparticles and their application as an adsorbent for glyphosate removal in an aqueous matrix. *J. Environ. Chem. Eng.*, *9*, 105178.
95. Mehndiratta, P., Jain, A., Srivastava, S., & Gupta, N., (**2013**). Environmental pollution and nanotechnology. *Environ. Pollut.*, *2*, 49.
96. Chen, M., Zeng, G., Xu, P., Zhang, Y., Jiang, D., & Zhou, S., (**2017**). Understanding enzymatic degradation of single-walled carbon nanotubes triggered by functionalization using molecular dynamics simulation. *Environ. Sci. Nano*, *4*, 720–727.
97. Gong, J. L., Wang, B., Zeng, G. M., Yang, C. P., Niu, C. G., Niu, Q. Y., Zhou, W. J., & Liang, Y., (**2009**). Removal of cationic dyes from aqueous solution using magnetic multi-wall carbon nanotube nanocomposite as adsorbent. *J. Hazard. Mater.*, *164*, 1517–1522.
98. Cai, C., Zhao, M., Yu, Z., Rong, H., & Zhang, C. (**2019**). Utilization of nanomaterials for in-situ remediation of heavy metal (loid) contaminated sediments: A review. *Sci. Total Environ.*, *662*, 205–217.
99. Chauhan, R., Yadav, H. O. S.,Sehrawat, N., (**2020**). Nanobioremediation: A new and versatile tool for sustainable environmental clean up-overview. *J. Mater. Environ. Sci*, *11*, 564–573.
100. Yan, H., Yang, X., Chen, J., Yin, C., Xiao, C., & Chen, H., (**2011**). Synergistic removal of aniline by carbon nanotubes and the enzymes of *Delftia* sp. XYJ6. *J. Environ. Sci.*, *23*, 1165–1170. doi: 10.1016/S1001-0742(10)60531-1.
101. Manickam, N., Reddy, M. K., Saini, H. S., & Shanker, R., (**2008**). Isolation of hexachlorocyclohexane-degrading *Sphingomonas* sp. by dehalogenase assay and

characterization of genes involved in γ-HCH degradation. *J. Appl. Microbiol.*, *104*, 952–960.

102. Singh, R., Manickam, N., Mudiam, M. K. R., Murthy, R. C., & Misra, V., (**2013**). An integrated (nano-bio) technique for degradation of γ-HCH contaminated soil. *J. Hazard. Mater.*, *258*, 35–41.
103. Hung, C. M., Huang, C. P., Chen, C. W., & Dong, C. D., (**2021**). Degradation of organic contaminants in marine sediments by peroxymonosulfate over $LaFeO_3$ nanoparticles supported on water caltrop shell-derived biochar and the associated microbial community responses. *J. Hazard. Mater.*, *420*, 126553.
104. Gholami, F., Shavandi, M., Dastgheib, S. M. M., & Amoozegar, M. A., (**2018**). Naphthalene remediation from groundwater by calcium peroxide (CaO_2) nanoparticles in permeable reactive barrier (PRB). *Chemosphere*, *212*, 105–113.
105. Rajput, V. D., Minkina, T., Feizi, M., Kumari, A., Khan, M., Mandzhieva, S., Sushkova, S., et al., (**2021**). Effects of silicon and silicon-based nanoparticles on rhizosphere microbiome, plant stress and growth. *Biology (Basel).*, *10*, 791.
106. Rizwan, M., Ali, S., Ur Rehman, M. Z., Malik, S., Adrees, M., Qayyum, M. F., Alamri, S. A., et al., (**2019**). Effect of foliar applications of silicon and titanium dioxide nanoparticles on growth, oxidative stress, and cadmium accumulation by rice (*Oryza sativa*). *Acta Physiol. Plant.*, *41*, 1–12.
107. Rajput, V., Minkina, T., Mazarji, M., Shende, S., Sushkova, S., Mandzhieva, S., Burachevskaya, M., et al., (**2020**). Accumulation of nanoparticles in the soil-plant systems and their effects on human health. *Ann. Agric. Sci.*, *65*, 137–143.

CHAPTER 11

Nanobioremediation of Metal and Salt Contaminated Soils

SARA ZAFAR, MUHAMMAD KAMRAN KHAN, NAEEM IQBAL, and SHAGUFTA PERVEEN

Government College University, Faisalabad, Pakistan

ABSTRACT

The global population is growing quickly and faces many challenges due to human overexploitation and environmental degradation. The employment of eco-friendly, sustainable, and financially feasible technology for environmental remediation and clean energy supplies for long-term growth is the 21st century's most difficult problem. The use of biosynthetic nanoparticles to remove toxins from a polluted environment is a new, rapidly growing technique. The current work speculates about nanoparticle manufacturing by plants, bacteria, yeast, and fungi, which are all evolving as nano-factories and could be used in environmental cleanup. Most biogenic nanoparticles that have been examined have shown to be quite effective. The biosynthetic route to nanoparticle manufacturing could be more successful and safer than previous approaches. Nanomaterials have distinct physical and chemical characteristics, which has aroused the interest of scientists and researchers working in a range of environmental sectors, such as bioremediation. Pollution of stream water, groundwater, and soil are on the rise as a result of growing industrialization, urbanization, and modern farming methods. The elimination of pollutants is the most difficult task for researchers. Natural resources have been exploited to meet human requirements for energy generation and other needs, resulting in water quality degradation and pollution, as well as an ecological imbalance. Even though current treatment methods

Nano-Bioremediation for Water and Soil Treatment: An Eco-Friendly Approach.
Vishnu D. Rajput, Arpna Kumari, and Tatiana M. Minkina (Eds.)

are effective, they have several drawbacks that make remediation operations more difficult. Nanoremediation uses nanomaterials and plants, referred to as phyto-nanoremediation, animals, referred to as zoo-nanoremediation, and bacteria, referred to as microbial nanoremediation.

11.1 INTRODUCTION

With each passing year, environmental pollution becomes a more serious concern as global industry advances. Environmental remediation is a step to take to safeguard our soil and water resources, and there are numerous things that individuals at all levels of power may do to alleviate the harm done to the environment. Given the continuing nature of the problem, it is critical to be aware of new strategies in the field of environmental remediation [141].

Rapid urbanization and industrial expansion have resulted in the constant addition of numerous wastes and harmful chemicals to the environment, contaminating it to the point of no return. It is a global concern since developing efficient cleanup procedures has proven difficult.

Bioremediation is a 70-year-old approach to removing contaminants from the environment and industrial pollution based on the natural phenomenon of waste product degradation by microorganisms. Nanotechnology is currently bridging technological gaps to develop more effective and efficient nanoparticles for application in bioremediation using biological systems. Bacterial, fungal, and plant systems are being employed in research to generate nanoparticles with unique adaptive qualities capable of detoxifying a wide range of pollutants, such as hydrocarbons and heavy metals. The phrase "nanobioremediation" refers to a new type of bioremediation that has a strong connection to nanotechnology [142].

Bioremediation is updated to nano-bioremediation using various types of metal nanoparticles (NPs) in combination with various bacteria, which provides ecologically friendly ways for cleansing dangerous ecological toxins. Nanobioremediation has gotten a lot of interest recently because of its high efficacy in creating a sustainable environment. Traditional bioremediation methods use microbes, plants, or derived products to treat environmental toxins, but they have a fundamental drawback: they are ineffective in the presence of large amounts of pollutants. Nanobioremediation also, is effective, quick, cheap, and efficient, and it reduces the detrimental effects of toxins present at extreme concentrations. Biogenic or manufactured NPs, each having their physiochemical characteristics [143]. The effective life, transport, and destiny of NPs have an impact on the efficacy of decontamination in soil ecosystems [144].

Industrialization and large-scale anthropogenic activities have resulted in environmental contamination. Thermal power plants, coal mines, cement, sponge iron, steel, ferroalloys, petroleum, and chemicals are all highly polluting industries that emit dust, smoke, fumes, and hazardous gas emissions. Furthermore, a vast number of highly toxic organic and inorganic substances are emitted, inflicting lasting damage to our ecology and ecosystem and frequently surpassing the carrying capacity of the environment. As a result, to limit negative consequences, suitable and effective pollution control measures should be implemented [145].

Traditional pollutant removal technologies have severe problems in terms of cost and secondary pollution generation therefore developing simple and environmentally acceptable approaches becomes a priority. For the removal of various toxins, new methods are being investigated, with nanotechnology showing enormous promise [140].

Heavy metal pollution in wastewater, groundwater, lakes, and streams is currently causing major long-term health effects in humans. Because of their non-specificity, inefficiency, and high cost, traditional procedures have become inappropriate for treatment as a result of industrialization.

Few studies have been published that combine biological approaches, biophysical approaches, biochemical methods, physiochemical methods, and nano-based physiochemical methods are some of the additional remediation alternatives to solve these challenges. This chapter examines common treatment approaches physical, chemical, physiochemical, and biological procedures, with a focus on bioremediation, its mechanism, and nanotechnology's applicability in bioremediation.

11.2 OVERVIEW OF NANOREMEDIATION

Nanoparticles made of metal oxides are excellent for the eradication of heavy metals and organic pollutants from water using nano-adsorbents in recent investigations. However, due to their potential dominance over conventional water management approaches, polymer templated nanoparticles and functionalized nanoparticles for heavy metal removal is attracting a lot of interest. Furthermore, numerous nano-based materials have been examined in depth in this chapter, presenting an outline of nanobioremediation for heavy metal pollution exclusion from ground water and industrial effluents. In the future, the proposed nano-bio-approach will provide the best result in terms of efficiency and cost, the most effective and sustainable treatment technology, as judged by the socioeconomic conditions of developing countries [5].

In terms of environmental science, there are a variety of ecologically friendly uses of nanomaterials (NMs), including materials that bring clean water from contaminated water sources in both large-scale and portable applications, as well as remediation, which detect and remove environmental contaminants (waste and harmful material) [1, 2]. The process of breaking down environmental toxins into less harmful forms using biological agents such as bacteria, fungi, protists, or their enzymes is known as "bioremediation." [3].

Bioremediation has several advantages over traditional treatments, including cost, high competence, biosorbent regeneration, reduced chemical and biological slurry, metal selectivity, no additional nutrient requirements, and the possibility of metal recovery [4].

Bioremediation is known as natural attenuation or intrinsic bioremediation when it happens spontaneously, and biostimulated bioremediation when it is induced by the addition of fertilizers to increase bioavailability within the medium. Bioventing, bioleaching, bioreactor, bioaugmentation, composting, biostimulating, land farming, phytoremediation, and rhizofiltration are some of the most prevalent bioremediation processes [6].

Polluted soil and groundwater, particularly in industrial and municipal regions, is a pervasive challenge that poses serious health hazards and environmental and human health problems [7, 8].

The treatment of contaminated soil, groundwater, wastewater, and several research has been conducted on landfill leachate [9, 10]. Groundwater and soil remediation may be divided into ex-situ and in-situ, depending on where the remediation takes place. Ex-situ remediation is the process of recovering polluted subterranean soil or groundwater and treating it on-site or transporting it to another area for treatment [11].

When polluted soil or underground water is treated straight in the subsurface, this is known as in situ remediations. Because in situ remediation is less expensive than ex-situ remediation, it is commonly chosen [8, 12]. The cost of removing and restoring contaminated soil, according to [13], is prohibitively costly (on the order of $3 million/ha), posing a serious concern for growing countries in terms of environmentally friendly sustainability [13].

The basic goal of remediation of soil and groundwater methods is to reduce contamination to a point that is both desirable and safe. The cleanup of soil and groundwater has been completed by using physical, chemical, and biological approaches. Several criteria, including soil features and contaminants, as well as the type of the selected and designed remediation technique, play a crucial impact in the choice of the appropriate groundwater and soil treatment [14].

Contaminants are removed from extracted groundwater via ex-situ treatments such as carbon adsorption, air stripping, chemical oxidation/precipitation, or biological reactors in classical approaches such as pump-and-treat. These approaches, on the other hand, are linked with significant operational costs and the creation of contaminated waste [10]. Emerging remediation approaches, such as surfactant enhanced remediation (SER), have been shown to be successful for groundwater and soil contaminated with organic pollutants in the form of dense nonaqueous phase liquids (DNAPLs). These technologies, however, are not without risk; as the interfacial force of DNAPLs reduces, unrestricted vertical movement may occur [15].

Nanotechnology in current days is being considered in a growing number of sectors. NPs exhibit a variety of important and promising qualities as a result of their many functions [16, 19]. Multidisciplinary domains such as molecular level manufacture concepts and engineering are used to create NPs. Nanotechnology, in general, is a method for creating particles with a range of 1–100 nanometers, studying the physical occurrences associated with such particles, and using it in a variety of fields [10]. Chemical, electrical, medical, and biotechnology industries all use nanotechnology. There have been various projects to use nanotechnology for environmental protection purposes, such as water and wastewater treatment, pollution control, and soil and groundwater remediation, while many firms manufacture and employ nanomaterials in diverse ways [20].

In recent years, nanoremediation technologies for contaminated locations have been deployed (2009 till now). The majority of nanoremediation technology evaluations are bench-scale applications, with a little field-scale application [8]. The major benefits of using nanoremediation for soil and groundwater remediation, particularly for large sites, include reduced cost and cleanup time, complete destruction of some toxins without the need for dirty soil disposal, and no need to transport or pump groundwater [20, 21].

The utilization of reactive NPs in nanoremediation technologies allows for the conversion and detoxification of pollutants. Catalysis and chemical reduction are the two primary processes for NP remediation [20, 22]. Furthermore, adsorption is another removal process aided by NPs because NPs have large surface-area-to-mass ratios and a wide distribution of active sites, which increases adsorption capacity [23]. Because of their novel surface coating and small size, numerous manufacturing NPs have extremely practicable properties for in situ remediation functions. Furthermore, unlike microparticles, NPs may diffuse and enter minuscule pores in the subsurface and float in groundwater for extended periods of time; unlike microparticles, NPs can travel long distances and have a greater geographical distribution [20].

Because of their nanoscale features, the physical movement of NPs and/or transport in groundwater is dominated by random motion or Brownian movement rather than the wall effect [24]. As a consequence, NPs are not impacted by gravity sedimentation and remain floating in groundwater throughout the restoration process, unlike microscale particles, which are heavily influenced by gravity sedimentation due to their density and huge size, macroscale particles are not. As a result, NPs provide a functional treatment strategy that allows for direct inoculation into the subsurface where contaminants exist [20].

Nanoremediation for soil and groundwater has been demonstrated in studies [25, 28]. However, the environmental impacts of those NPs are yet unknown, and additional research is needed to better understand their environmental destiny and toxicity, as these concerns are critical for environmental conservation.

In the lab, nanomaterials for soil and groundwater cleanup have been thoroughly tested against a variety of contaminants, with promising results [29, 30]. Nanomaterials can have beneficial or bad effects on biological creatures, the atmosphere, the world, and the financial system, and each instance should be analyzed individually. Appropriate nanoremediation hazard reporting, field-scale validation of remediation findings, science-policy interface discussions, and appropriate market development initiatives are all strategies to boost the appeal and acceptance of nanoremediation technology [31].

Nanotechnology's emergence has been the topic of intense investigation in recent years, including all forms of life and intersecting with numerous disciplines of science [32]. Richard Feynman proposed the concept of nanotechnology in 1959 [33], and it is now one of the world's fastest expanding disciplines of scientific inquiry and technology expansion. The "Next Industrial Revolution" [34] is a term used to describe this sector. Green nanoparticle production technologies have become popular in recent years, generating a great deal of interest in the chemical, electrical, and biological sciences. There are numerous possible environmental benefits from nanotechnology. The three categories that can be divided are pollution prevention, treatment, and cleaning, sensing, and detection.

The nanotechnologies discussed here are focused on on-site cleanup and wastewater treatment. Many nanotechnologies for air remediation are also being developed, in addition to applications for soil, groundwater, and wastewater. Smaller particles make it easier to build smaller sensors that can be placed in more remote locations. Nanotechnology's ability to reduce pollution is still being developed, but it has the potential to speed up some of the most groundbreaking environmental breakthroughs [35]. Another study highlighted the purposes of nanotechnology in the management of contaminated water, including nanoscale filtration techniques, adsorption, and degradation of pollutants by nanoparticle catalysts [36]. Among the several uses of nanotechnology, effluent treatment with nanoparticles is one of the highly popular [37].

To address the issue more comprehensively, modern research has been dedicated to the advancement of cost-intensive, environmentally favorable plans like bioremediation with low adverse consequences. The issue of bioremediation has been significantly rebuilt with the incorporation of nanotechnology, dealing with items in the nanoscale range (any one dimension spanning between 1 and 100 nm) that exhibit specific traits due to their extremely small size [38]. Nanotechnology has had enormous consequences in the remediation of HM contamination. Nanobioremediation boosts the efficacy and potential of the decontamination process substantially. Various nanomaterials can be used with microorganisms and plants concurrently or sequentially or as nanocarriers for microbial biosorbents to accelerate HM elimination [39].

Nanomaterials synthesized chemically, on the other hand, can be hazardous to the environment and have negative side effects. Various plant extracts and bacteria can be employed to fabricate biogenic NPs to overcome the challenges associated with standard NP manufacturing. The bioinspired synthesis of diverse nanomaterials requires less money and energy, and the resulting particles are safe for the environment.

11.3 TYPES OF NPS FOR SOIL CLEAN UP: APPLICATION AND METHODS

Our knowledge of ENM-soil interactions is limited, and due to the system's complexity, there is still much to learn before we fully comprehend the manmade nanoparticle behavior [57–59].

Soil research over the last 15 years has been encouraging in transcendental topics such as increasing fertility, reducing deterioration, attenuation, or degradation of pollutants, and developing nutrient and pollution sensors [61, 62].

Controlling the number of active compounds taken up consistently during crop growth, preventing overdose, and reducing input and waste are all important goals, according to the creation of nanocomposites and nanoencapsules [63]. Fertilizers [64, 65], herbicides [66, 67], plaguicides [68, 69], or growth promoters [70], and by removing the need for contaminated soil treatment and disposal, their rationalization and control of the amount of application should assist minimize the overall costs of cleaning up severely contaminated areas [71].

Nanoparticles or nanomaterials are used to clean up or remediate the environment, such as soil, groundwater, sediment, air, and wastewater is known as nano-remediation [72] (Table 11.1). Nano-remediation, for example, uses techniques like nanomembrane filtration, photocatalysis, and adsorption to clean wastewater [73].

There has been a lot of research on the advantages of nanoremediation or nanotechnology for environment cleanup [74–80], including heavy metal removal from soils utilizing plants [81–83], wastewater remediation [84, 85], and pesticide degradation [86–88] in soil and water. Table 11.1 shows some of the ways nanoparticles can be employed to improve soil and air quality [89].

TABLE 11.1 Nanoparticles for the Remediation of Contaminants from the Air

Nanoparticles	Pollutants to be Targeted	Effects or Observations	References
Silica nanoparticles (SiNPs)	Atmospheric lead (Pb)	The higher Pb uptake by SiNPs was attributed to the SiNPs having a large surface area and negative-charged groups.	[97]
$Zn_{12}O_{12}$ nanocage	Carbon disulfide (CS2)	The adsorption energy of CS2 per molecule dropped as the number of CS2 molecules increased, which could be related to steric revulsion between the CS2 molecules.	[98]
Aligned carbon nanotube	Aerosols	The filtration performing of the unique filters demonstrated that the filtration efficiency and pressure drop both increased significantly as the number of carbon nanotube layers grew.	[99]

11.4 NANOBIOREMEDIATION: A PROMISING APPROACH TOWARD SOIL CLEANUP

One of the most popular strategies to improve the long-term viability of polluted-land crop cultivation is to use nanoremediation with nanomaterials. These strategies include:

- Agro-biotechnology;
- Rhizospheric engineering or root biology;
- Molecular biology;
- Nanobiotechnology can be employed to produce viable crops in such areas [90–95].

The link between cultivating edible plants on contaminated grounds and the food chain is a huge problem and a major concern. As a result, any potential health risks associated with the biomagnification of contaminants by plants in the food chain should be avoided [95].

This is the conundrum: how can we grow food plants in contaminated environments with soils while also ensuring that the harvested plants are safe for human consumption? This may necessitate strategies such as:

- Pollutant breeding through the selection of low accumulator cultivars;
- Lowering pollutant bioavailability in soils;
- Limiting pollution translocation and uptake to the edible sections [95, 96].

It's worth noting that phytoremediation, biofortification (raising the nutritious content of agricultural goods), and safe phytoproducts (enhancing the nutrient content of agricultural products) are all advantages of revitalizing damaged soils (removal of pollutants from the environment) (Table 11.2).

Larger quantities of metalloids (particularly As) and metals (Cu, Cd, Pb, Zn, etc.) have been released into the environment as a result of industrial activity, causing substantial ecosystem damage [40]. Furthermore, metals and metalloids pose a risk to health because a large number of them are poisonous even in minute amounts, and some are even hazardous (e.g., As) and carcinogenic [41]. In contrast to organic pollutants, which disintegrate into innocuous tiny molecules, inorganic pollutants are resistant to numerous biological reactions, making them extremely challenging to eliminate from soils [42].

TABLE 11.2 Nanoparticle Application in the Remediation of Contaminated Soil

Nanoparticles	Degradation	Application	References
Ni/Fe Bimetallic	Decabromodiphenyl ether (BDE209)	At room temperature, Ni/Fe bimetallic nanoparticles decompose BDE209 on land, with a removal effectiveness of 72%.	[100]
Nano-Zerovalent Iron (nZVI)	Organochlorine insecticides (DDT)	In spiked soil, 1 g nZVI kg^1 was effective for the breakdown of dichloro-diphenyl trichloroethane (DDT), but a larger dose was used for the treatment of older pollutants in soil.	[101]
Goethite and hematite nanomaterials (nHA)	Cadmium and lead	Pb/Cd phosphate (e.g., hydroxypyromorphite-like mineral) was generated in large amounts by (nHA), which was able to diminish water-soluble, bio-accessibly, and phytoavailable Pb/Cd.	–

Several findings have found that 52,000 to 1,12,000 tons of As are emitted into the environment each year [40, 43, 44]. Contact with As can induce a variety of diseases, including lung cancer, skin cancer, bladder cancer, and others [45, 46]. As has been identified as a major source of toxic pollution because it is exceedingly harmful to humans, plants, and animals. Unlike the metals previously mentioned, it is a contaminant that is anionic. The simultaneous treatment of soils degraded by anionic and cationic pollutants is a difficulty in this context [47]; that is, remediation is much more complicated when metals are present.

Traditional remediation methods aim to elevate the pH of cationic metal-contaminated soils and thereby stabilize them, among other things. This method, however, results in the solubilization of As [48]. The disparity between the immobilization and mobilization of contaminants is the focus of the various remediation strategies [49].

From numerous perspectives, the soil ecosystem is a complicated system. The soil matrix, for example, is a tri-phasic system with phases or states that include solid, liquid, and gas. There are also biotic and abiotic interactions, the former including multicellular animals like nematodes, arachnids, mites, and earthworms, and the latter involving microorganisms like bacteria, viruses, fungi, protozoa, amoebas, and so on. Alkali and transition metals are examples of inorganic elements, in particular, are involved in abiotic interactions [50].

Plant growth and agroecosystems are dependent on soil microorganisms (they are involved in soil ecosystems and organic waste decomposition, and Carbon, nitrogen, phosphorus, and potassium are the cornerstones to driving nutrient cycle in soils, biogeochemistry cycling, to name a few). NPs are widely used for items with an extensive range of manufacturing, commercial, pharmaceutical, and farming purposes [51, 52] will result in a rise in their intensity in the soil, as well as their ecological and ecological risks [53]. Some metal oxides or metal NPs are extremely hazardous to soil microorganisms and have a major impact on a wide range of soil microbial species [54, 52].

However, when researching the effects of metal or metal oxide nanoparticles on soil microbial metabolism and essential ecological activities, Real-time polymerase chain reaction (Real-Time PCR), also known as quantitative polymerase chain reaction(qPCR), was used to measure changes in the abundances of bacteria, eukaryotes, and ammonia-oxidizing bacteria [52]. AgNP amendments (at 0.1, 1.0, and 10 mg kg^{-1} soil) were found to diminish soil microbial metabolic activity, nitrification capacity, and the number of bacterial and ammonia-oxidizing bacteria. In contrast, FeO-NPs were reported to have beneficial impacts on soil microbial metabolic activity (at 1 and 10 mg kg^{-1} soil) and soil nitrification capacity (at 1 and 10 mg kg^{-1} soil) [52].

Micronutrients like Cu, Mn, and Zn, as well as non-nutrient metals like Hg, Ni, U, Cd, and Cr, are accumulated in larger proportions by fungi and yeast than the nutritious needs. It has long been established that fungal biomass can be used as a biosorbent to extract hazardous heavy metals and radionuclides derived from polluted trash.

Because of the presence of numerous functional groups, such as hydroxyl, carboxyl, sulfhydryl, phosphate, and amino groups, fungal cell walls, and processes play a key role in heavy metal sequestration [55]. Even in the absence of physiological pH, temperature, and nutrition availability, fungal biomass can ingest substantial amounts of dangerous metals from an aqueous solution via adsorption or other mechanisms [56].

11.5 MECHANISM OF REMEDIATION

1. **Nanoscale Zero-Valent Iron, Carbon Nanotubes:** The formulation, characterization, and usage of nanoscale zero-valent iron (NZVI) stopped on the carbon nanotube (CNT) complex (NZVI/

CNT) for the integration of Se(IV) in water. The structural study demonstrated that NZVI were evenly halted on CNT surfaces, implying that NZVI oxidation and accumulation were reduced. Because of the good synergetic impact between CNT adsorption and NZVI reduction, the NZVI/CNT showed substantially greater efficiency on Se(IV) sequestration than bare NZVI. X-ray absorption fine structure (XAFS) studies showed that NZVI/CNT reduced Se(IV) nearly entirely into Se(0)/Se(-II), while limited reduction happened on NZVI with a trace of Se(IV) adsorbed on the corrosion products.

Furthermore, due to the buffering action of CNT, the involvement of pH in Se(IV) sequestration on NZVI/CNT was smaller than on NZVI. Furthermore, because CNT indicated significant adsorption for HA, which limits possible reactivity, CNT immobilization could diminish the inhibitory action of humic acid (HA) on the elimination of Se(IV) by NZVI. The high performance of NZVI/CNT makes it a potent substance for Se(IV) detoxification in wastewater [102].

In order to activate peroxides and produce free radicals for the remediation of organic contamination that chlorination and nitroxides have been unable to remove, nanoscale zero-valent iron (nZVI), a low-toxicity, environmentally friendly transition metal with strong electron-feeding and reducing abilities, has been widely used [103].

Unfortunately, nZVI's instability has limited its advancement, such as its inclination to agglomerate and oxidize. As a result, improving nZVI's dispersion and oxidation endurance is crucial. Carbon nanotubes (CNTs) are a form of material that are hollow tubes with thermal stability and strong electrical conductivity. They can both activate persulfate and act as carbon carriers for nZVI by immediately oxidizing adsorbed water, and to remove phenolics and dyes, hydroxyl radicals are generated [104].

2. **Metal and Magnetic Nanoparticles:** Magnetic nanoparticles have earned a lot of interest because of their unique properties and broad range of uses, which include water and wastewater treatment. Because of the negative consequences on individuals and the environment, the emission of heavy metals and other toxins into water supplies is a problem. Aminated or amino-functionalized magnetic nanoparticles are created by implanting amino functional groups

onto magnetic nanoparticles, and they outperform exposed magnetic nanoparticles for water treatment purposes [105].

Approximately 663 million populations in the world today do not have safe drinking water. Heavy metal poisoning is a key cause of the global drinking water disaster, which affects millions of people around the world. Lead (Pb), cadmium (Cd), arsenic (As), mercury (Hg), and chromium (Cr) are highly hazardous and non-biodegradable heavy metal contaminants. Their presence in the body above legal thresholds might harm important organs like the kidney, liver, and brain, as well as the reproductive and nervous systems [106].

These heavy metals are extensively utilized in a variety of industries, involving metal finishing, iron and steel, metallurgy, mining, and battery manufacturing, and they are distributed into a variety of water sources [107]. Solvent extraction, reverse osmosis, membrane partition, ion exchange, and other procedures have been described for removing metal ions from wastewater [106].

Nanomaterials (NMs) are increasingly being promoted as a viable, cheaper, and ecologically pleasant alternative to conventional treatment materials in both reserves' protection and environmental remediation [108]. Nano-bioremediation is the method of eliminating heavy metals from contaminated areas by using nanomaterials produced by plants, fungi, and bacteria using nanotechnology. The green nanoparticle synthesis technique has made great progress, resulting in novel materials that are environmentally friendly, cost-effective, and stable. Although nanoparticles can be produced using a variety of traditional methods, the biological approach is preferable due to its synthesis ease, low toxicity, control of size properties, cheaper, and environmental friendliness [106]. Adsorption has already been proven to be a winning strategy for contaminant removal in the development of several heavy metal cleanup systems [109].

The absorbent's applicability is crucial to the adsorption process. Because metal ions have a strong attraction for the adsorbent, the effluent produced is of higher quality than the remainder of the operations. For remediation applications, various adsorbents have been developed, including zeolites, activated carbons, chelating chemicals, and nanoparticles. High adsorption capability, quick kinetics, and low expense are the most important selection criteria for adsorbents.

As technology progresses, nanoparticles are being used for heavy metal cleanup due to their larger surface area and low cost of manufacturing. They have also proven to be effective adsorbents. Among all known nanomaterials, metal oxide-based nanocomposites and magnetic nanoparticles made of iron oxide have been widely used to remediate heavy metal ions. Superpara-magnetism is a unique feature of nanosized iron oxide particles that distinguishes them from other nanoparticles [106].

11.6 INTEGRATION OF NANOREMEDIATION WITH OTHER TECHNOLOGIES

For soil and groundwater remedies, several investigators have concentrated on the usage and improvement of nanoremediation skills [10]. Nanoremediation is a technology that is regarded to be environmentally benign. As a result, it's thought to be a viable option for traditional site cleanup technologies [10, 110, 111].

Because of its low environmental impact, quick kinetics, cost-effectiveness, and non-toxic nature, nZVI injection is well-suited to soil restoration. In the 1990s, the first synthesis and application of nZVI were described [112].

Fe^{2+} and Fe^{3+} were used to make iron nanoscale particles ranging in size from 10 to 100 nm [113]. Many pollutants were removed from water using nZVI, primarily organic compounds with halogens, which are generally found in soil and groundwater. For the first time, they demonstrated that nZVI can be used to detoxify and convert a variety of environmental toxins involving chlorinated organic solvents, polychlorinated biphenyls, and organochlorine pesticides [114].

Another study [115] used nZVI to remove nitrate from porous media and customized nZVI with Cu in an up-flow packaged sand column with a multilayer system. The results showed that a 10 cm layer of nZVI/sand provided the ideal conditions for high nitrate removal, with a nitrate removal efficacy of around 97%. However, when using Cu-altered nZVI/sand, the best results were obtained when a double 5 cm layer was utilized, resulting in 100% nitrate elimination. The findings show that a single layer of nZVI or a multilayer of Cu-modified nZVI could attain significant nitrate elimination.

In a subsequent study, Xue et al. assessed the efficiency of rhamnolipid modified nZVI (R-nZVI) in the immobilizing lead (Pb) and cadmium (Cd) in river deposits [116].

11.7 HEAVY METAL REMOVAL USING BIOGENIC NANOPARTICLES

Water contamination caused by heavy metals is becoming a serious global concern, necessitating effective treatment solutions. As heavy metal contamination enters the food web, it impacts both plants and fauna. Nanotechnology advancements and the production of novel nanomaterials have aroused the curiosity of researchers all around the world. Due to their unique qualities, such as thermal and chemical stability and high surface area, both carbon- and metal-based nanomaterials have proven to be effective adsorbents for heavy metal cleaning. Thanks to a unique green technique for nanomaterials synthesis, nanomaterials production may now be done in an ecologically responsible, low-cost, and user-friendly manner [110].

In environmental research, the coupling of nanoparticles and living entities improves measurement precision, bioremediation efficiency, and broader biochemical activity, thanks to recent advances in nanotechnology. Bioremediation using nanoparticles has little risk of genetic leakage into the environment and can enhance the biochemical process with new functions and features. Nano-remediation is a relatively young field that has been used in 44 clean-up sites around the world since 2009 [117].

Because of their high selectivity, nanoparticles can treat a wide range of pollutants without any drawbacks or limits. Furthermore, nanomaterials can be produced using "green" methods. For its non-toxic result and clean and eco-friendly style, green technology is a widely recognized technique for bioremediation. While there are several methods for producing nanoparticles, in comparison to other methods, such as the sol-gel technique and chemical synthesis, biological nanoparticle production is the most suitable and environmentally friendly [118].

Metals (Cu, Cd, Pb, Zn, etc.) and metalloids (particularly As) in massive quantities have been discharged into the environment as a result of industrial activity, causing considerable ecosystem harm [119]. Metals and metalloids also constitute a concern to human health because many of them are poisonous even in low quantities, and some are even carcinogenic (e.g., As) and mutagenic [120]. Inorganic pollutants, unlike organic pollutants, which degrade into harmless little molecules, resist numerous biological processes, making them extremely difficult to remove from soils [121].

Inorganic pollutants, unlike organic pollutants, which degrade into harmless little molecules, resist numerous biological processes, making them extremely difficult to remove from soils [121–123]. Remediation is significantly more complicated by the presence of both anionic and cationic pollutants at the same time [124–132]. Nanoparticles have a larger total surface area, a greater density of reactive sites on their covers, and/or a greater intrinsic reactivity of the surface sites, all of which contribute to their increased reactivity [131–134].

More nanocompounds, have also been examined as potential replacements for nZVI, while the majority of the research has concentrated on water cleanup [135–138]. Regarding this, graphene oxide nanoparticles (nGOx) have been employed to remove metals and other hazardous materials from the aquatic environment [124, 139]. However, the capability of nGOx for soil remediation has still to be proven.

Soil nanoremediation with Fe-nanoparticles (nZVI) decreases the accessibility of As and metals in contaminated lands at the same time, as demonstrated in modern pilot-scale in situ experiments [129] and hybrid soil-cleaning methods [130]. The Fe° core of nZVI is encased in an oxide/hydroxide shell that stiffens as iron oxidation continues. Additionally, nZVI has a larger surface area, which means they have a great-level adsorption capacity and reactivity [131, 132]. Nanoparticles have a larger total surface area, a greater density of reactive sites on their covers, and/or a greater intrinsic reactivity of the surface sites, all of which contribute to their increased reactivity [133]. The following are some of the metal-nZVI interactions: reduction, adsorption, oxidation/re-oxidation, co-precipitation, and precipitation [134].

More nanocompounds, such as Fe-oxides [135, 136], have also been examined as potential replacements for nZVI, while the majority of the research has concentrated on water cleanup [137, 138]. Regarding this, graphene oxide nanoparticles (nGOx) have been employed to remove metals and other hazardous materials from the aquatic environment [124, 139]. However, the capability of nGOx for soil remediation has still to be proven.

11.8 NANOBIOREMEDIATION OF SALT AFFECTED SOILS

Engineered nanomaterials (ENMs) and nanotechnology are frequently linked to their tiny size, great surface area, and adequate responsiveness and adaptability.

Researchers have discovered that incorporating ENMs into the standard in situ methods can facilitate for the exclusion of many pollutants at once, enhancing the collective soil remediation method due to the inherent advantages of these technologies. ENMs are thus great tools for removing persistent pollutants from complicated environmental media like soil.

This soil remediation process influences the inherent properties of polluted soil samples' efficiency and acidity concentrations, as well as the presence of organic matter.

Despite the evident advantages of ENMs for soil remediation, various ecological concerns for health and safety should be investigated.

The presence of ENMs will unavoidably have an impact on the native soil ecosystem, with potential properties including changes in plant seed germination, plant root, and shoot development, soil microorganism growth and metabolism, and even the existence of individual invertebrate animals residing within the soil, such as snails, earthworms, and other insects.

Though the potential detrimental impacts of ENMs must be noted, it should be stressed that the use of ENMs may have beneficial effects on soil ecology under specific conditions.

Carbon and cellulose NMs, as well as metal and metal oxide NMs, have all been shown to operate as nutrient stimulants or to increase seed, root, and leaf nutrient distribution in soil. As a result, these materials are expected to boost crop yields and speed up agricultural activities while decreasing the prevalence of harmful soil contaminants. Nanotechnology can also be utilized in sensors to monitor plant and soil health, as well as water quality for crop health. Biosensors based on nanotechnology are still in their infancy, but they could be utilized to identify extensive fertilizers, herbicides, pesticides, insecticides, diseases, moisture, and soil pH all factors to consider [146].

The unrestrained discharge of nanoparticles into the soil is likely to harm the soil microbiota and, in some situations, harm crucial abiotic parameters such as soil fertility and toxicity. Soil fungi and bacteria have exhibited a range of responses to nanoparticles in previous research. When stress is created from various causes, the enzymatic activity of microorganisms has been stated to be impacted [147].

Zinc oxides (ZnO) and zinc sulfates ($ZnSO_4{\bullet}H_2O$ or $ZnSO_4{\bullet}7H_2O$) are commonly used as zinc fertilizers to address soil zinc deficiency. Zinc breakdown and bioavailability in calcium carbonate soils may be enhanced by using zinc oxide nanoparticles as zinc composts. Zinc oxide nanoparticles have a better antibacterial effect than large zinc particles because their tiny size (less than 100 nm) and high surface-to-volume ratio allow for improved contact with microscopic organisms [148].

ZnO nanoparticles made from Moringa oleifera leaf extract have been utilized as antimicrobials alongside *Staphylococcus aureus*, *Bacillus subtilis*, *Pseudomonas aeruginosa*, *Proteus mirabilis*, *Escherichia coli*, *Candida albicans*, and *Candida tropicalis* bacterial strains. Zinc oxide nanoparticles were tested against *Aspergillus flavus* and *Aspergillus niger*, and a most extreme zone of inhibition was observed. Parthenium-synthesized zinc oxide nanoparticles have proven to be effective antifungal mediators that are also ecologically friendly [148].

The leading receptor for nanoparticles is soil. The performance of nanoparticles in soil and their risk assessment in arable soil ecosystems or other real-world settings are currently a hot topic. Organic nanoparticles, both as primary particles and as accumulate, abound in soil, which is the natural matrix. Nanoparticles introduced into the soil exaggeratedly may have a substantial influence since they are very resilient to degradation and can concentrate in the soil.

Nanoparticles influence several infinitesimal soil parameters. One of the most pressing issues in the field of sustainable soil use is the preservation of soil microbial biomass and diversity. The influence of nanoparticles on soil is determined by their quantity, soil type, and enzymatic activity. The pursuit of dehydrogenase enzymes is reduced at high concentrations of nanoparticles. Another detrimental consequence of nanoparticles is their impact on Soil self-cleaning rates and nutrient balance, which serve as the foundation for maintaining plant nutrition and soil fertility enhancement activities. It is critical to investigate the presence of nanoparticles in soil and their influence. Soil properties such as pH, texture, structure, and organic matter influence the soil microbial population and the ability of contaminants to kill microorganisms [146].

Nanoparticles may influence the movement of contaminants in the soil. As a result, it is vital to assess the toxicity of various forms of NPs in soils. Particle size dissemination and organic matter content alter microbial communities in polluted soils. The effect of nanoparticles can be altered by intentionally influencing soil characteristics, makeup, and texture using diverse substances. The effects of CeO_2 nanoparticles on plants in biochar-amended soil are limited. Biochar is a minor soil alteration intended to increase soil fertility and productivity. The interaction of nano-biochar-modified soil is poorly understood [147].

The soil pH is an important component that has a direct impact on soil productiveness and health. It indicates whether the soil is acidic or alkaline. The accessibility of nutrients for plants is affected by the soil pH. It has been established that the nature of a soil's acidity or basicity is determined by its composition. Plants thrive best in a pH range of 5–7. The buildup of many types of nanoparticles, such as Zn, Ag, Au, Cu, and others, has been reported to affect the pH of the soil. The pH of the soil has been discovered to influence the harmfulness of NPson microbes and nematodes [146].

The relationship between organic matter and soil pH on ZnO nanoparticle toxicity in *Folsomia candida*. The outcome of Zn nanoparticles on the bacterial population was shown to be more harmful than that of dissolved Zn. CuO nanoparticles have recently been discovered to be capable of altering the pH of paddy soil. CuO nanoparticles raise the soil pH, which influences its properties. The soil pH may also affect the insects' consumption of silver nanoparticles produced in soil. Nanomaterials' effects are strongly prejudiced by the changes in form that they go through in nature. AgNM is naturally oxidized, and the release of silver particles may be the cause of the fatal action. The pH of the soil affects both the oxidation of AgNMs and the release of particles [149].

The effect of ZnO nanoparticles, a study on the spread of *Folsomia candida* discovered that soil pH affected how harmful metallic nanoparticles increased as the pH dropped. For all forms of Zn, the effect of Zn risk on *Eisenia fetida* proliferation is altered by pH, indicating the influence of pH on Zn crumbling [149].

Particularly in agriculture, researchers are using nanomaterials to cut back on the use of plant protection agents, reduce nutrient losses during fertilization, and increase yields through improved nutrient management. Disease diagnosis and treatment, in addition to the use of nano-tools and technologies, nanoparticles, and even nano-capsules, are examples of how to improve plant nutrient uptake. The nanoparticles made by biological processes, in particular, can be employed to minimize plant tissue damage.

High salt (NaCl) levels in soil, in particular, impede stomatal conductance, photosynthesis, growth, and transpiration. Osmotic stress is caused by salinity, this lessens leaf water potential and turgor pressure. As a result of ion toxicity, salinity increases the number of ROS in the plant cell, disrupting ion homeostasis. Because it disrupts nutrient uptake, causing the cell's membrane and ultrastructure to disintegrate. Salt toxicity causes an ionic imbalance in addition to oxidative and osmotic stress by causing significant ion buildup (Na+, Cl-) and constraining K^+ and Ca^{2+} uptake. These effects in an ionic in equity in addition to oxidative and osmotic stress.

There's a relationship between salt stress tolerance and plant root morphology, and NPs may be able to change root structure to improve plant stress resistance. Iron oxides have been found to offer a rich iron supply for plants over a wide pH range (i.e., 3–10) as it progressively releases Fe. Under Fe_2O_3 treatment, the enzymatic system, nutritional content, and salinity stress of salt-stressed peppermint were all enhanced, as were the morpho-physiological parameters. Plants handled with NPs produce more secondary metabolites, indicating improved plant health and absorption efficiency, according to GC/MS analyzes. This shows that NPs may be a viable solution for enhancing plant development and physiology in challenging environments. Different NPs, such as ZnO, SiO_2, Fe_2O_3, and TiO_2, can alleviate salt-produced stress. Among the investigated NPs, ZnO has proved to be remarkably effective in giving salt stress tolerance [146].

The usage of ZnO NPs increased the amount of Zn in the and this important micronutrient required for tryptophan biosynthesis. Tryptophan is a well-known IAA precursor; this attribute of Zn may aid ZnO NPs in becoming superior to other NPs and promoting development and physiology more effectively. Elemental analysis shows that salt has a bad impact on the crucial element absorption in active cells, whereas NPs administration increases the amount of Ca, K, and C in the cell significantly. Under salt stress, ZnO NPs were observed to have the greatest ability to boost growth features in linseed [150]. Thus, the use of NPs was observed to be advantageous in both stress-free and salt-stressed circumstances.

Advanced microscopic studies back up the favorable effect of NPs on root cell feasibility. Amongst the several NPs, ZnO showed to be the most effective. The use of NPs to improve root cell viability is consistent with previous findings [150]. Abiotic stress causes cell destruction in Lycopersicon esculentum roots, according to their research; salt stressed *Linum* sp. produced comparable results in the recent study. NPs increased cell viability and root growth, resulting in a healthier root system and improved mineral absorption.

Linum thrives in the existence of NPs, which helps them to cope with a stressful environment. Many academics are working hard to produce new cost-efficient fertilizers to feed the world's rising population; NPs could be one such possibility. Plants supplemented with NPs overcome their deficit and become more productive. Because they are linked to insoluble soil components, nutrients like Zn are difficult to reach in soil.

The most widespread practice of providing nutrients to plants is the nutrient application in the soil. The quality of the soil, salt content, pH, soil texture, and how long the nanoparticles will be able to release the agrochemicals are all important considerations when delivering nanoparticles by soil application. Mineral nutrient adsorption is influenced by the existence of negative soil elements in the soil. Most agricultural soil has a higher cation exchange capacity than anion exchange capacity. The most extensively used method is the use of silica nanoparticles in soil, which involves making a solution of the NPs and applying it directly to the soil. The concentration of silica nanoparticles changes dependent on the water requirements of the plants and the kind of soil. About 25 mg/L, 50 mg/L, and 100 mg/L are the most widely utilized concentrations [152].

11.9 CONCLUSION

Existing technologies in a variety of fields, including pollution control, could be revolutionized by nanotechnology. Nanotechnology has a big role to play in the advancement of innovative goods that can replace conventional production processes with better performance and potentially save money and the environment.

Reduced material utilization is also advantageous. Furthermore, nanotechnology can manage and expand production activities in a more viable manner, ultimately bringing them as near to zero emissions as possible. While great emphasis has been placed on the development of nanomaterials and their potential benefits in water treatment techniques, the reviewed literature shows that worries about their potential human and environmental harm have also grown.

Nanotechnology has the power to solve a diversity of pollution-related issues, including heavy metal contamination, the negative impacts of chemical pollutants, oil pollution, etc. Nanotechnology can deliver environmentally friendly options to environmental management that do not affect the environment. Various plants, fungi, and bacteria have been found as "hyper accumulators," which are capable of accumulating exceptionally high amounts of metals. Plants, fungi, and microbes like these could be efficient in heavy metal pollution bioremediation. Other environmental contaminants can be removed using nanomaterials in various forms.

Nanoparticles (nano-scale particles or NSPs) derived from plants, fungus, and bacteria have been used to eliminate heavy metals from polluted sites.

In extremely polluted situations, plant, fungal, and bacterium nanoparticles can aid in the detoxification and soil, water, and ecosystems bioremediation. Nanotechnology will be modified and adapted in the upcoming future to expand the condition and duration of bioremediation.

The role of NMs in waste and hazardous matter degradation, and how they might minimize the expense of waste and toxic material deprivation, is the topic of this study.

NMs not just help microorganisms decompose waste and hazardous materials more efficiently, but they also help them break down waste and deadly compounds that are damaging to them. This also indicates that heavy metals may be removed from polluted soil using phytoremediation.

As an outcome of the preceding discussion, it can be decided that, like its applications in a variety of other research, it has enormous applicability in bioremediation. Due to their enormous potential, it is projected that their use will skyrocket soon, and they will perform a crucial part in long-term development.

Furthermore, polluted lands provide a significant hazard to the environment, particularly toxic chemicals, However, these degraded soils could provide a tremendous opportunity for many cropping in food production and biorefineries for the bioeconomy. More study is required on:

- The biotoxicity of nanomaterials/nanoparticles;
- The research of local environmental factors and their influence on NP/nanomaterial destiny, transit, and conversion; and
- The antagonistic properties of NPs and microbial activities to cope with nano-remediation.

KEYWORDS

- **biogenic nanoparticles**
- **nanobioremediation**
- **nanoremediation**
- **nanotechnology**
- **remediation**
- **salinity**
- **soil clean up**
- **soil pollution**

REFERENCES

1. Schrick, B., Hydutsky, B. W., Blough, J. L., & Mallouk, T. E., (2004). Delivery vehicles for zerovalent metal nanoparticles in soil and groundwater. *Chemistry of Materials, 16*(11), 2187–2193.
2. Liu, M., Wang, Z., Zong, S., Chen, H., Zhu, D., Wu, L., & Cui, Y., (2014). SERS detection and removal of mercury(II)/silver(I) using oligonucleotide-functionalized core/shell magnetic silica sphere@ Au nanoparticles. *ACS Applied Materials & Interfaces, 6*(10), 7371–7379.
3. Van Dillewijn, P., Caballero, A., Paz, J. A., González-Pérez, M. M., Oliva, J. M., & Ramos, J. L., (2007). Bioremediation of 2, 4, and 6-trinitrotoluene under field conditions. *Environmental Science & Technology, 41*(4), 1378–1383.
4. Kratochvil, D., & Volesky, B., (1998). Advances in the biosorption of heavy metals. *Trends in Biotechnology, 16*(7), 291–300.
5. Kumar, S. R., & Gopinath, P., (2017). Nano-bioremediation applications of nanotechnology for bioremediation. *Handbook of Advanced Industrial and Hazardous Wastes Management*, 27–48.
6. Li, Y. Y., & Li, B., (2011). Study on fungi-bacteria consortium bioremediation of petroleum contaminated mangrove sediments amended with mixed biosurfactants. In: *Advanced Materials Research* (Vol. 183, pp. 1163–1167). Trans Tech Publications Ltd.
7. Cheng, M., Zeng, G., Huang, D., Lai, C., Xu, P., Zhang, C., & Liu, Y., (2016). Hydroxyl radicals based advanced oxidation processes (AOPs) for remediation of soils contaminated with organic compounds: A review. *Chemical Engineering Journal, 284*, 582–598.
8. Thomé, A., Reddy, K. R., Reginatto, C., & Cecchin, I., (2015). Review of nanotechnology for soil and groundwater remediation: Brazilian perspectives. *Water, Air, & Soil Pollution, 226*(4), 1–20.
9. Mateas, D. J., Tick, G. R., & Carroll, K. C., (2017). In situ, stabilization of NAPL contaminant source-zones as a remediation technique to reduce mass discharge and flux to groundwater. *Journal of Contaminant Hydrology, 204*, 40–56.
10. Rajan, C. S., (2011). Nanotechnology in groundwater remediation. *International Journal of Environmental Science and Development, 2*(3), 182.
11. Tsai, T. T., Kao, C. M., Yeh, T. Y., & Lee, M. S., (2008). Chemical oxidation of chlorinated solvents in contaminated groundwater. *Practice Periodical of Hazardous, Toxic, and Radioactive Waste Management, 12*(2), 116–126.
12. Honetschlägerová, L., Martinec, M., & Škarohlíd, R., (2019). Coupling in situ chemical oxidation with bioremediation of chloroethenes: A review. *Reviews in Environmental Science and Bio/Technology, 18*(4), 699–714.
13. Chaney, R. L., Reeves, P. G., Ryan, J. A., Simmons, R. W., Welch, R. M., & Angle, J. S., (2004). An improved understanding of soil Cd risk to humans and low cost methods to phytoextract Cd from contaminated soils to prevent soil Cd risks. *Biometals, 17*(5), 549–553.
14. Liu, J. W., Wei, K. H., Xu, S. W., Cui, J., Ma, J., Xiao, X. L., & He, X. S., (2021). Surfactant-enhanced remediation of oil-contaminated soil and groundwater: A review. *Science of the Total Environment, 756*, 144142.
15. Yang, C., Offiong, N. A., Zhang, C., Liu, F., & Dong, J., (2021). Mechanisms of irreversible density modification using colloidal biliquid aphron for dense nonaqueous

phase liquids in contaminated aquifer remediation. *Journal of Hazardous Materials, 415*, 125667.
16. Gupta, V. K., Agarwal, S., Sadegh, H., Ali, G. A., Bharti, A. K., & Makhlouf, A. S. H., (2017). Facile route synthesis of novel graphene oxide-β-cyclodextrin nanocomposite and its application as adsorbent for removal of toxic bisphenol A from the aqueous phase. *Journal of Molecular Liquids, 237*, 466–472.
17. Sadegh, H., Ali, G. A., Makhlouf, A. S. H., Chong, K. F., Alharbi, N. S., Agarwal, S., & Gupta, V. K., (2018). MWCNTs-Fe_3O_4 nanocomposite for Hg(II) high adsorption efficiency. *Journal of Molecular Liquids, 258*, 345–353.
18. Lee, S. P., Ali, G. A., Algarni, H., & Chong, K. F., (2019). Flake size-dependent adsorption of graphene oxide aerogel. *Journal of Molecular Liquids, 277*, 175–180.
19. Arabi, S. M. S., Lalehloo, R. S., Olyai, M. R. T. B., Ali, G. A., & Sadegh, H., (2019). Removal of Congo red azo dye from aqueous solution by ZnO nanoparticles loaded on multiwall carbon nanotubes. *Physica E: Low-Dimensional Systems and Nanostructures, 106*, 150–155.
20. Karn, B., Kuiken, T., & Otto, M., (2009). Nanotechnology and in situ remediation: A review of the benefits and potential risks. *Environmental Health Perspectives, 117*(12), 1813–1831.
21. Alazaiza, M. Y., Albahnasawi, A., Ali, G. A., Bashir, M. J., Copty, N. K., Amr, S. S. A., & Al Maskari, T., (2021). Recent advances of nanoremediation technologies for soil and groundwater remediation: A review. *Water, 13*(16), 2186.
22. Akharame, M. O., Fatoki, O. S., & Opeolu, B. O., (2019). Regeneration and reuse of polymeric nanocomposites in wastewater remediation: The future of economic water management. *Polymer Bulletin, 76*(2), 647–681.
23. Chaturvedi, S., & Dave, P. N., (2019). Water purification using nanotechnology is an emerging opportunity. *Chemical Methodologies, 3*(1), 115–144.
24. Kumar, A., Joshi, H., & Kumar, A., (2021). Remediation of arsenic by metal/metal oxide based nanocomposites/nanohybrids: Contamination scenario in groundwater, practical challenges, and future perspectives. *Separation & Purification Reviews, 50*(3), 283–314.
25. Blundell, S. P., & Owens, G., (2021). Evaluation of enhancement techniques for the dechlorination of DDT by nanoscale zero-valent iron. *Chemosphere, 264*, 128324.
26. Chen, J., Dong, H., Tian, R., Li, R., & Xie, Q., (2020). Remediation of trichloroethylene-contaminated groundwater by sulfide-modified nanoscale zero-valent iron supported on biochar: Investigation of critical factors. *Water, Air, & Soil Pollution, 231*(8), 1–13.
27. Gil-Díaz, M., Álvarez, M. A., Alonso, J., & Lobo, M. C., (2020). Effectiveness of nanoscale zero-valent iron for the immobilization of Cu and/or Ni in water and soil samples. *Scientific Reports, 10*(1), 1–10.
28. Mpouras, T., Polydera, A., Dermatas, D., Verdone, N., & Vilardi, G., (2021). Multi wall carbon nanotubes application for treatment of Cr (VI)-contaminated groundwater; modeling of batch & column experiments. *Chemosphere, 269*, 128749.
29. Galdames, A., Ruiz-Rubio, L., Orueta, M., Sánchez-Arzalluz, M., & Vilas-Vilela, J. L., (2020). Zero-valent iron nanoparticles for soil and groundwater remediation. *International Journal of Environmental Research and Public Health, 17*(16), 5817.
30. Araújo, R., Castro, A. C. M., & Fiúza, A., (2015). The use of nanoparticles in soil and water remediation processes. *Materials Today: Proceedings, 2*(1), 315–320.

31. Mukhopadhyay, R., Sarkar, B., Khan, E., Alessi, D. S., Biswas, J. K., Manjaiah, K. M., & Ok, Y. S., (2021). Nanomaterials for sustainable remediation of chemical contaminants in water and soil. *Critical Reviews in Environmental Science and Technology*, 1–50.
32. Baker, S., & Satish, S., (2012). Endophytes: Toward a vision in the synthesis of nanoparticle for future therapeutic agents. *Int. J. Bio-Inorg. Hybd. Nanomat, 1*(2), 67–77.
33. Feynman, R., (1959). *There's Plenty of Room at the Bottom. An Invitation to Enter a New Field of Science.* Lecture given at the annual meeting of the American Physical Society, California Institute of Technology.
34. Roco, M. C., & Bainbridge, W. S., (2005). Societal implications of nanoscience and nanotechnology: Maximizing human benefit. *Journal of Nanoparticle Research, 7*(1), 1–13.
35. Wiek, A., Foley, R. W., & Guston, D. H., (2012). Nanotechnology for sustainability: What does nanotechnology offer to address complex sustainability problems? In: *Nanotechnology for Sustainable Development* (pp. 371–390). Springer, Cham.
36. Smith, A., (2006). Opinion: Nanotech–the way forward for clean water? *Filtration & Separation, 43*(8), 32–33.
37. Ghosh, S., & Webster, T. J., (2021). Nanotechnology for water processing. In: *The Future of Effluent Treatment Plants* (pp. 335–360). Elsevier.
38. Das, S., Chakraborty, J., Chatterjee, S., & Kumar, H., (2018). Prospects of biosynthesized nanomaterials for the remediation of organic and inorganic environmental contaminants. *Environmental Science: Nano, 5*(12), 2784–2808.
39. Abdi, O., & Kazemi, M., (2015). A review study of biosorption of heavy metals and comparison between different biosorbents. *J. Mater. Environ. Sci., 6*(5), 1386–1399.
40. Nriagu, J. O., & Pacyna, J. M., (1988). Quantitative assessment of worldwide contamination of air, water, and soils by trace metals. *Nature, 333*(6169), 134–139.
41. Dixit, R., Malaviya, D., Pandiyan, K., Singh, U. B., Sahu, A., Shukla, R., & Paul, D., (2015). Bioremediation of heavy metals from soil and aquatic environment: An overview of principles and criteria of fundamental processes. *Sustainability, 7*(2), 2189–2212.
42. Fulekar, M. H., Singh, A., & Bhaduri, A. M., (2009). Genetic engineering strategies for enhancing phytoremediation of heavy metals. *African Journal of Biotechnology, 8*(4).
43. Huang, Q., Liu, Q., Lin, L., Li, F. J., Han, Y., & Song, Z. G., (2018). Reduction of arsenic toxicity in two rice cultivar seedlings by different nanoparticles. *Ecotoxicology and Environmental Safety, 159*, 261–271.
44. Shibayama, A., Takasaki, Y., William, T., Yamatodani, A., Higuchi, Y., Sunagawa, S., & Ono, E., (2010). Treatment of smelting residue for arsenic removal and recovery of copper using pyro–hydrometallurgical process. *Journal of Hazardous Materials, 181*(1–3), 1016–1023.
45. Gopalakrishnan, A., Krishnan, R., Thangavel, S., Venugopal, G., & Kim, S. J., (2015). Removal of heavy metal ions from pharma-effluents using graphene-oxide nanosorbents and study of their adsorption kinetics. *Journal of Industrial and Engineering Chemistry, 30*, 14–19.
46. Wu, Y., Ma, X., Feng, M., & Liu, M., (2008). The behavior of chromium and arsenic on activated carbon. *Journal of Hazardous Materials, 159*(2–3), 380–384.
47. Beiyuan, J., Awad, Y. M., Beckers, F., Tsang, D. C., Ok, Y. S., & Rinklebe, J., (2017). Mobility and phytoavailability of As and Pb in contaminated soil using pine sawdust biochar under systematic change of redox conditions. *Chemosphere, 178*, 110–118.

48. Beesley, L., & Marmiroli, M., (2011). The immobilization and retention of soluble arsenic, cadmium, and zinc by biochar. *Environmental Pollution, 159*(2), 474–480.
49. Bolan, N., Kunhikrishnan, A., Thangarajan, R., Kumpiene, J., Park, J., Makino, T., & Scheckel, K., (2014). Remediation of heavy metal (loid) contaminated soils–to mobilize or to immobilize? *Journal of Hazardous Materials, 266*, 141–166.
50. Medina-Pérez, G., Fernández-Luqueño, F., Vazquez-Nuñez, E., López-Valdez, F., Prieto-Mendez, J., Madariaga-Navarrete, A., & Miranda-Arámbula, M., (2019). Remediating polluted soils using nanotechnologies: Environmental benefits and risks. *Polish Journal of Environmental Studies, 28*(3).
51. Fernández-Luqueño, F., López-Valdez, F., Pérez-Morales, C., García-Mayagoitia, S., Sarabia-Castillo, C. R., & Pérez-Ríos, S. R., (2017). Enhancing decontamination of PAHs-polluted soils: Role of organic and mineral amendments. In: *Enhancing Cleanup of Environmental Pollutants* (pp. 339–368). Springer, Cham.
52. He, S., Feng, Y., Ni, J., Sun, Y., Xue, L., Feng, Y., & Yang, L., (2016). Different responses of soil microbial metabolic activity to silver and iron oxide nanoparticles. *Chemosphere, 147*, 195–202.
53. Xu, C., Peng, C., Sun, L., Zhang, S., Huang, H., Chen, Y., & Shi, J., (2015). Distinctive effects of TiO_2 and CuO nanoparticles on soil microbes and their community structures in flooded paddy soil. *Soil Biology and Biochemistry, 86*, 24–33.
54. Fernández-Luqueño, F., López-Valdez, F., Sarabia-Castillo, C. R., García-Mayagoitia, S., & Pérez-Ríos, S. R., (2017). Bioremediation of polycyclic aromatic hydrocarbons-polluted soils at laboratory and field scale: A review of the literature on plants and microorganisms. *Enhancing Cleanup of Environmental Pollutants*, 43–64.
55. Fourest, E., & Volesky, B., (1997). Alginate properties and heavy metal biosorption by marine algae. *Applied Biochemistry and Biotechnology, 67*(3), 215–226.
56. Ayangbenro, A. S., & Babalola, O. O., (2017). A new strategy for heavy metal polluted environments: A review of microbial biosorbents. *International Journal of Environmental Research and Public Health, 14*(1), 94.
57. Bawa, R., Bawa, S. R., Maebius, S. B., Flynn, T., & Wei, C., (2005). Protecting new ideas and inventions in nanomedicine with patents. *Nanomedicine: Nanotechnology, Biology and Medicine, 1*(2), 150–158.
58. Keller, A. A., McFerran, S., Lazareva, A., & Suh, S., (2013). Global life cycle releases of engineered nanomaterials. *Journal of Nanoparticle Research, 15*(6), 1–17.
59. Parisi, C., Vigani, M., & Rodríguez-Cerezo, E., (2015). Agricultural nanotechnologies: What are the current possibilities? *Nano Today, 10*(2), 124–127.
60. Takeuchi, M. T., Kojima, M., & Luetzow, M., (2014). State of the art on the initiatives and activities relevant to risk assessment and risk management of nanotechnologies in the food and agriculture sectors. *Food Research International, 64*, 976–981.
61. Aragay, G., Pino, F., & Merkoçi, A., (2012). Nanomaterials for sensing and destroying pesticides. *Chemical Reviews, 112*(10), 5317–5338.
62. Yao, J., Yang, M., & Duan, Y., (2014). Chemistry, biology, and medicine of fluorescent nanomaterials and related systems: New insights into biosensing, bioimaging, genomics, diagnostics, and therapy. *Chemical Reviews, 114*(12), 6130–6178.
63. Chen, H., & Yada, R., (2011). Nanotechnologies in agriculture: New tools for sustainable development. *Trends in Food Science & Technology, 22*(11), 585–594.

64. Corradini, E., De Moura, M. R., & Mattoso, L. H. C., (2010). A preliminary study of the incorparation of NPK fertilizer into chitosan nanoparticles. *Express Polymer Letters, 4*(8).
65. DeRosa, M. C., Monreal, C., Schnitzer, M., Walsh, R., & Sultan, Y., (2010). Nanotechnology in fertilizers. *Nature Nanotechnology, 5*(2), 91–91.
66. Grillo, R., De Melo, N. F., De Lima, R., Lourenço, R. W., Rosa, A. H., & Fraceto, L. F., (2010). Characterization of atrazine-loaded biodegradable poly (hydroxybutyrate-co-hydroxyvalerate) microspheres. *Journal of Polymers and the Environment, 18*(1), 26–32.
67. Goldwasser, Y., Eizenberg, H., Golan, S., & Kleifeld, Y., (2003). Control of *Orobanche crenata* and *Orobanche aegyptiaca* in parsley. *Crop Protection, 22*(2), 295–305.
68. Mishra, S., & Singh, H. B., (2015). Biosynthesized silver nanoparticles as a nanoweapon against phytopathogens: Exploring their scope and potential in agriculture. *Applied Microbiology and Biotechnology, 99*(3), 1097–1107.
69. Nuruzzaman, M. D., Rahman, M. M., Liu, Y., & Naidu, R., (2016). Nanoencapsulation, nano-guard for pesticides: A new window for safe application. *Journal of Agricultural and Food Chemistry, 64*(7), 1447–1483.
70. Peteu, S. F., Oancea, F., Sicuia, O. A., Constantinescu, F., & Dinu, S., (2010). Responsive polymers for crop protection. *Polymers, 2*(3), 229–251.
71. Mukhopadhyay, S. S., (2014). Nanotechnology in agriculture: Prospects and constraints. *Nanotechnology, Science and Applications, 7*, 63.
72. Hamza, R. A., Iorhemen, O. T., & Tay, J. H., (2016). Occurrence, impacts, and removal of emerging substances of concern from wastewater. *Environmental Technology & Innovation, 5*, 161–175.
73. Qu, X., Alvarez, P. J., & Li, Q., (2013). Applications of nanotechnology in water and wastewater treatment. *Water Research, 47*(12), 3931–3946.
74. Ingle, A. P., Seabra, A. B., Duran, N., & Rai, M., (2014). Nanoremediation: A new and emerging technology for the removal of toxic contaminants from the environment. In: *Microbial Biodegradation and Bioremediation* (pp. 233–250). Elsevier.
75. Araújo, R., Castro, A. C. M., & Fiúza, A., (2015). The use of nanoparticles in soil and water remediation processes. *Materials Today: Proceedings, 2*(1), 315–320.
76. Jain, R., Jordan, N., Schild, D., Van, H. E. D., Weiss, S., Franzen, C., & Lens, P. N., (2015). Adsorption of zinc by biogenic elemental selenium nanoparticles. *Chemical Engineering Journal, 260*, 855–863.
77. Fajardo, C., Gil-Díaz, M., Costa, G., Alonso, J., Guerrero, A. M., Nande, M., & Martín, M., (2015). The residual impact of aged nZVI on heavy metal-polluted soils. *Science of the Total Environment, 535*, 79–84.
78. Jain, R., Dominic, D., Jordan, N., Rene, E. R., Weiss, S., Van, H. E. D., & Lens, P. N., (2016). Higher Cd adsorption on biogenic elemental selenium nanoparticles. *Environmental Chemistry Letters, 14*(3), 381–386.
79. Gillies, G., Mackenzie, K., Kopinke, F. D., & Georgi, A., (2016). Fluorescence labelling as a tool for zeolite particle tracking in nanoremediation approaches. *Science of the Total Environment, 550*, 820–826.
80. Gil-Díaz, M., Diez-Pascual, S., González, A., Alonso, J., Rodríguez-Valdés, E., Gallego, J. R., & Lobo, M. C., (2016). A nanoremediation strategy for the recovery of an As-polluted soil. *Chemosphere, 149*, 137–145.

81. Ghormade, V., Deshpande, M. V., & Paknikar, K. M., (2011). Perspectives for nano-biotechnology enabled protection and nutrition of plants. *Biotechnology Advances, 29*(6), 792–803.
82. Arruda, S. C. C., Silva, A. L. D., Galazzi, R. M., Azevedo, R. A., & Arruda, M. A. Z., (2015). Nanoparticles applied to plant science: A review. *Talanta, 131*, 693–705.
83. Gil-Díaz, M., González, A., Alonso, J., & Lobo, M. C., (2016). Evaluation of the stability of a nanoremediation strategy using barley plants. *Journal of Environmental Management, 165*, 150–158.
84. Hamza, R. A., Iorhemen, O. T., & Tay, J. H., (2016). Occurrence, impacts, and removal of emerging substances of concern from wastewater. *Environmental Technology & Innovation, 5*, 161–175.
85. Peeters, K., Lespes, G., Zuliani, T., Ščančar, J., & Milačič, R., (2016). The fate of iron nanoparticles in environmental waters treated with nanoscale zero-valent iron, FeONPs, and Fe_3O_4NPs. *Water Research, 94*, 315–327.
86. El-Temsah, Y. S., & Joner, E. J., (2013). Effects of nano-sized zero-valent iron (nZVI) on DDT degradation in soil and its toxicity to collembola and ostracods. *Chemosphere, 92*(1), 131–137.
87. El-Temsah, Y. S., Sevcu, A., Bobcikova, K., Cernik, M., & Joner, E. J., (2016). DDT degradation efficiency and ecotoxicological effects of two types of nano-sized zero-valent iron (nZVI) in water and soil. *Chemosphere, 144*, 2221–2228.
88. Gomes, H. I., Fan, G., Mateus, E. P., Dias-Ferreira, C., & Ribeiro, A. B., (2014). Assessment of combined electro–nanoremediation of molinate contaminated soil. *Science of the Total Environment, 493*, 178–184.
89. Ibrahim, R. K., Hayyan, M., AlSaadi, M. A., Hayyan, A., & Ibrahim, S., (2016). Environmental application of nanotechnology: Air, soil, and water. *Environmental Science and Pollution Research, 23*(14), 13754–13788.
90. Germaine, K. J., Keogh, E., Ryan, D., & Dowling, D. N., (2009). Bacterial endophyte-mediated naphthalene phytoprotection and phytoremediation. *FEMS Microbiology Letters, 296*(2), 226–234.
91. Hur, M., Kim, Y., Song, H. R., Kim, J. M., Choi, Y. I., & Yi, H., (2011). Effect of genetically modified poplars on soil microbial communities during the phytoremediation of waste mine tailings. *Applied and Environmental Microbiology, 77*(21), 7611–7619.
92. Houben, D., Evrard, L., & Sonnet, P., (2013). Beneficial effects of biochar application to contaminated soils on the bioavailability of Cd, Pb, and Zn and the biomass production of rapeseed (*Brassica napus* L.*). Biomass and Bioenergy, 57*, 196–204.
93. Meister, R., Rajani, M. S., Ruzicka, D., & Schachtman, D. P., (2014). Challenges of modifying root traits in crops for agriculture. *Trends in Plant Science, 19*(12), 779–788.
94. Abhilash, P. C., & Dubey, R. K., (2015). Root System Engineering: Prospects and Promises. *Trends in Plant Science*, *20*(7), 408–409.
95. Abhilash, P. C., Tripathi, V., Edrisi, S. A., Dubey, R. K., Bakshi, M., Dubey, P. K., & Ebbs, S. D., (2016). Sustainability of crop production from polluted lands. *Energy, Ecology and Environment, 1*(1), 54–65.
96. Ye-Tao, T. A. N. G., Teng-Hao-Bo, D. E. N. G., Qi-Hang, W. U., Shi-Zhong, W. A. N. G., Rong-Liang, Q. I. U., Ze-Bin, W. E. I., & Morel, J. L., (2012). Designing cropping systems for metal-contaminated sites: A review. *Pedosphere, 22*(4), 470–488.

97. Yang, X., Shen, Z., Zhang, B., Yang, J., Hong, W. X., Zhuang, Z., & Liu, J., (2013). Silica nanoparticles capture atmospheric lead: Implications in the treatment of environmental heavy metal pollution. *Chemosphere, 90*(2), 653–656.
98. Ghenaatian, H. R., Baei, M. T., & Hashemian, S., (2013). $Zn_{12}O_{12}$ nano-cage as a promising adsorbent for CS_2 capture. *Superlattices and Microstructures, 58*, 198–204.
99. Yildiz, O., & Bradford, P. D., (2013). Aligned carbon nanotube sheet high efficiency particulate air filters. *Carbon, 64*, 295–304.
100. Xie, Y., Fang, Z., Cheng, W., Tsang, P. E., & Zhao, D., (2014). Remediation of polybrominated diphenyl ethers in soil using Ni/Fe bimetallic nanoparticles: Influencing factors, kinetics, and mechanism. *Science of the Total Environment, 485*, 363–370.
101. El-Temsah, Y. S., & Joner, E. J., (2013). Effects of nano-sized zero-valent iron (nZVI) on DDT degradation in soil and its toxicity to collembola and ostracods. *Chemosphere, 92*(1), 131–137.
102. Sheng, G., Alsaedi, A., Shammakh, W., Monaquel, S., Sheng, J., Wang, X., & Huang, Y., (2016). Enhanced sequestration of selenite in water by nanoscale zero valent iron immobilization on carbon nanotubes by a combined batch, XPS, and XAFS investigation. *Carbon, 99*, 123–130.
103. He, D., Ma, J., Collins, R. N., & Waite, T. D., (2016). Effect of structural transformation of nanoparticulate zero-valent iron on generation of reactive oxygen species. *Environmental Science & Technology, 50*(7), 3820–3828.
104. Duan, X., Sun, H., Kang, J., Wang, Y., Indrawirawan, S., & Wang, S., (2015). Insights into heterogeneous catalysis of persulfate activation on dimensional-structured nanocarbons. *ACS Catalysis, 5*(8), 4629–4636.
105. You, J., Wang, L., Zhao, Y., & Bao, W., (2021). A review of amino-functionalized magnetic nanoparticles for water treatment: Features and prospects. *Journal of Cleaner Production, 281*, 124668.
106. Khan, S., & Kumari, S., (2021). Interaction of heavy metal ions with nanomaterials: A remediation process. *Research Anthology on Synthesis, Characterization, and Applications of Nanomaterials*, 1170–1183.
107. Dai, J., Xu, M., Chen, J., Yang, X., & Ke, Z., (2007). PCDD/F, PAH, and heavy metals in the sewage sludge from six wastewater treatment plants in Beijing, China. *Chemosphere, 66*(2), 353–361.
108. Zare, K., Gupta, V. K., Moradi, O., Makhlouf, A. S. H., Sillanpää, M., Nadagouda, M. N., & Kazemi, M., (2015). A comparative study on the basis of adsorption capacity between CNTs and activated carbon as adsorbents for removal of noxious synthetic dyes: A review. *Journal of Nanostructure in Chemistry, 5*(2), 227–236.
109. Mandal, S., Tripathy, S., Padhi, T., Sahu, M. K., & Patel, R. K., (2013). Removal efficiency of fluoride by novel Mg-Cr-Cl layered double hydroxide by batch process from water. *Journal of Environmental Sciences, 25*(5), 993–1000.
110. Chaturvedi, S., & Dave, P. N., (2019). Water purification using nanotechnology is an emerging opportunity. *Chemical Methodologies, 3*(1), 115–144.
111. Benamar, A., Mahjoubi, F., Ali, G. A., Kzaiber, F., & Oussama, A., (2020). A chemometric method for contamination sources identification along the Oum Er Rbia river (Morocco). *Bulgarian Chemical Communications, 52*, 159–171.
112. Mondal, A., Dubey, B. K., Arora, M., & Mumford, K., (2021). Porous media transport of iron nanoparticles for site remediation application: A review of lab scale column study,

transport modelling and field-scale application. *Journal of Hazardous Materials, 403*, 123443.
113. Karn, B., Kuiken, T., & Otto, M., (2009). Nanotechnology and in situ remediation: A review of the benefits and potential risks. *Environmental Health Perspectives, 117*(12), 1813–1831.
114. Alazaiza, M. Y., Albahnasawi, A., Ali, G. A., Bashir, M. J., Copty, N. K., Amr, S. S. A., & Al Maskari, T., (2021). Recent advances of nanoremediation technologies for soil and groundwater remediation: A review. *Water, 13*(16), 2186.
115. Shubair, T., Eljamal, O., Khalil, A. M., & Matsunaga, N., (2018). Multilayer system of nanoscale zero valent iron and nano-Fe/Cu particles for nitrate removal in porous media. *Separation and Purification Technology, 193*, 242–254.
116. Xue, W., Huang, D., Zeng, G., Wan, J., Zhang, C., Xu, R., & Deng, R., (2018). Nanoscale zero-valent iron coated with rhamnolipid as an effective stabilizer for immobilization of Cd and Pb in river sediments. *Journal of Hazardous Materials, 341*, 381–389.
117. Gholami-Shabani, M., Gholami-Shabani, Z., Shams-Ghahfarokhi, M., & Razzaghi-Abyaneh, M., (2018). Application of nanotechnology in mycoremediation: Current status and future prospects. In: *Fungal Nanobionics: Principles and Applications* (pp. 89–116). Springer, Singapore.
118. Mishra, A., Kumari, M., Pandey, S., Chaudhry, V., Gupta, K. C., & Nautiyal, C. S., (2014). Biocatalytic and antimicrobial activities of gold nanoparticles synthesized by *Trichoderma* sp. *Bioresource Technology, 166*, 235–242.
119. Nriagu, J. O., & Pacyna, J. M., (1988). Quantitative assessment of worldwide contamination of air, water, and soils by trace metals. *Nature, 333*(6169), 134–139.
120. Dixit, R., Malaviya, D., Pandiyan, K., Singh, U. B., Sahu, A., Shukla, R., & Paul, D., (2015). Bioremediation of heavy metals from soil and aquatic environment: An overview of principles and criteria of fundamental processes. *Sustainability, 7*(2), 2189–2212.
121. Fulekar, M. H., Singh, A., & Bhaduri, A. M., (2009). Genetic engineering strategies for enhancing phytoremediation of heavy metals. *African Journal of Biotechnology, 8*(4).
122. Huang, Q., Liu, Q., Lin, L., Li, F. J., Han, Y., & Song, Z. G., (2018). Reduction of arsenic toxicity in two rice cultivar seedlings by different nanoparticles. *Ecotoxicology and Environmental Safety, 159*, 261–271.
123. Shibayama, A., Takasaki, Y., William, T., Yamatodani, A., Higuchi, Y., Sunagawa, S., & Ono, E., (2010). Treatment of smelting residue for arsenic removal and recovery of copper using pyro–hydrometallurgical process. *Journal of Hazardous Materials, 181*(1–3), 1016–1023.
124. Gopalakrishnan, A., Krishnan, R., Thangavel, S., Venugopal, G., & Kim, S. J., (2015). Removal of heavy metal ions from pharma-effluents using graphene-oxide nanosorbents and study of their adsorption kinetics. *Journal of Industrial and Engineering Chemistry, 30*, 14–19.
125. Wu, Y., Ma, X., Feng, M., & Liu, M., (2008). The behavior of chromium and arsenic on activated carbon. *Journal of Hazardous Materials, 159*(2–3), 380–384.
126. Beiyuan, J., Awad, Y. M., Beckers, F., Tsang, D. C., Ok, Y. S., & Rinklebe, J., (2017). Mobility and phytoavailability of As and Pb in contaminated soil using pine sawdust biochar under systematic change of redox conditions. *Chemosphere, 178*, 110–118.
127. Beesley, L., & Marmiroli, M., (2011). The immobilization and retention of soluble arsenic, cadmium, and zinc by biochar. *Environmental Pollution, 159*(2), 474–480.

128. Bolan, N., Kunhikrishnan, A., Thangarajan, R., Kumpiene, J., Park, J., Makino, T., & Scheckel, K., (2014). Remediation of heavy metal (loid) contaminated soils–to mobilize or to immobilize? *Journal of Hazardous Materials, 266*, 141–166.
129. Gil-Díaz, M., Rodríguez-Valdés, E., Alonso, J., Baragaño, D., Gallego, J. R., & Lobo, M. C., (2019). Nanoremediation and long-term monitoring of brownfield soil highly polluted with As and Hg. *Science of the Total Environment, 675*, 165–175.
130. Boente, C., Sierra, C., Martínez-Blanco, D., Menéndez-Aguado, J. M., & Gallego, J. R., (2018). Nanoscale zero-valent iron-assisted soil washing for the removal of potentially toxic elements. *Journal of Hazardous Materials, 350*, 55–65.
131. Emadi, M., Savasari, M., Bahmanyar, M. A., & Biparva, P., (2019). Application of stabilized zero valent iron nanoparticles for immobilization of lead in three contrasting spiked soils. *Research on Chemical Intermediates, 45*(9), 4261–4274.
132. Tafazoli, M., Hojjati, S. M., Biparva, P., Kooch, Y., & Lamersdorf, N., (2017). Reduction of soil heavy metal bioavailability by nanoparticles and cellulosic wastes improved the biomass of tree seedlings. *Journal of Plant Nutrition and Soil Science, 180*(6), 683–693.
133. Mar Gil-Díaz, M., Pérez-Sanz, A., Angeles, V. M., & Carmen, L. M., (2014). Immobilization of Pb and Zn in soils using stabilized zero-valent iron nanoparticles: Effects on soil properties. *Clean–Soil, Air, Water, 42*(12), 1776–1784.
134. Xue, W., Huang, D., Zeng, G., Wan, J., Cheng, M., Zhang, C., & Li, J., (2018). Performance and toxicity assessment of nanoscale zero valent iron particles in the remediation of contaminated soil: A review. *Chemosphere, 210*, 1145–1156.
135. Shipley, H. J., Engates, K. E., & Guettner, A. M., (2011). Study of iron oxide nanoparticles in soil for remediation of arsenic. *Journal of Nanoparticle Research, 13*(6), 2387–2397.
136. Chen, Y. H., & Li, F. A., (2010). Kinetic study on removal of copper(II) using goethite and hematite nano-photocatalysts. *Journal of Colloid and Interface Science, 347*(2), 277–281.
137. Giraldo, L., Erto, A., & Moreno-Piraján, J. C., (2013). Magnetite nanoparticles for removal of heavy metals from aqueous solutions: Synthesis and characterization. *Adsorption, 19*(2), 465–474.
138. Rahimi, S., Moattari, R. M., Rajabi, L., Derakhshan, A. A., & Keyhani, M., (2015). Iron oxide/hydroxide (α, γ-FeOOH) nanoparticles as high potential adsorbents for lead removal from polluted aquatic media. *Journal of Industrial and Engineering Chemistry, 23*, 33–43.
139. Srivastava, S., (2013). Sorption of divalent metal ions from aqueous solution by oxidized carbon nanotubes and nanocages: A review. *Advanced Materials Letters, 4*(1), 2–8.
140. Misra, M., Chattopadhyay, S., Sachan, A., & Sachan, S. G., (2022). Microbially synthesized nanoparticles and their applications in environmental clean-up. *Environmental Technology Reviews, 11*(1), 18–32.
141. Mohanty, C., Satpathy, S. S., & Mohanty, S., (2022). An eco-friendly approach for the eradication of heavy metal contaminants by nano-bioremediation. In: *Research Anthology on Emerging Techniques in Environmental Remediation* (pp. 543–559). IGI Global.
142. Choudhary, R., Hiti-Bandaralage, J. C. A., Ahlawat, J., Gaur, N., & Diwan, B., (2022). Nanobioremediation: An introduction. In: *Nano-Bioremediation: Fundamentals and Applications* (pp. 263–279). Elsevier.
143. Banerjee, M., Mukherjee, S., & Mukherjee, S., (2022). Facets of nanoparticle-microbe interactions and their roles in nanobioremediation of environmental pollutants:

Biochemical, molecular, and technological perspectives. In: *Nano-Bioremediation: Fundamentals and Applications* (pp. 111–145). Elsevier.
144. De Oca-Vásquez, G. M., Menezes, D. B., Vega-Baudrit, J. R., Ferreira, L. F. R., Bharagava, R. N., Iqbal, H. M., & Bilal, M., (2022). Fate and transport of nanoparticles used in the remediation of polluted soil. In: *Nano-Bioremediation: Fundamentals and Applications* (pp. 541–562). Elsevier.
145. Chatterjee, S., Kumari, S., Rath, S., & Das, S., (2022). Prospects and scope of microbial bioremediation for the restoration of the contaminated sites. In: *Microbial Biodegradation and Bioremediation* (pp. 3–31). Elsevier.
146. Shang, Y., Hasan, M., Ahammed, G. J., Li, M., Yin, H., & Zhou, J., (2019). Applications of nanotechnology in plant growth and crop protection: A review. *Molecules, 24*(14), 2558.
147. Perveen, S., Sara, Z., & Naeem, I., (2020). Applications of bionanocomposites in agriculture. In Mahmod, Z., (ed.), *Bionanocomposites: Green Synthesis and Applications, Micro and Nano Technologies* (pp. 485–504). Elsevier. https://doi.org/10.1016/B978-0-12-816751-9.00018-0.
148. Sirelkhatim, A., Mahmud, S., Seeni, A., Kaus, N. H. M., Ann, L. C., Bakhori, S. K. M., & Mohamad, D., (2015). Review on zinc oxide nanoparticles: Antibacterial activity and toxicity mechanism. *Nano-Micro Letters, 7*(3), 219–242.
149. Shi, J., Ye, J., Fang, H., Zhang, S., & Xu, C., (2018). Effects of copper oxide nanoparticles on paddy soil properties and components. *Nanomaterials, 8*(10), 839.
150. Faizan, M., Bhat, J. A., Chen, C., Alyemeni, M. N., Wijaya, L., Ahmad, P., & Yu, F., (2021). Zinc oxide nanoparticles (ZnO-NPs) induce salt tolerance by improving the antioxidant system and photosynthetic machinery in tomatoes. *Plant Physiology and Biochemistry, 161*, 122–130.
151. Rajput, V. D., Minkina, T., Feizi, M., Kumari, A., Khan, M., Mandzhieva, S., & Choudhary, R., (2021). Effects of silicon and silicon-based nanoparticles on rhizosphere microbiome, plant stress and growth. *Biology, 10*(8), 791.
152. Rajput, V. D., Singh, A., Minkina, T., Rawat, S., Mandzhieva, S., Sushkova, S., & Upadhyay, S. K., (2021). Nano-enabled products: Challenges and opportunities for sustainable agriculture. *Plants, 10*(12), 2727.

CHAPTER 12

Nanotechnological Approaches for Restoring Metalloid Contaminated Soil

SHAGUFTA PERVEEN, ABIDA PARVEEN, IQBAL HUSSAIN, RIZWAN RASHEED, SAQIB MAHMOOD, MUHAMMAD ARSLAN ASHRAF, and AMARA HASSAN

Department of Botany, Government College University, Faisalabad, Pakistan

ABSTRACT

The global degradation of water, land, and atmosphere caused by liberating hazardous substances from ongoing human activity is becoming a severe issue. This presents a number of ecological and health-related problems that make it more difficult for standard treatment solutions to be applied. The threats to human health are linked to the entry of heavy metals from contaminated soil into the food chain. Nanotechnology provides a beneficial impact on restoring metal-contaminated soil. Several biological, physical, and chemical techniques have been developed to cope with metal-contaminated soil. Physical remediation involves surface capping, vapor extraction, thermal treatment (electrical and steam-based heating), and electro-kinetic process, while chemical remediation works as solidification and stabilization, soil washing and flushing, nanotechnology, vitrification, and chelation. In addition, biological remediation involves bioventing or bio-ventilation, bio-augmentation, bio-stimulation, vermiremediation, and phytoremediation. Above-discussed techniques restore metal-contaminated soil. It also brings both traditional and advanced techniques in order to compare, understand, and apply these strategies effectively. This study sheds light on the advancement in nanotechnology and its novel roles in monitoring and

Nano-Bioremediation for Water and Soil Treatment: An Eco-Friendly Approach.
Vishnu D. Rajput, Arpna Kumari, and Tatiana M. Minkina (Eds.)

treating toxicants with low cost, low energy, and high efficiency. The key aspect of the present study is to briefly highlight the uses, efficiency, and advantages of nanotechnology. In crux, nanotechnology is better than others due to its high competency rate, eco-friendly nature, potentially reduced waste material, and economic impact as well.

12.1 INTRODUCTION

From a biological point of view, heavy metals are a class of metals and metalloids that could damage plants and harm human health even in minute quantities. While chemically speaking, heavy metals refer to have an atomic mass >20 and specific gravity >5, for example, cadmium, mercury, copper, arsenic, lead, chromium, nickel, and zinc [1]. Heavy metal contamination is increasing day by day due to rapid industrialization, destruction of the natural ecosystems, and overpopulation that, ultimately poses toxic effects on food security and the environment [2, 3]. Heavy metal contamination is permanent and persistent; that not only enters the food chain and harms human health but also destroys the atmosphere and water bodies [4]. The issue of soil contamination with heavy metal stress has caught the public attention and become an environmental issue worldwide [5]. Soil pollution due to heavy metal stress covered 500 mha of land [6]. According to an estimate, the economic impact of heavy metal pollution in the soil is US $10 billion per year [7].

Nanotechnology is capable of making an impact on environmental issues, like suburbanization, energy and resource constrictions, viable use of resources, and accretion of pesticides and fertilizers [8, 9]. Nanotechnology enhances food quality and production globally in an environment friendly way because it provides advanced solutions to remediate water and soils [10–13]. The materials that are of nano-sized dimensions are termed nanomaterials (NMs) [14]. Nanotechnology has become a hot issue for research in the field of environmental xenobiotics mitigation [15]. It manipulates the atoms and molecules to make up the materials in the nanometer range [16–18]. By using nanoparticles, nanotechnology revolutionizes environmental remediation and has become the most promising approach in the field of remediation [19]. The size of the nano-sized metal-based particles (e.g., zero-valent metal, metal oxides, metal-containing, clay minerals, graphene, etc.) ranges between 1 and 100 nm [20]. Nanotechnology successfully provides ideas for the identification and generation of nano-adsorbents for environmental remediation [21]. NMs are excessively used as catalysts and adsorbents in remediation due to their unique features, huge specific areas, slight modifications in temperature, defensible pore size, a shorter space of interparticle diffusion, etc. [6, 22–24]. Water and industrial waste contain metallic contaminants that could be treated by zero-valent nanoparticles. Nanoparticles are commonly known for the treatment of halogenated hydrocarbons, radionuclides, and organic compounds. The remediation efficiency of hexavalent chromium and Pb(II) is 30 times less than nanoscale zero-valent iron. Fe-oxide-based nanomaterial is also capable of removing arsenic (As-V and As-III). It is beneficial as by applying a magnetic field, iron oxide adsorbent could be easily isolated [25]. For the removal of some other metals, such as lead, chromium, cadmium, etc., nano-alumina particles are used, and they are considered beneficial due to their low cost and high stability [26]. Due to their distinctive features, nanoparticles have the ability to remove metal and other pollutants from wastewater [27–29]. Nanoparticles have multiple active sites, which they can easily remove contaminants [29].

Nanotechnology is superior to all other metal remediation methods as it has excellent features for multi-functional procedures that could improve the check and balance of pollution, remediation efficiency, and solve all the discussed problems. It improves the remediation system as it inhibits the formation of secondary by-products, decomposes toxicants via zero waste operations, and prevents the soil from further contamination by altering the liable phase of pollutants to the non-labile phase. To monitor the emission of toxicants is very challenging for the management of environmental remediation. Several techniques, including plasmon resonance (SPR) [30–32], HPLC [33], and GC-MS [34], are extensively used to detect and monitor pollution. Due to high cost and time consumption, they are considered unsuitable for environmental detection [35]. Nano-technological advancement influences the monitoring and sensing of the environment by producing a variety of nanoparticles to remediate environmental contaminants [36–38]. Finally, a versatile and vibrant system is introduced by nanotechnology. It includes the latest technology that can sense and monitor many toxicants in a variety of environmental media [19].

12.2 SOURCES OF HEAVY METALS

Due to some natural processes and anthropogenic activities, toxic heavy metals become a part of the agro-ecosystem [39]. Naturally occurring heavy metals are inherited by soil from parent material, as the increased level of these metals is present in parent material, which has a negative impact on plants and organisms [2]. Heavy metals originate from parent material, as Earth's total crust is comprised of ingenious rocks and sedimentary rocks in 95 and 5% ratios, respectively [3]. Basaltic rocks have high concentrations of copper, cadmium, nickel, and cobalt, whereas shales are rich in Pb, Cu, Zn, Mn, and Cd. Some of the procedures by which heavy metals enter the soil are meteoric, biogenic, terrestrial, and volcanic eruption [40]. Due to anthropogenic processes, heavy metals get into the soil and disturb nature's gradually occurring geochemical cycle [39]. Some anthropogenic processes, such as mining and smelting [41], fossil fuel combustion [40], dumping municipal wastes [42], pesticides [43], sewage irrigation [44], and fertilizers [45] are major sources of increasing heavy metal concentration in the agricultural soil environment.

Heavy metals are very toxic, and even their low concentration is hazardous for life because they are non-biodegradable [46, 47]. There are three groups of metals; the first group (mercury, cadmium, and lead) is toxic at minimum concentrations. The second group (bismuth, indium, arsenic, thallium, and antimony) is slightly toxic, while the third group (zinc, cobalt, copper, iron, and selenium) is only toxic above a certain concentration [48]. Due to metal contamination, environmental toxicity is increasing. Naturally, heavy metals accumulate from wind-blown soil debris, forest fires, volcanic eruptions, biogenic processes, and marine salt [49]. Human activities are also one of the major sources of heavy metals, such as mining pesticides, fertilizers, and herbicides use, and crop field irrigation with industrial and sewage water [50].

Physicochemical and biological interfacial interactions influenced the conversion of metals and metalloids in the atmosphere and are considered significant in the rhizosphere [51, 52]. Intense biological processes are brought on by a plant's interaction with bacteria in the rhizosphere. It is more difficult to distinguish in the rhizosphere between the impacts of root growth activity and the impacts of microbes on the composition of metals and metalloids. Similar to a plant root, microorganisms take up and adsorb metals and metalloids and then mobilize them through microbial excretions and exudates [53].

12.3 NANOTECHNOLOGY FOR DETECTION AND REMEDIATION OF ENVIRONMENTAL POLLUTANTS

In today's industrialized world, detection and removal of environmental pollutants is an effective way but has become a challenge [54]. The devices to sense, diagnose, and remediate on-site applications might provide close monitoring of environmental conditions [55], While from all approaches, nanomaterials are auspicious to detect and remediate environmental toxicity [56, 57].

12.4 NANOTECHNOLOGY FOR SOIL REMEDIATION

Environmental pollution is the worst thing worldwide nowadays [58]. Soil becomes contaminated by toxic pollutants at a concentration that is harmful to human health or the environment [59]. Nanotechnology-based approaches could take part in new cost-effective methods for metal remediation [60, 61]. The purification of contaminants from nano-remediation methods comprises the application of reactive nanomaterials. They showed desirable characteristics and suitability for both in situ and ex situ applications [62]. NPs occupy less space in the subsurface bec, are suspended in underground water, and accomplish a large area as compared to macro-particles [63].

12.4.1 REMEDIATION OF HEAVY METAL POLLUTED SOILS

To solve the soil contamination issues, several approaches and techniques have been adopted. Generally, remediation techniques are divided into two groups: in situ remediation and ex situ remediation [64]. The in-situ remedy entails treating the pollutant at its original location, while the ex situ remedy involves the digging and removal of pollutants elsewhere from the original place. Overall, in situ, remediation provides many technical, economic, and ecological benefits as compared to ex situ remediation [65]. However, the selection of most suitable remedy to remove soil pollution depends upon the features of the site, form of removable pollutant, the level of toxicity, and ultimately, the use of contaminated medium [66]. Examples of ex-situ nanotechnology are dendritic polymers, SAMMS (self-assembled monolayers on mesoporous substrates), and dendrimer-enhanced ultrafiltration to remove Cu(II) from water [67] and soil washing to remove Pb(II) [68] contamination. The noble metals supported by alumina or nonionic amphiphilic polyurethane used in situ cleanup are currently of the utmost interest [69]. A common type of in situ remedy is a permeable reactive barrier (PRB) [70]:

1. **Physical Remediation:** This process refers to the control or removal of pollutants from the soil by physical means [71, 72] Following technologies are used in physical remediation:

i. **Surface Capping:** It includes the covering of soil with the material having low permeability. The soil covering prevents people from the risk of exposure to the polluted site and prevents the water passage, which ultimately controls the movement of contaminants [73]. This could be used for civil purposes like parking lots [74]. In the procedure of soil replacement, several techniques, surface capping, landfilling, and encapsulation, are used to cover the polluted soil with healthy soil. This process helps to dilute the pollutants and enhance the environmental capability of soil. In the isolation method, polluted soil is isolated by installing some boundaries or barriers so that its dispersal can be prevented. Physical barriers are composed of impermeable materials, such as steel, cement, bentonite, and grout, which are used for capping. The soil isolation method is not a direct remediation strategy, but it could minimize the relocation of heavy metals into the groundwater [75].

ii. **Vapor Extraction:** This method uses in-situ removal of organic pollutants, both volatile and sub-volatile, from the soil. The vapors are divided into the soil gaps, and extraction walls are placed in the ground. Once the fumes have been handled, they should dissipate. The effectiveness of this method depends on the toxicants' solubility and vapor pressure [76].

2. **Electro-Kinetic Remediation:** Through electrical adsorption, it is utilized to remove both organic and inorganic soil in situ. An increase in the transport of cations from the soil to the cathode and anions to the anode through the created electrical field results from the application of a low-density electrical current in the soil electrodes [74, 77]. This method is better suited to partially and fully saturated soils as well as soils with low electrical conductivity [78].

3. **Thermal Treatment:** Of contaminated soil, several heating methodologies such as conductive heating, electrical resistive heating, and steam-based heating are used. The application of thermal treatment to treat toxic soil is performed on the basis of instability of pollutants by heating the subsurface [65]. This technology vanishes the pollutant in an efficient way by high vapor pressure such as Hg, but the properties of soil are greatly affected [79]. The removal of Hg from the soil is done by thermal treatment at 600 °C, but at this temperature, the mineral composition and physiological and chemical

properties of the soil are altered at a drastic level [80]. Heavy metal remediation through thermal procedure could cause repartitioning of heavy metals, that's why acid washing or chemical extraction was recommended before thermal treatment for the removal of Hg [81].

12.4.2 CHEMICAL REMEDIATION

The most popular method for removing contaminants is chemical remediation, which comprises absorption, retaliation, load transfer, oxidation, reduction, or a combination of these [65, 82]. The solidification or stabilization, vitrification, flushing and washing of soil, and electro-kinetics are considered major remediation strategies [75]. To improve the quality of soil and immobility of heavy metals in soil, many cost-effective and eco-friendly waste resources have been testified, such as lime-based agents [83], calcined oyster shells [84], eggshells [85], waste mussel shells [86], and calcined cockle shell [87]. Some chemical technologies are given below:

1. **Solidification and Stabilization:** This, which releases or immobilizes toxicants by in-situ or ex-situ processes using chemical mediators that change mobile contaminants into immobile ones, is also referred to as chemical immobilization. Contaminants cannot be removed with this method; it can only stop their migration. It is employed to remove severely poisonous, radioactive, or metallic pollutants. The technique's disadvantage is that it lessens the soil's capacity to support plants and microorganisms [88].

   In the solidification process, cement, gray steering wheel, and thermoplastics are used to encapsulate the contaminant in a solid form. The resultant solidified block is waterproof and doesn't allow the movement of toxicants. The main advantage of this process is that no gradual monitoring is required as with the passage of time, solid matrix can undergo weathering [74]. The stabilization technique immobilizes the contaminants by adding chemicals into the soil, which maximizes the physic-chemical reactions with contaminants and reduces their mobility [74, 89–91]. The addition of bonemeal, made up of fine, poorly crystalline apatite and $Ca_{10}(PO_4)6OH_2$, could minimize the mobility and bioavailability of heavy metals [92].

2. **Soil Washing:** This is an ex-situ technology in which contaminants are extracted from the environment by using aqueous solutions. The

mined soil is homogenized with extraction solution, and then it is agitated and washed. The clean soil could be reseated at its origin. This process could also be performed in in-situ (soil flushing) [93].

3. **Nanotechnology:** The nano-scaled zero-valent metals like iron, nickel, palladium, etc., are extensively used to remove toxic substances from contaminated soil. This technology is very suitable for both in-situ and ex-situ implementation due to the small size and high surface area of nanoparticles. Nanomaterial reduces the inorganic contaminants like chromium and arsenic [59, 94]. Among all nanomaterials, the use of zero-valent iron particles is advantageous because they are cheap and less toxic [95]. Still, further investigations are required to check their mobility, reactivity, toxicity, and reaction time [96, 97] (Figure 12.1).

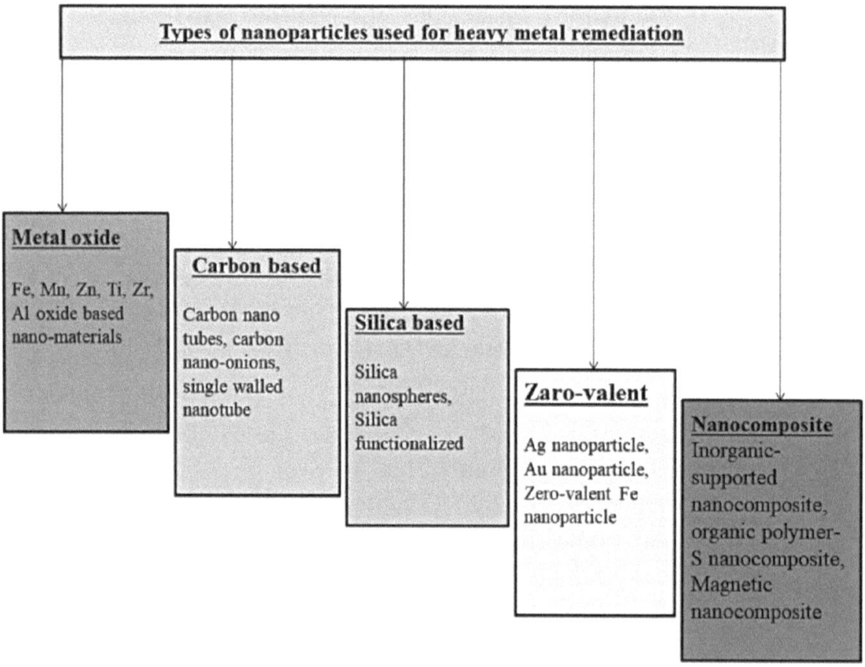

FIGURE 12.1 Different types of nanoparticles are used for heavy metal remediation.

4. **Vitrification:** This needs thermal energy of about 1,400–2,000°C, which could be attained by mixing the glass-forming precursors with polluted soil, heating this mixture until it converts into liquid,

and obtaining an amorphous homogeneous glass after cooling [71]. Chemical bonding and encapsulation are two major interactions with glass matrix by which heavy metals could become immobilized [98]. In vitrification, the heating temperature is very significant. The recapitulation of toxicants and their leaching ability may be improved by adding the efficient additives in the vitrification [99]. The washing and flushing of soil with the help of water or any other suitable washing solution removes the toxicants from polluted soil [2]. Some washing agents, such as water [100], saponin [101], organic acid [102], chelating agents [103], surfactants [104], and low-molecular-weight organic acids [105], have been used to attain the optimal remediation of toxic metals from the medium.

5. **Chelation:** Some chelating agents, such as ethylene diamine tetra acetic acid (EDTA), are considered a very effective chelating agent to vanish heavy metals from contaminated soils [106]. EDTA is beneficial to use as it provides low biodegradability and efficient removal of heavy metals [107]. Some other remediation include electro-migration, electro-osmosis, electrophoresis, and electrolysis, which is a new technique to recover metal-polluted soil [75]. Chelators are used to enhance the activity of electric kinetic remediation (EK). Several chelators along with EK were applied to enhance the working ability of EK to remove metal-contaminated soil [108].

12.4.3 BIOLOGICAL REMEDIATION

The removal of environmental pollutants with the help of living organisms is referred to as biological remediation [109]. As compared to other techniques, bioremediation is relatively cheap and environment friendly and do not produce a large amount of toxic waste [48]. It involves the adsorption, transformation, or degradation of contaminants [65]. The following are basic biological techniques:

1. **Bioventing or Bio-Ventilation:** Bioventing involves the addition of oxygen in the soil voids to stimulate the growth of microorganisms. The involvement of oxygen in microbial metabolism acts as a substrate to initiate the mechanism of biodegradation [77, 110]. The disadvantage of this technique is that it is dependent on the ability of air movement in the soil, that's why particle size and permeability

are considered significant for the proper functioning of this technique [111].

2. **Bio-Augmentation:** During this technique, genetically modified microorganisms are used to remove contaminants from soil. These microbes enhance the degradation rate of contaminants [112]. These microorganisms are selected on the base of their physiology and metabolic ability to degrade the contaminants [77].

3. **Bio-Stimulation:** This involves the adjustment of various environmental attributes, such as nutrients, enhances the solubility of some contaminants, the addition of bio-surfactants or biopolymers, the addition of oxygen, temperature, and humidity control to provide optimal conditions for the deterioration of microorganisms [113]. It could be done in a controlled way as it changes the surface loads of the soil particles and minimizes the interaction of the soil with the contaminants [114, 115].

4. **Vermiremediation:** Worms are utilized in this method to remove pollutants from the soil. The worms have the capacity to ingest and digest organic pollutants, affecting the soil's composition, biomass, and microbial activities. Worms carry co-metabolism with soil microorganisms and increase mineral and nutrient content, which promotes the microbiological activity of the polluted site [116].

5. **Phytoremediation:** It is the process of cleaning soil or sediment with plants and plant components that are associated with a microbial population [77]. Eventually referred to as "green technology" and gained widespread acceptance. It is employed to weaken, eliminate, or stabilize organic and inorganic pollutants [65], further subdivided into: phytoextration [72], phytodegradation [117], phytovolatization [77], phytostabilization [117], and rhizodegradation [72].

Bacteria, fungi, yeast, and algae are microorganisms that are frequently used for metal remediation [118]. A detailed illustration of heavy metal remediation includes the following: *Sporosarcina ginsengisoli* [119], Pseudomonas putida [120], and Bacillus subtilis [121]. For successful bioremediation by microorganisms, a group of bacterial strains is used instead of a single strain culture. Kang et al. [122] examine the effects of *Viridibacillus arenosi*, *Sporosarcina soli*, *E. cloacae*, and *E. cloacae* (bacterial mixtures) to recover polluted soil. Consequently, it was noted that the bacterial mixture was effective and competent to remediate polluted soil. Microorganisms help to remove toxicants from soil via precipitation, biosorption, and by some enzymes [48]. The combination of plants and microorganisms in a remedy to remove metals from polluted soil provides efficient results [123].

Mycorrhizal fungi are extensively used in remediation techniques, as they could apply some methods, for example, acidification, immobilization, and modification of root exudates, hyphal segregation, and chemical precipitation [124]. Metal-tolerant arbuscular mycorrhizal fungi separated from contaminated soil works better to cope with metal removal than that isolated from unpolluted soils [125] (Figure 12.2).

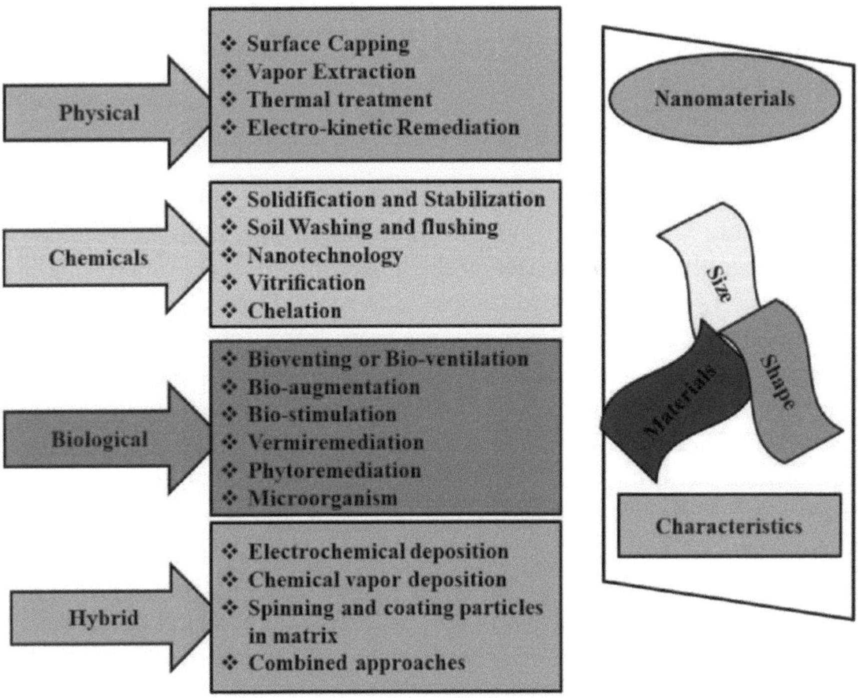

FIGURE 12.2 Several approaches, along with their techniques used to remove metals.

12.4.4 PLANT SPECIES FOR NANO-PHYTOREMEDIATION

The usage of plants to remediate metal-polluted soil is known as phytoremediation [126]. It is an appropriate method to use when contaminates are spread on a vast area and occupy the plant root zone [127]. Phytoremediation basically consists of phytoextraction, phyto-stabilization, phytovolatilization, and phyto-degradation [128]. Phytoextraction provides long-term clean-up as it involves the absorption/uptake of toxicants from water and soil via plant root and their transfer to the shoot or any other removable part of the plant [129]. The mobility and availability of heavy metals could be minimized by phytostabilization [130]. Phytovolatilization refers to the process in which plants absorb and convert toxic metals into volatile forms and then release them into the surroundings. Phytodegradation is a process in which the metabolic processes of plants transform or breakdown down metal contaminants from the soils [128]. For phytoremediation, a plant species must have the following characteristics: fast-growing ability, an efficient root system, high plant biomass, tolerance to high metal concentrations, and ability to accumulate high concentrations of metals [131] (Table 12.1).

12.5 CONCLUSION AND FUTURE PROSPECTS

By introducing significant quantities of dangerous toxicants that contaminate soil, water, and the atmosphere and ultimately endanger human health, human activities are upsetting the ecosystem's delicate equilibrium. In an effort to choose a treatment strategy that is consistent with removing all the waste products left over from the Industrial Revolution, this chapter simply contrasts the application of beneficial nanotechnology to conventional technologies in environmental remediation. It has been shown that nanotechnology offers unique properties for sophisticated, strong, and multi-functional treatment methods that can accelerate the rehabilitation of polluted soil and get around all of the aforementioned obstacles. It has been demonstrated that several organic, inorganic, metal-based, zero-valent, and silica-based nanoparticles are used to remove pollutants from soil. In brief, nanotechnology is capable of improving the environmental remediation process as it inhibits the formation of secondary by-products.

Its major concern is global food safety and human health. In the future, it may have a substantial impact on agricultural and farming development. The ultimate goal is to increase the agricultural output and decrease the input (fertilizers).

TABLE 12.1 Types of Nanoparticles and Their Physiological Effects on Several Crops under Different Metal Stresses and Growth Media

Sl. No.	Type of Nanoparticle	Crop	Metal Stress	Media	Physiological Effects	References
1.	FeO	Wheat	Cadmium	Soil	Increased the growth, nutrients, and antioxidant enzymes in wheat while decreasing the uptake of Cd in root and shoot.	[132]
2.	ZnO	Rice	Cadmium	Hydroponic	Improves plant growth and development while reducing the toxic metal ions in roots and shoots.	[133]
3.	AgNPs	Moringa	Cd and Pb	Soil	Mitigate oxidative stress by increasing antioxidant activities in plants. Reduces the uptake of Cd and Pb.	[134]
4.	Si	Coriander	Pb	Soil	Inhibits ROS production and increases plant growth, development, and yield.	[135]
5.	Cu	Wheat	Cr	Soil	Decreased Cr toxicity and oxidative stress while increasing plant growth, biomass, and yield.	[136]
6.	FeO	Tomato	Cr	Soil	Improved the photosynthetic activity of plants and ultimately improved growth and yield.	[137]
7.	TiO_2	Maize	Cd	Soil	Decreased root shoot concentration of Cd and mitigates Cd toxicity.	[138]
8.	Cu	Wheat	Cd	Soil	Increase total chlorophyll content and improve plant growth.	[139]
9.	MgO	Rice	As	Soil	Significantly reduces As toxicity and improves the rice plant growth, cellular activities, and antioxidant contents.	[140]
10.	Zinc oxide	Wheat	Cd	Soil	Increased plant growth and biomass, photosynthesis efficiency, soil moisture content, and reduced Cd accumulation from soil.	[141]

LANGUAGE PROOF

The authors are thankful to the Institute of Faisalabad Learning and Development for their input in English language improvement.

CONFLICT OF INTEREST

Authors have no conflict.

KEYWORDS

- **bioremediation**
- **decontamination**
- **heavy metals**
- **nanocomposite**
- **nanomaterial**
- **nano-phytoremediation**
- **phytoremediation**
- **soil pollution**

REFERENCES

1. Rascio, N., & Navari-Izzo, F., (2011). Heavy metal hyperaccumulating plants: How and why do they do it? And what makes them so interesting? *Journal of Plant Science, 180*(2), 169–181.
2. Li, C., et al., (2019). *A Review on Heavy Metals Contamination in Soil: Effects, Sources, and Remediation Techniques, 28*(4), 380–394.
3. Sarwar, N., et al., (2017). *Phytoremediation Strategies for Soils Contaminated with Heavy Metals: Modifications and Future Perspectives, 171,* 710–721.
4. Kankia, H., & Abdulhamid, Y., (2014). Determination of accumulated heavy metals in benthic invertebrates found in Ajiwa Dam, Katsina State, Northern Nigeria. *Journal of Archives of Applied Science Research, 6*(6), 80–87.
5. Hu, B., et al., (2017). *Application of Portable XRF and VNIR Sensors for Rapid Assessment of Soil Heavy Metal Pollution, 12*(2), e0172438.
6. Gong, X., et al., (2018). *Remediation of Contaminated Soils by Biotechnology with Nanomaterials: Bio-behavior, Applications, and Perspectives, 38*(3), 455–468.

7. He, Z., et al., (2015). *Heavy Metal Contamination of Soils: Sources, Indicators and Assessment, 9*, 17–18.
8. Chen, H., & Yada, R., (2011). Nanotechnologies in agriculture: New tools for sustainable development. *Trends in Food Science & Technology, 22*(11), 585–594.
9. Parisi, C., Vigani, M., & Rodríguez-Cerezo, E., (2015). Agricultural nanotechnologies: What are the current possibilities? *Journal of Nanotoday, 10*(2), 124–127.
10. Biswal, S. K., et al., (2012). *Applications of Nanotechnology in Agriculture and Food Sciences, 2*(1), 21–36.
11. Khot, L. R., et al., (2012). *Applications of Nanomaterials in Agricultural Production and Crop Protection: A Review, 35,* 64–70.
12. Sekhon, B. S., (2014). Nanotechnology in Agri-food production: An overview. *Journal of Nanotechnology Science and Applications, 7,* 31.
13. Sonkaria, S., et al., (2012). *Nanotechnology and its Impact on Food and Nutrition: A Review, 4*(1), 8–18.
14. Muthukumaran, P., et al., (2022). *Nanotechnological Approaches as a Promising Way for Heavy Metal Mitigation in an Aqueous System, 62*(3, 4), 376–394.
15. Mapelli, F., et al., (2017). *Biotechnologies for Marine Oil Spill Cleanup: Indissoluble Ties with Microorganisms, 35*(9), 860–870.
16. Rawtani, D., et al., (2018). *Nanotechnology-Based Recent Approaches for Sensing and Remediation of Pesticides, 206,* 749–762.
17. Khatri, N., Tyagi, S., & Rawtani, D. (2017). Recent strategies for the removal of iron from water: A review. *Journal of Water Process Engineering, 19,* 291–304.
18. Tharmavaram, M., Rawtani, D., & Pandey, G., (2017). Fabrication routes for one-dimensional nanostructures via block copolymers. *Journal of Nano Convergence, 4*(1), 1–13.
19. Ibrahim, R. K., et al., (2016). *Environmental Application of Nanotechnology: Air, Soil, and Water, 23*(14), 13754–13788.
20. Chen, Y., et al., (2019). *Modification, Application, and Reaction Mechanisms of Nano-Sized Iron Sulfide Particles for Pollutant Removal from Soil and Water: A Review, 362,* 144–159.
21. Wu, Y., et al., (2019). *Environmental Remediation of Heavy Metal Ions by Novel-Nanomaterials: A Review, 246,* 608–620.
22. Ahmed, S. F., et al., (2022). *Green Approaches in Synthesizing Nanomaterials for Environmental Nanobioremediation: Technological Advancements, Applications, Benefits and Challenges, 204,* 111967.
23. Dai, S., et al., (2019). *Preparation of Core-Shell Structure* Fe_3O_4*@ C@* MnO_2 *Nanoparticles for Efficient Elimination of U (VI) and Eu (III) Ions, 685,* 986–996.
24. Singh, J., et al., (2018). *'Green' Synthesis of Metals and Their Oxide Nanoparticles: Applications for Environmental Remediation, 16*(1), 1–24.
25. Singh, D. V., et al., (2020). *W*onders of nanotechnology for remediation of polluted aquatic environs. In: *Fresh Water Pollution Dynamics and Remediation* (pp. 319–339). Springer.
26. Pacheco, S., & Rodríguez, R., (2001). Adsorption properties of metal ions using alumina nano-particles in aqueous and alcoholic solutions. *Journal of Sol-Gel Science and Technology, 20*(3), 263–273.
27. Sharma, V. K., et al., (2009). *Silver Nanoparticles: Green Synthesis and Their Antimicrobial Activities, 145*(1, 2), 83–96.

28. Sharma, V., & Sharma, A., (2012). Nanotechnology: An emerging future trend in wastewater treatment with its innovative products and processes. *International Journal of Enhanced Research in Science Technology & Engineering, 1*(2), 1–8.
29. Hristovski, K., Baumgardner, A., & Westerhoff, P., (2007). Selecting metal oxide nanomaterials for arsenic removal in fixed bed columns: From nanopowders to aggregated nanoparticle media. *Journal of Hazardous Materials, 147*(1, 2), 265–274.
30. Homola, J., (2006). *Surface Plasmon Resonance Based Sensors* (Vol. 4). Springer Science & Business Media.
31. Shankaran, D. R., et al., (2007). *Recent Advancements in Surface Plasmon Resonance Immunosensors for Detection of Small Molecules of Biomedical, Food and Environmental Interest, 121*(1), 158–177.
32. Salah, T. A., et al., (2014). *Development of Nano-Hydroxyapatite/Chitosan Composite for Cadmium Ions Removal in Wastewater Treatment, 45*(4), 1571–1577.
33. Shintani, H., (2014). Toxic compounds analysis with high performance liquid chromatography detected by electrochemical detector (ECD). *International Journal of Clinical Pharmacology & Toxicology (IJCPT), 3*(3), 121–27.
34. Tranchida, P. Q., et al., (2015). *Flow-modulated Comprehensive Two-Dimensional Gas Chromatography Combined with a High-Resolution Time-of-Flight Mass Spectrometer: A Proof-of-Principle Study, 87*(5), 2925–2930.
35. Su, S., et al., (2012). *Nanomaterials-Based Sensors for Applications in Environmental Monitoring, 22*(35), 18101–18110.
36. Andreescu, S., et al., (2009). *JEM Spotlight: Applications of Advanced Nanomaterials for Environmental Monitoring, 11*(1), 27–40.
37. Theron, J., Walker, J. A., & Cloete, T. E., (2008). Nanotechnology and water treatment: Applications and emerging opportunities. *Critical Reviews in Microbiology, 34*(1), 43–69.
38. Theron, J., Eugene, C. T., & De Kwaadsteniet, M., (2010). Current molecular and emerging nanobiotechnology approaches for the detection of microbial pathogens. *Crit. Rev. Microbiol. 36*(4), 318–339.
39. Dixit, R., et al., (2015). *Bioremediation of Heavy Metals from Soil and Aquatic Environment: An Overview of Principles and Criteria of Fundamental Processes, 7*(2), 2189–2212.
40. Muradoglu, F., et al., (2015). *Cadmium Toxicity Affects Chlorophyll A and B Content, Antioxidant Enzyme Activities and Mineral Nutrient Accumulation in Strawberry, 48*(1), 1–7.
41. Chen, Z., et al., (2015). *Cadmium (Cd) Localization in Tissues of Cotton (Gossypium hirsutum L.), and its Phytoremediation Potential for Cd-Contaminated Soils, 95*(6), 784–789.
42. Khan, I., et al., (2016). *Comparative Analysis of Heavy Metal Profile of Brassica Campestris (L.) and Raphanus Sativus (L.) Irrigated with Municipal Waste Water of Sargodha City, 6,* 1–4.
43. Ogunlade, M., & Agbeniyi, S., (2011). Impact of pesticides use on heavy metals pollution in cocoa soils of Cross River State, Nigeria. *African Journal of Agricultural Research, 6*(16), 3725–3728.
44. Sun, W. H., Jiang, Y. X., & Li, X., (2013). Research of the evaluation of heavy-metal pollution in rice by sewage irrigation. In: *Applied Mechanics and Materials*. Trans Tech Publ.

45. Atafar, Z., et al., (2010). *Effect of Fertilizer Application on Soil Heavy Metal Concentration, 160*(1), 83–89.
46. Brodin, M., et al., (2017). *Lignocellulosics as Sustainable Resources for Production of Bioplastics: A Review, 162,* 646–664.
47. Ferrey, M. L., et al., (2018). *Pharmaceuticals and Other Anthropogenic Chemicals in Atmospheric Particulates and Precipitation, 612,* 1488–1497.
48. Ojuederie, O. B., & Babalola, O. O., (2017). Microbial and plant-assisted bioremediation of heavy metal polluted environments: A review. *Int. J. Environ. Res. Public Health, 14*(12), 1504.
49. Muhammad, S., Shah, M. T., & Khan, S., (2011). Health risk assessment of heavy metals and their source apportionment in drinking water of Kohistan region, northern Pakistan. *Microchemical Journal, 98*(2), 334–343.
50. Zaynab, M., et al., (2022). *Health and Environmental Effects of Heavy Metals, 34*(1), 101653.
51. Huang, P., & Germida, J. J., (2002). Chemical and biological processes in the rhizosphere: Metal pollutants. *5th International Symposium ISMOM, 8,* 381–438.
52. Gobran, G., et al., (2005). Contribution of rhizospheric processes to mineral weathering in forest soils. Huang, P. M., & Gobran, G. R., (eds.), *Biogeochemistry of Trace Elements in the Rhizosphere* (pp. 3–28). Elsevier, Amsterdam.
53. Huang, P. M., (2008). Impacts of physicochemical-biological interactions on metal and metalloid transformations in soils: An overview. *Biophysico-Chemical Processes of Heavy Metals and Metalloids in Soil Environments, 1,* 3–52.
54. Liu, Y., et al., (2011). *Nanoparticle-Based Strategies for Detection and Remediation of Environmental Pollutants, 136*(5), 872–877.
55. Ghormade, V., Deshpande, M. V., & Paknikar, K. M., (2011). Perspectives for nano-biotechnology enabled protection and nutrition of plants. *Biotechnol Adv., 29*(6), 792–803.
56. Baruah, S., & Dutta, J., (2009). Nanotechnology applications in pollution sensing and degradation in agriculture: A review. *Environmental Chemistry Letters, 7*(3), 191–204.
57. Zhang, L., & Fang, M., (2010). Nanomaterials in pollution trace detection and environmental improvement. *Journal of Nanotoday, 5*(2), 128–142.
58. Das, S., et al., (2015). *Recent Trends in Nanomaterials Applications in Environmental Monitoring and Remediation, 22*(23), 18333–18344.
59. Thomé, A., et al., (2015). *Review of Nanotechnology for Soil and Groundwater Remediation: Brazilian Perspectives, 226*(4), 1–20.
60. Gupta, V. K., et al., (2013). *Adsorptive Removal of Dyes from Aqueous Solution Onto Carbon Nanotubes: A Review, 193,* 24–34.
61. Park, H., et al., (2013). *Surface Modification of TiO_2 Photocatalyst for Environmental Applications, 15,* 1–20.
62. Reddy, D. H. K., & Lee, S. M., (2013). Application of magnetic chitosan composites for the removal of toxic metal and dyes from aqueous solutions. *Advances in Colloid and Interface Science, 201,* 68–93.
63. Li, Q., et al., (2016). *Decontaminating Soil Organic Pollutants with Manufactured Nanoparticles, 23*(12), 11533–11548.
64. Gomes, H. I., Dias-Ferreira, C., & Ribeiro, A. B., (2013). Overview of in situ and ex situ remediation technologies for PCB-contaminated soils and sediments and obstacles for full-scale application. *Science of the Total Environment, 445,* 237–260.

65. Song, B., et al., (2017). *Evaluation Methods for Assessing Effectiveness of in Situ Remediation of Soil and Sediment Contaminated with Organic Pollutants and Heavy Metals, 105,* 43–55.
66. Mulligan, C., Yong, R., & Gibbs, B., (2001). Remediation technologies for metal-contaminated soils and groundwater: An evaluation. *Journal of Engineering Geology, 60*(1–4), 193–207.
67. Prabhakar, V., et al., (2013). *Nanotechnology, Future Tools for Water Remediation, 3*(7), 54–59.
68. Diallo, M. S., et al., (2005). *Dendrimer Enhanced Ultrafiltration. 1. Recovery of Cu(II) from Aqueous Solutions Using PAMAM Dendrimers with Ethylene Diamine Core and Terminal NH_2 groups, 39*(5), 1366–1377.
69. Hoffmann, M. R., et al., (1995). *Environmental Applications of Semiconductor Photocatalysis, 95*(1), 69–96.
70. Rajan, C., (2011). Nanotechnology in groundwater remediation. *International Journal of Environmental Science and Development, 2*(3), 182.
71. Yao, Z., et al., (2012). *Review on Remediation Technologies of Soil Contaminated by Heavy Metals, 16,* 722–729.
72. Thomé, A., et al., (2018). Remediation technologies applied in polluted soils: New perspectives in this field. In: *The International Congress on Environmental Geotechnics*. Springer.
73. Tomaszewski, J. E., et al., (2006). Treatment and containment of contaminated sediments. In: *Assessment and Remediation of Contaminated Sediments* (pp. 137–178). Springer.
74. Liu, L., et al., (2018). *Remediation Techniques for Heavy Metal-Contaminated Soils: Principles and Applicability, 633,* 206–219.
75. Jankaitė, A., & Vasarevičius, S., (2005). Sunkiaisiais metalais užterštų dirvožemių atkūrimo būdai. *Journal of Environmental Engineering and Landscape Management, 13*(2), 109–113.
76. Albergaria, J. T., Maria Da, C. M., & Delerue-Matos, C., (2012). Remediation of sandy soils contaminated with hydrocarbons and halogenated hydrocarbons by soil vapor extraction. *Journal of Environmental Management, 104,* 195–201.
77. Lim, M. W., Von, L. E., & Poh, P. E., (2016). A comprehensive guide of remediation technologies for oil contaminated soil—Present works and future directions. *Journal of Marine Pollution Bulletin, 109*(1), 14–45.
78. Reddy, K. R., (2013). Electrokinetic remediation of soils at complex contaminated sites: Technology status, challenges, and opportunities. *Coupled Phenomenain Environmental Geotechnics,* 131–147. Taylor and Francis Group.
79. Chang, T., & Yen, J., (2006). On-site mercury-contaminated soil remediation by using thermal desorption technology. *Journal of Hazard Mater., 128*(2, 3), 208–217.
80. Roh, Y., et al., (2000). *Thermal-treated Soil for Mercury Removal: Soil and Phytotoxicity Tests*. Wiley Online Library.
81. Huang, Y. T., Hseu, Z. Y., & Hsi, H. C., (2011). Influences of thermal decontamination on mercury removal, soil properties, and repartitioning of coexisting heavy metals. *Journal of Chemosphere, 84*(9), 1244–1249.
82. Hu, X. J., et al., (2011). *Adsorption of Chromium (VI) by Ethylenediamine-Modified Cross-Linked Magnetic Chitosan Resin: Isotherms, Kinetics, and Thermodynamics, 185*(1), 306–314.

83. Lim, J. E., et al., (2013). *Effects of Lime-Based Waste Materials on Immobilization and Phytoavailability of Cadmium and Lead in Contaminated Soil, 41*(12), 1235–1241.
84. Ok, Y. S., et al., (2010). *Effects of Natural and Calcined Oyster Shells on Cd and Pb Immobilization in Contaminated Soils, 61*(6), 1301–1308.
85. Soares, M. A., Quina, M. J., & Quinta-Ferreira, R. M., (2015). Immobilization of lead and zinc in contaminated soil using compost derived from industrial eggshell. *Journal of Environmental Management, 164,* 137–145.
86. Otero, M., et al., (2015). *Cr (VI) Sorption/Desorption on Untreated and Mussel-Shell-Treated Soil Materials: Fractionation and Effects of pH and Chromium Concentration, 6*(2), 373–382.
87. Islam, M. N., et al., (2017). *Heavy Metal Stabilization in Contaminated Soil by Treatment with Calcined Cockle Shell, 24*(8), 7177–7183.
88. Tajudin, S. A., Azmi, M. M., &. Nabila, A., (2016). Stabilization/solidification remediation method for contaminated soil: A review. In: *IOP Conference Series: Materials Science and Engineering*. IOP Publishing.
89. Xie, T., et al., (2015). *Characteristics and Applications of Biochar for Environmental Remediation: A Review, 45*(9), 939–969.
90. Reddy, K. R., et al., (2015). *Effects of Biochar Amendment on Geotechnical Properties of Landfill Cover Soil, 33*(6), 524–532.
91. Seshadri, B., et al., (2017). *Potential Value of Phosphate Compounds in Enhancing Immobilization and Reducing Bioavailability of Mixed Heavy Metal Contaminants in Shooting Range Soil, 184,* 197–206.
92. Hodson, M. E., et al., (2000). *Bonemeal Additions as a Remediation Treatment for Metal Contaminated Soil, 34*(16), 3501–3507.
93. Morillo, E., & Villaverde, J., (2017). Advanced technologies for the remediation of pesticide-contaminated soils. *Science of the Total Environment, 586,* 576–597.
94. Crane, R. A., & Scott, T. B., (2012). Nanoscale zero-valent iron: Future prospects for an emerging water treatment technology. *Journal of Hazardous Materials, 211,* 112–125.
95. Yan, W., et al., (2013). *Iron Nanoparticles for Environmental Clean-Up: Recent Developments and Future Outlook, 15*(1), 63–77.
96. Reddy, K. R., et al., (2012). *Transport and Reactivity of Lactate-Modified Nanoscale iron Particles in PCP-Contaminated Soils, 16*(1), 68–74.
97. Lefevre, E., et al., (2016). *A Review of the Environmental Implications of in Situ Remediation by Nanoscale Zero Valent Iron (nZVI): Behavior, Transport, and Impacts on Microbial Communities, 565,* 889–901.
98. Navarro, A., (2012). Effect of sludge amendment on remediation of metal contaminated soils. *Journal of Minerals, 2*(4), 473–492.
99. Guo, G., et al., (2006). *Availability and Assessment of Fixing Additives for the In Situ Remediation of Heavy Metal Contaminated Soils: A Review, 116*(1), 513–528.
100. Dermont, G., et al., (2008). *Soil Washing for Metal Removal: A Review of Physical/Chemical Technologies and Field Applications, 152*(1), 1–31.
101. Maity, J. P., et al., (2013). *Evaluation of Remediation Process with Soapberry Derived Saponin for Removal of Heavy Metals from Contaminated Soils in Hai-Pu, Taiwan, 25*(6), 1180–1185.
102. Kim, J. O., Lee, Y. W., & Chung, J., (2013). The role of organic acids in the mobilization of heavy metals from soil. *KSCE Journal of Civil Engineering, 17*(7), 1596–1602.

103. Wei, J., Tao, T., & Zhi-Ming, L., (2011). Removal of heavy metal from contaminated soil with chelating agents. *Open Journal of Soil Science, 2011.*
104. Sun, H., et al., (2011). Study on surfactants remediation in heavy metals contaminated soils. In: *2011 International Symposium on Water Resource and Environmental Protection*. IEEE.
105. Almaroai, Y. A., et al., (2012). *Effects of Synthetic Chelators and Low-Molecular-Weight Organic Acids on Chromium, Copper, and Arsenic Uptake and Translocation in Maize (Zea mays L.), 177*(11), 655–663.
106. Leštan, D., Luo, C. L., & Li, X. D., (2008). The use of chelating agents in the remediation of metal-contaminated soils: A review. *Environmental Pollution, 153*(1), 3–13.
107. Qiao, J., et al., (2017). *EDTA-Assisted Leaching of Pb and Cd from Contaminated Soil, 167,* 422–428.
108. Song, Y., et al., (2016). *Effect of EDTA, EDDS, NTA, and Citric Acid on Electrokinetic Remediation of As, Cd, Cr, Cu, Ni, Pb, and Zn Contaminated Dredged Marine Sediment, 23*(11), 10577–10586.
109. Ayangbenro, A. S., & Babalola, O. O., (2017). A new strategy for heavy metal polluted environments: A review of microbial biosorbents. *IJERPH, 14*(1), 94.
110. Karamalidis, A., et al., (2010). *Laboratory Scale Bioremediation of Petroleum-Contaminated Soil by Indigenous Microorganisms and Added Pseudomonas Aeruginosa Strain Spet, 101*(16), 6545–6552.
111. Thomé, A., et al., (2014). *Bioventing in a Residual Clayey Soil Contaminated with a Blend of Biodiesel and Diesel Oil, 140*(11), 06014005.
112. Abdulsalam, S., et al., (2011). *Comparison of Biostimulation and Bioaugmentation for Remediation of Soil Contaminated with Spent Motor Oil, 8*(1), 187–194.
113. Abed, R. M., et al., (2015). *Effect of Biostimulation, Temperature and Salinity on Respiration Activities and Bacterial Community Composition in an oil Polluted Desert Soil, 98,* 43–52.
114. Cecchin, I., et al., (2017). *Nanobioremediation: Integration of Nanoparticles and Bioremediation for Sustainable Remediation of Chlorinated Organic Contaminants in Soils, 119,* 419–428.
115. Thomé, A., et al., (2017). *Biostimulation and Rainfall Infiltration: Influence on Retention of Biodiesel in Residual Clayey Soil, 24*(10), 9594–9604.
116. Rodriguez-Campos, J., et al., (2014). *Potential of Earthworms to Accelerate Removal of Organic Contaminants from Soil: A Review, 79,* 10–25.
117. Germida, J., Frick, C., & Farrell, R., (2002). Phytoremediation of oil-contaminated soils. In: *Developments in Soil Science* (pp. 169–186). Elsevier.
118. Coelho, L. M., et al., (2015). *Bioremediation of Polluted Waters Using Microorganisms, 10,* 60770.
119. Achal, V., et al., (2012). *Biomineralization Based Remediation of As(III) Contaminated Soil by Sporosarcina Ginsengisoli, 201,* 178–184.
120. Balamurugan, D., Udayasooriyan, C., & Kamaladevi, B., (2014). Chromium (VI) reduction by *Pseudomonas putida* and *Bacillus subtilis* isolated from contaminated soils. *International Journal of Environmental Sciences, 5*(3), 522–529.
121. Imam, S. A., et al., (2016). *Comparative Study of Heavy Metal Bioremediation in Soil by Bacillus Subtilis and Saccharomyces Cerevisiae, 9.*
122. Kang, C. H., Kwon, Y. J., & So, J. S. (2016). Bioremediation of heavy metals by using bacterial mixtures. *Journal of Ecological Engineering, 89,* 64–69.

123. Vangronsveld, J., et al., (2009). *Phytoremediation of Contaminated Soils and Groundwater: Lessons from the Field, 16*(7), 765–794.
124. Hristozkova, M., et al., (2017). *Mycorrhizal Fungi and Microalgae Modulate Antioxidant Capacity of Basil Plants, 57*(4).
125. Cornejo, P., et al., (2013). *Copper Compartmentalization in Spores as a Survival Strategy of Arbuscular Mycorrhizal Fungi in Cu-Polluted Environments, 57,* 925–928.
126. Ali, H., Khan, E., & Sajad, M. A., (2013). Phytoremediation of heavy metals—Concepts and applications. *Journal of Chemosphere, 91*(7), 869–881.
127. Chibuike, G. U., & Obiora, S. C., (2014). Heavy metal polluted soils: Effect on plants and bioremediation methods. *Applied and Environmental Soil Science, 2014.*
128. Kong, Z., & Glick, B. R., (2017). The role of plant growth-promoting bacteria in metal phytoremediation. *J. Adv. Microb. Physiol., 71,* 97–132.
129. Bhargava, A., et al., (2012). *Approaches for Enhanced Phytoextraction of Heavy Metals, 105,* 103–120.
130. Radziemska, M., et al., (2018). *Concept of Aided Phytostabilization of Contaminated Soils in Postindustrial Areas, 15*(1), 24.
131. Jabeen, R., Ahmad, A., & Iqbal, M., (2009). Phytoremediation of heavy metals: Physiological and molecular mechanisms. *Journal of the Botanical Review, 75*(4), 339–364.
132. Manzoor, N., et al., (2021). *Iron Oxide Nanoparticles Ameliorated the Cadmium and Salinity Stresses in Wheat Plants, Facilitating Photosynthetic Pigments and Restricting Cadmium Uptake, 769,* 145221.
133. Ma, C., et al., (2021). *Graphitic Carbon Nitride (C3N4) Reduces Cadmium and Arsenic Phytotoxicity and Accumulation in Rice (Oryza Sativa L.), 11*(4), 839.
134. Azeez, L., et al., (2019). *Zero-Valent Silver Nanoparticles Attenuate Cd and Pb Toxicities on Moringa Oleifera via Immobilization and Induction of Phytochemicals, 139,* 283–292.
135. Fatemi, H., et al., (2021). *Foliar Application of Silicon Nanoparticles Affected the Growth, Vitamin C, Flavonoid, and Antioxidant Enzyme Activities of Coriander (Coriandrum Sativum L.) Plants Grown in Lead (Pb)-Spiked Soil, 28*(2), 1417–1425.
136. Noman, M., et al., (2020). *Biogenic Copper Nanoparticles Synthesized by Using a Copper-Resistant Strain Shigella Flexneri SNT22 Reduced the Translocation of Cadmium from Soil to Wheat Plants, 398,* 123175.
137. Brasili, E., et al., (2020). *Remediation of Hexavalent Chromium Contaminated Water Through Zero-Valent Iron Nanoparticles and Effects on Tomato Plant Growth Performance, 10*(1), 1–11.
138. Lian, J., et al., (2020). *Foliar Spray of TiO_2 Nanoparticles Prevails Over Root Application in Reducing Cd Accumulation and Mitigating Cd-Induced Phytotoxicity in Maize (Zea Mays L.), 239,* 124794.
139. Noman, M., et al., (2020). *Green Copper Nanoparticles from a Native Klebsiella Pneumoniae Strain Alleviated Oxidative Stress Impairment of Wheat Plants by Reducing the Chromium Bioavailability and Increasing the Growth, 192,* 110303.
140. Ahmed, T., et al., (2021). *Green Magnesium Oxide Nanoparticles-Based Modulation of Cellular Oxidative Repair Mechanisms to Reduce Arsenic Uptake and Translocation in Rice (Oryza sativa L.) Plants, 288,* 117785.

141. Khan, Z. S., et al., (2019). *The Accumulation of Cadmium in Wheat (Triticum Aestivum) as Influenced by Zinc Oxide Nanoparticles and Soil Moisture Conditions, 26*(19), 19859–19870.

CHAPTER 13

Removal of Dyes by Nano-Bioremediation: Importance and Future Aspects

LAKHA V. CHOPDA[1] and PRAGNESH N. DAVE[2]

[1]*Government Engineering College, Bhuj, Gujarat, India*

[2]*Department of Chemistry, Sardar Patel University, Vallabh Vidyanagar, Gujarat, India*

ABSTRACT

Nowadays, global concern is for the removal of heavy metals, hydrocarbons, and radioactive wastes from soil and water. These are dangerous for the environment and human health. Removal of these contaminants from polluted soils and water is one of the major challenging tasks. The textile industry is spread worldwide and consumes more water. Therefore, water pollution by dyes is significantly increased and causes undesirable effects. Various methods and techniques are available for the clean-up of pollutants from water bodies. Nano-bioremediation is an emerging technology for the remediation of pollutants with the aid of biosynthetic nanoparticles. The application of biogenic nanoparticles for the removal of a wide range of pollutants is growing rapidly as they are non-toxic and easy to prepare using biological entities. This present chapter discusses the application of biogenic nanoparticles synthesized from yeast, fungi, bacteria, and plants as a nanocatalyst for the remediation of dyes from the aqueous system.

13.1 INTRODUCTION

Water is an essential compound on Earth that is needed to sustain the lives of creatures. Anthropogenic activities such as agriculture, washing, and industry require water for human development. It is difficult to imagine human life

Nano-Bioremediation for Water and Soil Treatment: An Eco-Friendly Approach.
Vishnu D. Rajput, Arpna Kumari, and Tatiana M. Minkina (Eds.)

without the use of water. However, these anthropogenic activities have led to water pollution [1]. Providing safe water that is free from contaminants is a significant obligation. Water can become contaminated by various pollutants [2], such as heavy metals, dyes, pesticides, pharmaceuticals, persistent organic pollutants, microorganisms, etc., which is a growing concern today. Water pollution causes several problems for human health, creates environmental issues, and ultimately affects socio-economic development. The quality of water in rivers all over the world has worsened due to water pollution.

The textile industry is a leading sector in terms of providing employment. Dyes are organic compounds applied to transmit color to textiles, food, leather, paper, and several other materials. The textile industry requires a significant amount of water, especially for dyeing and washing of the material. The solubility of various types of dyes, such as reactive, disperse, azo, direct, acid, and basic, in water is high, making it difficult to remove. The untreated discharge of the textile industry is the main cause of water pollution. Dyes contribute to causing cancer along with other pollutants [1].

The removal of dyes from the water system proceeds by physical and chemical methods [3]. The adsorption is the main physical process for the removal of dyes [4]. It is the simplest and economical viable method for removal of dyes from the water. The activated carbon is the promising adsorbent use to remove not only dyes but other contaminates from the water system [5, 6]. The other carbon materials such as plant, agriculture waste, polymer, carbon nanotube and graphene/graphene oxide/reduced graphene and non-carbon material like zeolite, clay, metal nanoparticles, flay ash, etc., are used to remove the dyes from aqueous system [7, 8]. The adsorption capacity of dye relies on the interaction between adsorbent and dye molecule. The adsorption of dye from the water on these materials occurs mostly through electrostatic interaction, hydrogen bonding, and chemical interaction depends on the nature of adsorbent and dye [9]. The coagulation technique consists of addition of aluminum or iron salts into dye containing water system [10, 11]. The aluminum or iron salts converts into their respective metal hydroxide. The form metal hydroxide chemically interacts with dye molecule and makes the water free from dye. The electro-coagulation is also an effective method for the decontamination of dye from the water [3]. The key difference between coagulation and electro-coagulation is that former involves the direct addition of metal salts while latter bears the oxidation of metal either of aluminum or iron. The oxidize metals contribute to form metal hydroxide and reaming removal of dyes process is similar to that of coagulation method. The advanced oxidation process by using of hydrogen peroxide, Fenton's reagent and ozonization are the chemical method for removal of dyes [3]. These methods work on the formation of powerful hydroxyl radical that eventually lead the decomposition of the dye molecule into carbon dioxide by oxidation process.

The pollutants persisted in a water longer time without of any treatment. The decontamination of water is the most essentiality for enhancing the quality of water. The various methods mentioned above to remediate dyes possessing some merits and demerits [12]. The application of metal nanoparticles and metal oxide nanoparticles them self in the field of water remediation found considerable position due to their consisting of unique physical and chemical property compared to their counter bulk material. The synthesis of metal nanoparticles and metal oxide nanoparticles by chemical method is not environment eco-friendly. The produced biogenic nanoparticles both metal and metal oxide nanoparticles, are considered as less toxic and cheap compared to produce by chemical methods. The chemical synthesis requires the external stabilizing agent to control the size of nanoparticles which is greatly nullified by biosynthetic nanoparticles as all necessary features present in the biological sources.

Among the existence method, bioremediation is the uppermost technology for water purification [13] as it originates from bio-source or material synthesis derives from bio-source (biomolecule or biopolymer) or using microorganisms (bacterial, algae, fungi, and viruses). This chapter provides an account on the biogenic nanoparticles as a nanocatalyst for the decontamination of dyes from the water.

13.2 BIOGENIC NANOPARTICLES FOR WATER REMEDIATION

Nanotechnology is the popular research area since inception of well-known lecture ("There's Plenty of Room at the Bottom") presented by Richard P Feynman. The nanoparticles synthesis considers as the potential field of science. The nanosized materials in the range of 1–100 nm in all three dimensions possesses good properties like electronic, chemical, catalytic, magnetic, photo-fluorescent, and optical compare to their bulk materials [14] hence found them significant application in various fields such sensor, drug delivery, gene delivery, diagnosis, bio-imaging, treatment of cancer, stem cell therapy, catalyst, and many more [15, 16]. The top down and bottom-up approach are used to manufacture of nanoparticles [17, 18]. The top-down approach involves the utilize the bulk material for nanoparticles synthesis while simple precursor (atoms or molecule) employ for nanoparticles synthesis undergoes to bottom-up approach. For the example, synthesis of Fe-C-O-Mo alloy nano-rods synthesized from chemical decomposition of a screw is the example of top-down approach [19] and synthesis of metal oxide nanoparticles by sol-gel process is the example of bottom-up approach [20]. The mechanical grinding (ball milling), lithography, etching, erosion, laser ablation, sputtering is top-down methodologies while bottom approach entails sol-gel, chemical vapor deposition, physical vapor deposition, electrochemical deposition, hydrothermal method and atomic layer are methodologies for the nanoparticles synthesis [12, 21]. Conventionally, nanoparticles prepared by chemical and physical method [17] (illustrated in Figure 13.1) but high cost of used materials, attachment of environment hazardous and toxic chemicals and produced large amount waste at the end of process which are also sometimes create negative effect on the surroundings [12, 13]. To ascertain these problems, biosynthesis of nanoparticles is evolved that usage of microorganisms or plant extract or agriculture waste and constructed nanoparticles known as biogenic nanoparticles [1, 22]. As this pathway believes as green method as no involvement of toxic and hazardous chemicals, generated waste are biological compounds hence easily degraded out, no worse effect on environment, mild reaction conditions and easy to prepare from microorganism culture and plant extract hence biosynthesis pathway for nanoparticles is developed for the nanoparticles synthesis. The nanoparticles synthesis by biological source is the bottom-up approach and similar to chemical method. The metal salts reduced to nanosize by attendance of bimolecular particularly enzyme and others presence in the microorganisms (bacteria, algae, yeast, fungi, and viruses) and in the case of plant extract mediated nanoparticles synthesis in which phytochemicals existence in extract bears antioxidant or reducing properties accountable for reduction of metal salts into their respective nanoparticles as well additionally presence of bio-molecules in microorganisms and plant extract provide extra stability to the nanoparticles and many cases these biomolecules act as green capping agent for nanoparticles [12, 22]. Biogenic synthesized nanoparticles by intracellular and extracellular pathways, in the case of extracellular process, extract of microorganisms or plant materials or agriculture waste exploit for the synthesis whereas intracellular method is the in-cell synthesis method and separate procedure is needed to detach the nanoparticles [23]. The small and uniform particle size of nanoparticles is achieved by intracellular method. The aggregation of nanoparticles size is maintained by the extracellular method that resulted into the size and uniformity of nanoparticles.

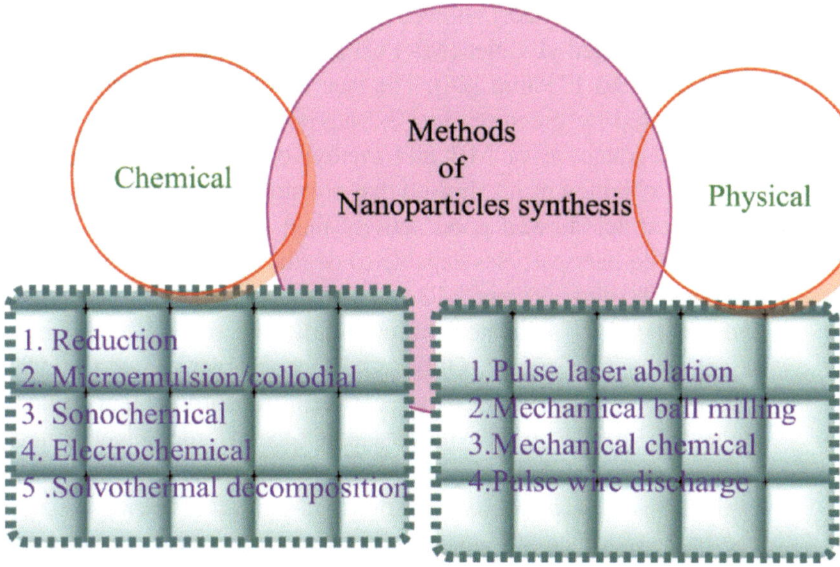

FIGURE 13.1 Physical and chemical methods for synthesis of nanoparticles (NPs).

13.3 CATALYTIC APPLICATION OF NANOPARTICLES

Water pollution with wide range of pollutants from organic to inorganic to microorganism increased due to rapid development in the whole sphere of life. The pollutants in water systems are directly contact to all living organism which created many health issues. For the better health life, it is necessary to remove the pollutants from the water system by using available efficient techniques. The nano-based materials play the significant role in the form of adsorbent and catalyst to remediate pollutants from the water system. The photo catalytic and some examples of advanced oxidation process (AOP) for the degradation of leading dyes are discussed.

1. **For Methylene Blue (MB) Dye:** The nanomaterials as a photo catalyst for the degradation of organic pollutants are widely studied. The SnO_2 NPs prepared by using 1, 2 and 4 wt.% of *C. aurantifolia* peel extract [24]. The TEM image revealed that as the percentage of extract increased resulted to decrease the diameter of nanoparticles which contributed to activity of catalyst. The 1, 2, and 4 wt.% of extract showed 64, 75, and 96% degradation efficiency of MB. The Ag NPs prepared by *C. pyrenoidosa* (algae) showed high degradation efficiency than bulk Ag powder and commercial Ag NPs [25]. The Cu NPs with average diameter of 45–62 nm synthesized by *Aloe vera*

leaves (aqueous extract) showed exhibited 93 and 96% degradation efficiency of methyl violet (MV) 6B and methylene blue (MB) dyes within 150 and 135 min [26]. The iron oxide and ironoxyhydroxide NPs prepared by green tea leaf extract acted as Fenton-like catalyst for the degradation of MB and methyl orange (MO) respectively [27]. The obtained result showed that prepared nanoparticles by green tea leaf extract showed good activity than synthesized iron NPs by $NaBH_4$. The nanoparticles capable to degrade 80% dye within 5 min and after 200 and 350 min, MB and MO are completely degraded out. The kinetic study indicated that MB followed second order kinetic model while in the case MO, first and second order displayed 0.9896 and 0.9623 values of regression coefficient repetitively that could be suggested MO accorded with first and second order kinetic model but with more can be inclined towards to second order. The Pd NPs produced using *gum olibanum* (Boswellia serrata*)* efficiently decolorized in the presence of reducing agent $NaBH_4$ in a very short reaction time [28]. The *Achillea millefolium* L. extract used to prepare Ag NPs which subsequently supported over peach kernel shell. The catalyst Ag/pech kernel shell completely decolorized MB and MO at 10 mg amount of nanoparticles in 50 and 48 second, respectively [29]. The spherical Ag NPs synthesized by employing *S. acuminata* fruit extract [30]. The prepared catalyst showed excellent degradation activity in the presence of $NaBH_4$ for four dyes such as MB, MO, phenol red (PR) and direct blue (DB) [24]. The Ag NPs prepared by extract of *Cicer arietinum* leaves rapidly degraded the Congo red (CR) and MB [31]. The nanoparticles with spherical morphology with two different size (58.5 and 78.13 nm) successfully degraded the MB under sunlight in 110 min [32]. The microwave assisted Ag NPs synthesis reported by using *B. sensitivum* leaf extract. The nanoparticles are spherical type of 19.06 nm in diameter. The Ag NPs showed remarkable catalytic activity towards MB and MO dyes degradation [33]. The leaf extract of *P. hydropiper* produced AgNPs in spherical nature with diameter of 60 nm [34]. The nanoparticles exhibited good catalytic activity towards the degradation of MB dye in the presence of $NaBH_4$ that convert MB into leucomethylene blue (colorless) product. Absence of AgNPs did not induce catalytic activity indicated that AgNPs pronounced effect on the degradation of MB. The reduced grapheme (rGO) prepared by fruit extract of *P. emblica* displayed 92 and 91% degradation efficiency for MO and MB in 90 min of contact of sunlight, respectively using

20 mg/L of catalyst amount [35]. The iron NPs was prepared by *L. speciosa* plant leaves. The nanoparticles are in spheroidal shape with diameters in the range of 50–100 nm was obtained by the utilization of citrate buffer [36]. The yielded material is Fe_3O_4 as confirmed by EDS analysis. The dyes MB, MO, Allura red (AR), brilliant blue (BB), and green S (GS) efficiently degraded out by Fe_3O_4. The SnO_2 (15–40 nm) of tetragonal morphology was prepared by *Erwinia herbicola* bacterium that later annealed at 425 K [37]. The nanoparticles degraded MB, MO, and eriochrome black T (EBT) in 93.3, 97.8, and 94.0% efficiency using 0.2 g of SnO_2 NPs with 120 min exposure of UV light. The degradation of MB by other nanoparticles is shown in Table 13.1. Table 13.1 indicates that MB degraded out efficiently under investigated conditions except Pd NPs prepared by *Andean blackberry* leaf extract that displayed moderate activity around 72% and ZnO NPs prepared by *Monsonia burkeana* exhibited 48% efficiency. The low activity of ZnO was attributed by poisonous of catalyst caused by deposition of extract over the surface of ZnO.

TABLE 13.1 The Catalytic Activity Outline of Nanoparticles

NPs	Reducing Agent	Shape and Size (nm)	Mode of Light	Dose of Material	Degradation Efficiency	References
SnO_2	*C. aurantifolia* peel extract (4 wt.%)	Hemispherical-5.77 (TEM)	UV	15 ppm	96% (120 min)	[24]
α-Fe_2O_3	*C. ramiflora* fruit extract	Spherical – 58.5 and 78.13 (SEM)	Solar	30 mg	94% (110 min)	[32]
Cu	*P. granatum* seeds extract	Semispherical – 40–80 (SEM)	Solar	100 ppm	87.11% (180 min)	[38]
Ag	*Lychee* (*Litci chinensis*) fruit	Spherical – 4–8 nm (HRTEM)	UV	4 mg	99.24% (11 min)	[39]
Ag	*G. arborea* fruit extract	Spherical – 17 (TEM)	$NaBH_4$	3 mL	100% (10 min)	[40]
ZnO	*Monsonia burkeana*	Hexagonal-5–15 (TEM)	UV	20 mg	48% (45 min)	[41]
Ag	*C. racemosa* (Algae)	Distorted spherical-25 (HRTEM)	$NaBH_4$	0.05 mL	100%	[42]
Fe_2O_3	Onion peel extract	Nanofiber-like-24–44 (SEM)	Normal source	3 mg	97% (30 min)	[43]
FeO	Garlic peel extract	Nanosheet-like – 29–32 (SEM)	Normal source	3 mg	90% (35 min)	[43]
Pd	Andean blackberry leaf extract	Decahedron-55–60 (TEM)	Solar/ UV	300 and 500 μL	>72%	[44]

2. **For Rhodamine B Dye:** *Delonix elata* leaf extract applied for the synthesis of SnO_2 NPs [45]. The SnO_2 NPs produced by three routes sonication, wet-chemical, and microwave. FESEM images revealed that cluster like foam morphology with small agglomeration of particles achieved by sonication and wet-chemical methods while microwave methods produced uniform particles. The surface area of nanoparticles is estimated and values of it: 101, 169, and 196 m^2/g as prepared by sonication, wet-chemical, and microwave methods, respectively and same the value of pore size and pore volume also enhanced. The reverse trend in the case of particle size determination has observed by DLS method. The 13.68, 17.40 and 18.7 nm particle size obtained by sonication, wet-chemical, and microwave methods, respectively. The photo catalytic activity by UV light irradiation showed that 82.3, 85.6, and 92.8% activity achieved by SnO_2 NPs prepared by sonication, wet-chemical, and microwave methods, respectively. The microwave assisted SnO_2 displayed high activity because of its high surface properties and small particle size compared to wet-chemical and microwave methods and same in the cases of catalytic activity of wet-chemical method compared to sonication method. The 1, 2, and 4 wt% leave extract of *C. sinensis* used to synthesize SnO_2 NPs [46]. The nanoparticles showed 6.91, 5.2, and 4.7 nm average particle size as determined by TEM. The photocatalytic activity of nanoparticles was investigated by UV and solar light irradiation. The UV light irradiation exhibited excellent catalytic activity while solar irradiation displayed negligible activity. The SnO_2 NPs showed 81% degradation for MO and 100% degradation efficiency for MB and rhodamine B (Rh-B) after 180 min exposure of UV light. The syntheses of Ag NPs govern by extract of *C. arvensis* leaves. The nanoparticles well deeply characterized by FT-IR, XRD, DLS, SEM, and TEM techniques. The high negative value of zeta potential contributed to high stability of nanoparticles. The Ag NPs displayed catalytic activity towards the three azo dyes (reactive black (RB)-5, MO, and direct yellow (DY)-42) in the presence of $NaBH_4$. The quasi-spherical Au NPs with average particle size of 72.32 nm prepared by soil fungus *Cladosporium oxysporum* AJP03 that signified excellent degradation activity towards Rh-B in 7 min using $NaBH_4$ [47]. The SnO_2/ZnO (2:1 molar ratio) nanoparticles

prepared by leaf extract of *Urtica dioica* [48]. The photocatalytic activity of nanoparticles evaluated for Rh B using UV irradiation. The SnO_2/ZnO displayed 91.9% degradation efficiency within 55 min which exceeded than SnO_2 (91.7% in 190 min) and ZnO (85% in 140 min). The catalytic activity of SnO_2/ZnO, SnO_2 and ZnO attributed their band gap value, the band value of SnO_2/ZnO is 3.44 eV, SnO_2 is 3.14 eV and ZnO is 3.32 eV. The high the band value of SnO_2/ZnO than SnO_2 and ZnO resulted into high degradation efficiency of SnO_2/ZnO. SnO_2 quantum dots (QDs) successfully prepared by flower extract of *Aparajitha* (*Clitoria ternatea*) of 4–10 nm size [49]. The SnO_2 quantum displayed higher surface area (78 m^2/g) than bulk SnO_2 (35.5 m^2/g). The high surface area of SnO_2 quantum dots resulted into high degradation activity of Rh B. The ZnO NPs were prepared and caped by glycine [50]. The synthesis outlined of nanoparticles preparation is shown in Figure 13.2. The morphology study showed that sample GZ1 (0.5 M glycine) produced few ZnO nanorods with high agglomeration, GZ2 displayed (2 M glycine) form flower bud-like ZnO nanostructures. The GZ3 (1M Glycine showed that ZnO nanorods arranged into bundles (agglomerated ZnO nanoroda are seen). The GZ5 revealed sheet-like structure intertwined each other and GZ5 showed thicker surface of ZnO nanosheets with thicker surface and less agglomeration. It shows that morphology depends on the synthesis parameters. The band gap of GZ NPs is around 2.98 ± 0.07 eV which is lower than bulk ZnO (~3.3 eV). The low value of band gap of GZ NPs is due to presence of various surface defects and these defects evaluated by photoluminescence (PL) spectroscopic. The photo catalytic activity revealed that GZ2 NPs displayed high activity (~80%) while other prepared nanoparticles displayed ~50% efficiency. The high activity of GZ2 is because of formation of reactive oxidative species (ROS) such as hydroxyl radicals ($OH^{\cdot}$), superoxide radicals (O_2^{-}) and holes (h^{+}) as their role is verified by performed radical scavenging experiment. The obtained results showed that presence of radical scavengers reduced photocatalytic activity of GZ2.

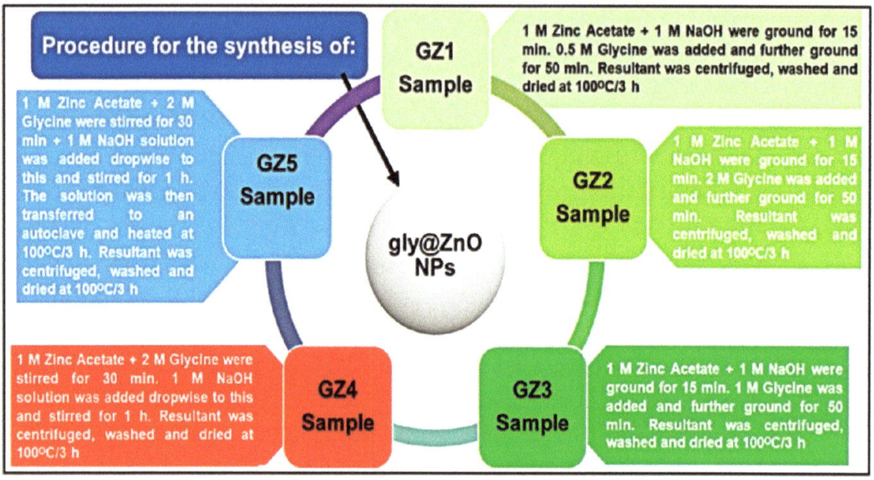

FIGURE 13.2 Procedure for synthesis of gly@ZnO NPs [50].

Source: Reprinted with permission from Ref. [50]. Copyright © 2020, Springer Science Business Media, LLC

3. **For Congo Red, EBT, and Methyl Orange:** The Mn oxide NPs prepared by bark extract of *C. verum* and sodium alginate [51]. The nanoparticles size less than 100 nm was obtained with spherical morphology. The 78.5% of Congo red dye degraded at pH 7.0, 0.06 g/L of MnNPs dose, 10 mg/L initial dye concentration with 60 min exposure of UV irradiation. The irregular hematite nanoparticles (α-Fe_2O_3 NPs) were prepared by using *P. granatum* seed extract [52]. The average particle size is 26.53 nm. The surface area estimate to 31.52 m^2/g and average pore diameter is 5.54 nm. The nanoparticles displayed ferro magnetic behavior. The catalytic activity showed that 89.42% of CR and 87.96% of bromophenol blue (BPB) was degraded out in 240 min contact of sun light and 0.05 g/L of amount of nanoparticles. The leaf extract of *A. gangeticus Linn* used to prepare Ag NPs [53]. The nanoparticles with 11–15 nm size was obtained with globular shape and polycrystalline in nature. In the presence of $NaBH_4$, Ag NPs exhibited good catalytic activity towards degradation of Congo red. Ag NPs prepared by *Ficus retusa* leaf extract [54]. The TEM image showed that 5–35 nm sizes of nanoparticles were obtained. The zeta potential of nanoparticles was – 39.1 that revealed of high stability of nanoparticles. The 86.05 and 80.11% catalytic activity was achieved by the $NaBH_4$ and light source, respectively in 60 and 120 min towards the degradation efficiency of EBT. The SnO_2

NPs were prepared by aqueous gelatin solution [55]. The TEM and FESEM images indicated that appearance of SnO_2 NPs is tetragonal rutile with mean particle size ~27 nm. The photo catalytic study using UV source showed that 35.9% of EBT dye degraded out. The chitosan biopolymer constructed spherical SnO_2 NPs [56]. The XRD profile indicated that produced nanoparticles have the tetragonal crystalline structure with average particle size found to be 25.6 nm. The photo catalytic activity indicated that 77% of EBT degraded by exposure of UV light using 21.1 mg of catalyst. The nano-scale dysprosium cerate ($Dy_2Ce_2O_7$) NPs were synthesized by using extract of *Ananas comosu* [57]. The amount of extract has pronounced effect on the shape, dimension, and catalytic activity. The nanoparticles prepared by 3 mL extract showed excellent stability and photocatalytic activity using UV irradiation towards the degradation of EBT. The *Ulva lactuca* (seaweed) used to generated spherical Ag NPs of 48.59 nm [58]. The negative value of zeta potential (–34 mV) suggested that high stability of nanoparticles. The Ag NPS possessed good photo catalytic activity towards MO degradation using solar irradiation. The Au and Ag NPs prepared by leaf extract of *Mussaenda glabrata* assisted by microwave [59]. The Au NPs are obtained in spherical and triangular geometry with average diameter estimated to 10.59 nm and Ag NPs are showed spherical geometry with average diameter estimated to 51.32 nm. Both catalysts effectively exhibited catalytic activity for the degradation of Rh B and MO in the presence of $NaBH_4$. The leaf extract of *Mussaenda erythrophylla* used to prepare Ag NPs [60]. The zeta potential value of – 47.7 mV signified the excellent stability of Ag NPs. The SEM images revealed that two different spherical size of nanoparticles (88 and 82 nm) were produced. The catalyst showed good catalytic activity towards the degradation of MO using $NaBH_4$. The silver nanocatalysts prepared employing *Trigonella foenum-graecum* seeds [61]. The AG NPs synthesized by using different amount of extract. The Ag NPs are obtained between 22 and 32 nm. The Ag NPs showed good catalytic activity towards the degradation of eosin Y, MO, and MB, respectively.

4. **For the Other Dyes:** Iron NPs were synthesized using *oolong* tea extracts [62]. The spherical nanoparticles are obtained of 40–50 nm of diameter. The nanoparticles applied for the degradation of malachite green (MG). The nanoparticles removed MG up to 75.5% followed by pseudo first-order kinetic model. The iron oxide NPs prepared by

the extract of green tea [63]. The iron oxide nanoparticles prepared by the different volume ratio of Fe(II) and tea extracts (1:1, 1:2, and 1:3). The spherical nanoparticles are achieved with 70–80 nm of diameter. The degradation activity of MG was undertaken, and iron oxide NPs showed 90.56% degradation activity of MG. The green tea extract synthesized iron NPs with 1:2 volume ratio of $FeSO_4$:tea extract was able to remove 96% of MG dye [64]. The highly crystalline ZnO and CuO NPs were prepared by using aqueous extract of fruit hull (Jaft) as reducing and stabilizing agent [65]. The FESEM images of both nanoparticles indicated that CuO NPs are in quasi-spherical shape while ZnO NPs are uniform spherical in nature. The average particle sizes of both nanoparticles estimated to 34 nm. The photo catalytic activities of both nanoparticles are carried out for basic violet 3 under visible light. The ZnO NPs showed high catalytic activity than CuO NPs. The degradation of basic violet 3 by both NPs followed the pseudo-first-order kinetics. The uniform spherical nature of ZnO could be the crucial factor for the high catalytic activity compared to CuO NPs. The zero-valent iron (ZVI) NPS prepared by green tea that subsequently supported over bentonite [66]. The SEM morphology indicated that nanoparticles are in irregular spherical shape and less than 50 nm average particle size of nanoparticles are observed as evaluated by AFM. Blue 238 dye degradation carried out using H_2O_2 (Fenton-like oxidation process). The zero-valent iron NPS and supported to bentonite showed 93.5 and 96.2% activity at $c(H_2O_2) = 5$ mmol/L, ZVI) or ρ(B-GT-ZVI) = 0.5 g/L, c(RB 238 dye) = 0.05 mmol/L, and pH = 2.5 at 180 min. Second-order kinetic model suited for RB 238 dye degradation and activation energy decreased from 38.22 kJ/mol for ZVI to 14.13 kJ/mol for B-ZVI showed that supported nanoparticles effectively decrease energy barrier than ZVI. The SnO_2 NPs using leaf extract of *Amaranthus tricolor* L.) was prepared [67]. The photo catalytic activity of bromophenol blue (BPB) was conducted in the presence and absence of H_2O_2. The photo catalytic activity monitored by UV-visible HPLC showed that superior decomposition activity of BPB shown by nanoparticles. The flower like (by SEM) and uniform spherical shape diameter ranging from 8 to 20 nm (by TEM) of SnO_2 NPs obtained as prepared by using *Pometia pinnata* leaf extract [68]. The dye BPB 99.93% was degraded by the nanoparticles under UV-visible light illumination. The *Streptomyces griseobrunneus* strain FSHH12 (based on the 16S rDNA gene sequence analysis) selected for the preparation of selenium nanoparticles (Se NPs) [69]. The photo catalytic degradation of bromothymol blue (BTB) using the purified Se NPs (64 μg/mL) was performed. The 62.3% dye degradation efficiency under UV illumination (15 W) obtained after 60 min. The catalytic activity of Se NPs was higher than bulk SeO_2 NPs (32.5%). The zero valent iron NPs (ZVI) using tea (*Camellia sinensis*) polyphenols was prepared [70]. The prepared catalyst tested for the degradation of BTB in the presence of H_2O_2. The degradation of dye increased with increased of concentration of ZVI. The prepared other two materials Fe-EDTA (ethylenediamine tetraacetate) and FE-EDDS ((S,S)-ethylenediamine-N,N-disuccinic acid) found to be less effective than ZVI. At 0.33 mM Fe concentration and 0.33 mM hydrogen peroxide concentration, initial rate constants for bromothymol blue oxidation were 0.1447, 0.0038, and 0.0148 for the catalysts ZVI, Fe-EDTA, and Fe-EDDS, respectively. The ZnO NPs synthesis by employing using leaf extract of *Plectranthus amboinicus* that displayed rod shape structure with an average size of 88 nm [71]. The band gap of ZnO NPs is estimated to 3.07 eV. The ZnO NPs prepared by leaf extract of *Plectranthus amboinicus* exhibited superior activity for the degradation methyl red (MR) than chemically synthesized ZnO NPs. The Au NPs prepared by *Plumeria alba* flower extract (PAFE) [72]. The 1 and 5% of PAFE employed for the synthesis of two different size 28 and 15.6 nm spherical nanoparticles. The small size nanoparticles showed good degradation activity towards the five dyes (MB, eosin Y (EY), MR, CR, and Evans blue (EB) in the presence of $NaBH_4$. The Pd NPs supported to anaerobic granular sludge (AGS) was synthesized and evaluated for azo dyes reduction [73]. The Pd NPs@AGS was *in-situ* prepared by using sodium formate, and M9 medium in serum bottles. The PD/AGS enhanced the reduction rate of azo dyes (Orange II, EB, and CR) in the presence hydrogen donors such as formate, formic acid, acetate, glucose, ethanol, and lactate. The activity trend of hydrogen donors is as followed: formic acid > formate > ethanol > glucose > lactate > acetate. The of Ag@AgCl (silver@silverchloride) NPs by using *Aquilaria agallocha* (AA) leaves juice [74]. The synthesized Ag@AgCl NPs were used for the degradation of Victoria Blue (VB) B using solar light irradiation. The 99.46% of Victoria Blue B dye was degraded out by Ag@AgCl NPs. After 5th cycle, no altered in the activity of Ag@AgCl NPs was observed.

13.4 CONCLUSION AND FUTURE PROSPECT

Green technology is highly competent to remove the different kind of pollutants from the water system. Dyes accumulate easily in the water. The various physical and chemical methods are available to treat dyes present in the water, but bioremediation assisted by biogenic nanoparticles as a nano-photocatalyst catalyst are rapidly in demand for the remediation of pollutants from the water system. The nanoparticles effectively adsorbed the light and enhanced the degradation efficiency of the dye molecules into lesser nontoxic compound. Biogenic nanoparticles can be deeply dictated in the role of photo catalyst for the water treatment in the near future. The synthesis procedure for biogenic nanoparticles should be in robustness *viz* not harsh reaction conditions involve in the preparation procedure, facile separation of nanoparticles from the system and no alter in the catalytic activity after many uses.

KEYWORDS

- **bioremediation**
- **catalytic application**
- **contamination**
- **green synthesis**
- **nanoparticles**
- **nanotechnology**
- **water remediation**

REFERENCES

1. Lellis, B., Fávaro-Polonio, C. Z., Pamphile, J. A., & Polonio, J. C., (2019). Effects of textile dyes on health and the environment and bioremediation potential of living organisms. *Biotechnol. Res. Innov., 3*, 275–290. https://doi.org/10.1016/j.biori.2019.09.001.
2. Schwarzenbach, R. P., Egli, T., Hofstetter, T. B., Von, G. U., & Wehrli, B., (2010). Global water pollution and human health. *Annu. Rev. Environ. Resour., 35*, 109–136. https://doi.org/10.1146/annurev-environ-100809-125342.
3. Sharma, S., & Kaur, A., (2018). Various methods for removal of dyes from industrial effluents-A review. *Indian J. Sci. Technol., 11*, 1–21. https://doi.org/10.17485/ijst/2018/v11i12/120847.

4. Kandisa, R. V., & Saibaba, K. V. N., (2016). Dye removal by adsorption: A review. *J. Bioremediation Biodegrad., 7,* 1000371. https://doi.org/10.4172/2155-6199.1000371.
5. Mezohegyi, G., Van, D. Z. F. P., Font, J., Fortuny, A., & Fabregat, A., (2012). Towards advanced aqueous dye removal processes: A short review on the versatile role of activated carbon. *J. Environ. Manage., 102,* 148–164. https://doi.org/10.1016/j.jenvman.2012.02.021.
6. Jeirani, Z., Niu, C. H., & Soltan, J., (2017). Adsorption of emerging pollutants on activated carbon. *Rev. Chem. Eng., 33,* 1–32. https://doi.org/10.1515/revce-2016-0027.
7. Soltani, A., Faramarzi, M., & Mousavi, P. S. A., (2021). A review on adsorbent parameters for removal of dye products from industrial wastewater. *Water Qual. Res. J., 56,* 181–193. https://doi.org/10.2166/wqrj.2021.023.
8. Homaeigohar, S., (2020). The nanosized dye adsorbents for water treatment. *Nanomaterials., 10,* 295. https://doi.org/10.3390/nano10020295.
9. Dutta, S., Gupta, B., Srivastava, S. K., & Gupta, A. K., (2021). Recent advances on the removal of dyes from wastewater using various adsorbents: A critical review. *Mater. Adv., 2,* 4497–4531. https://doi.org/10.1039/d1ma00354b.
10. Papić, S., Koprivanac, N., & Božić, A. L., (2000). Removal of reactive dyes from wastewater using Fe(III) coagulant. *Color. Technol., 116,* 352–358. https://doi.org/10.1111/j.1478-4408.2000.tb00013.x.
11. Yeheyes, K., Zewge, F., & Chandravanshi, B. S., (2007). Removal of dye from water by coagulation using aluminum sulfate and lime. *SINET Ethiop. J. Sci., 29,* 133–140. https://doi.org/10.4314/sinet.v29i2.18267.
12. Baig, N., Kammakakam, I., & Falath, W., (2021). Nanomaterials: A review of synthesis methods, properties, recent progress, and challenges. *Mater. Adv., 2,* 1821–1871. https://doi.org/10.1039/d0ma00807a.
13. Vallepu, N., Gaddam, S. A., Kotakadi, V. S., Goli, P. P., DVR, S. G., & Gudivada, S. (2021). Biogenic silver nanoparticles can be an effective and efficient water purification agents of future. *Inorganic and Nano-Metal Chemistry, 51*(9), 1195–1206.
14. Khan, I., Saeed, K., & Khan, I., (2019). Nanoparticles: Properties, applications, and toxicities. *Arab. J. Chem., 12,* 908–931. https://doi.org/10.1016/j.arabjc.2017.05.011.
15. Yadav, K. K., Singh, J. K., Gupta, N., & Kumar, V., (2017). A review of nano bioremediation technologies for environmental cleanup: A novel biological approach. *J. Mater. Environ. Sci., 8,* 740–757.
16. Li, X., Xu, H., Chen, Z. S., & Chen, G., (2011). Biosynthesis of nanoparticles by microorganisms and their applications. *J. Nanomater., 2011,* 1–16. https://doi.org/10.1155/2011/270974.
17. Satyanarayana, T., & Reddy, S., (2018). A Review on chemical and physical synthesis methods of nanomaterials. *Int. J. Res. Appl. Sci. Eng. Technol., 6,* 2885–2889. https://doi.org/10.22214/ijraset.2018.1396.
18. Ijaz, I., Gilani, E., Nazir, A., & Bukhari, A., (2020). Detail review on chemical, physical and green synthesis, classification, characterizations and applications of nanoparticles. *Green Chem. Lett. Rev., 13,* 223–245. https://doi.org/10.1080/17518253.2020.1802517.
19. Wu, P., Feng, L., Liang, Y., Zhang, X., Mahmoudi, B., & Kazemnejadi, M., (2020). Magnetic Fe-C-O-Mo alloy nano-rods prepared from chemical decomposition of a screw (a top-down approach): An efficient and cheap catalyst for the preparation of dihydropyridine and dihydropyrimidone derivatives. *Appl. Catal. Gen., 590,* 117301. https://doi.org/10.1016/j.apcata.2019.117301.

20. Singh, L. P., Bhattacharyya, S. K., Kumar, R., Mishra, G., Sharma, U., Singh, G., & Ahalawat, S., (2014). Sol-gel processing of silica nanoparticles and their applications. *Adv. Colloid Interface Sci., 214*, 17–37. https://doi.org/10.1016/j.cis.2014.10.007.
21. Iqbal, P., Preece, J. A., & Mendes, P. M., (2012). Nanotechnology: The "top-down" and "bottom-up" approaches. in: Gale, P. A., & Steed, J. W., (eds.), *Supramol. Chem.* (p. smc195). John Wiley & Sons, Ltd, Chichester, UK. https://doi.org/10.1002/9780470661345.smc195.
22. Gautam, P. K., Singh, A., Misra, K., Sahoo, A. K., & Samanta, S. K., (2019). Synthesis and applications of biogenic nanomaterials in drinking and wastewater treatment. *J. Environ. Manage., 231*, 734–748. https://doi.org/10.1016/j.jenvman.2018.10.104.
23. Kim, S., Kim, J., Kim, S., Lee, J., & Yoon, J., (2018). Electrochemical lithium recovery and organic pollutant removal from industrial wastewater of a battery recycling plant. *Environ. Sci. Water Res. Technol., 4*, 175–182. https://doi.org/10.1039/c7ew00454k.
24. Luque, P. A., Nava, O., Soto-Robles, C. A., Garrafa-Galvez, H. E., Martínez-Rosas, M. E., Chinchillas-Chinchillas, M. J., Vilchis-Nestor, A. R., & Castro-Beltrán, A., (2020). SnO_2 nanoparticles synthesized with *Citrus aurantifolia* and their performance in photocatalysis. *J. Mater. Sci. Mater. Electron., 31*, 16859–16866. https://doi.org/10.1007/s10854-020-04242-5.
25. Aziz, N., Faraz, M., Pandey, R., Shakir, M., Fatma, T., Varma, A., Barman, I., & Prasad, R., (2015). Facile algae-derived route to biogenic silver nanoparticles: Synthesis, antibacterial, and photocatalytic properties. *Langmuir, 31*, 11605–11612. https://doi.org/10.1021/acs.langmuir.5b03081.
26. Md. Ahmaruzzaman, (2022). Ecofriendly biosynthetic route for production of Cu nanoparticles and evaluation of their photocatalytic activities for degradation of organic compounds. *J. Iran. Chem. Soc., 19*, 645–654. https://doi.org/10.1007/s13738-021-02325-6.
27. Shahwan, T., Abu, S. S., Nairat, M., Boyacı, E., Eroğlu, A. E., Scott, T. B., & Hallam, K. R., (2011). Green synthesis of iron nanoparticles and their application as a Fenton-like catalyst for the degradation of aqueous cationic and anionic dyes. *Chem. Eng. J., 172*, 258–266. https://doi.org/10.1016/j.cej.2011.05.103.
28. Kora, A. J., & Rastogi, L., (2016). Catalytic degradation of anthropogenic dye pollutants using palladium nanoparticles synthesized by gum olibanum, a glucuronoarabinogalactan biopolymer. *Ind. Crops Prod., 81*, 1–10. https://doi.org/10.1016/j.indcrop.2015.11.055.
29. Khodadadi, B., Bordbar, M., & Nasrollahzadeh, M., (2017). *Achillea millefolium* L. extract mediated green synthesis of waste peach kernel shell supported silver nanoparticles: Application of the nanoparticles for catalytic reduction of a variety of dyes in water. *J. Colloid Interface Sci., 493*, 85–93. https://doi.org/10.1016/j.jcis.2017.01.012.
30. Bogireddy, N. K. R., Kiran, K. H. A., & Mandal, B. K., (2016). Biofabricated silver nanoparticles as green catalyst in the degradation of different textile dyes. *J. Environ. Chem. Eng., 4*, 56–64. https://doi.org/10.1016/j.jece.2015.11.004.
31. Arya, G., Sharma, N., Ahmed, J., Gupta, N., Kumar, A., Chandra, R., & Nimesh, S., (2017). Degradation of anthropogenic pollutant and organic dyes by biosynthesized silver nano-catalyst from *Cicer arietinum* leaves. *J. Photochem. Photobiol. B., 174*, 90–96. https://doi.org/10.1016/j.jphotobiol.2017.07.019.
32. Bishnoi, S., Kumar, A., & Selvaraj, R., (2018). Facile synthesis of magnetic iron oxide nanoparticles using inedible *Cynometra ramiflora* fruit extract waste and their

photocatalytic degradation of methylene blue dye. *Mater. Res. Bull., 97*, 121–127. https://doi.org/10.1016/j.materresbull.2017.08.040.

33. Joseph, S., & Mathew, B., (2015). Microwave-assisted green synthesis of silver nanoparticles and the study on catalytic activity in the degradation of dyes. *J. Mol. Liq., 204*, 184–191. https://doi.org/10.1016/j.molliq.2015.01.027.
34. Bonnia, N. N., Kamaruddin, M. S., Nawawi, M. H., Ratim, S., Azlina, H. N., & Ali, E. S., (2016). Green biosynthesis of silver nanoparticles using '*Polygonum hydropiper*' and study its catalytic degradation of methylene blue. *Procedia Chem., 19*, 594–602. https://doi.org/10.1016/j.proche.2016.03.058.
35. Parthipan, P., Cheng, L., Rajasekar, A., Govarthanan, M., & Subramania, A., (2021). Biologically reduced graphene oxide as a green and easily available photocatalyst for degradation of organic dyes. *Environ. Res., 196*, 110983. https://doi.org/10.1016/j.envres.2021.110983.
36. Garole, V. J., Choudhary, B. C., Tetgure, S. R., Garole, D. J., & Borse, A. U., (2018). Detoxification of toxic dyes using biosynthesized iron nanoparticles by photo-Fenton processes. *Int. J. Environ. Sci. Technol., 15*, 1649–1656. https://doi.org/10.1007/s13762-017-1510-0.
37. Srivastava, N., & Mukhopadhyay, M., (2014). Biosynthesis of SnO_2 nanoparticles using bacterium *Erwinia herbicola* and their photocatalytic activity for degradation of dyes. *Ind. Eng. Chem. Res., 53*, 13971–13979. https://doi.org/10.1021/ie5020052.
38. Nazar, N., Bibi, I., Kamal, S., Iqbal, M., Nouren, S., Jilani, K., Umair, M., & Ata, S., (2018). Cu nanoparticles synthesis using biological molecule of *P. granatum* seeds extract as reducing and capping agent: Growth mechanism and photo-catalytic activity. *Int. J. Biol. Macromol., 106*, 1203–1210. https://doi.org/10.1016/j.ijbiomac.2017.08.126.
39. Khan, A. U., Yuan, Q., Wei, Y., Ul H. Khan, Z., Tahir, K., Khan, S. U., Ahmad, A., et al., (2016). Ultra-efficient photocatalytic deprivation of methylene blue and biological activities of biogenic silver nanoparticles. *J. Photochem. Photobiol. B., 159*, 49–58. https://doi.org/10.1016/j.jphotobiol.2016.03.017.
40. Saha, J., Begum, A., Mukherjee, A., & Kumar, S., (2017). A novel green synthesis of silver nanoparticles and their catalytic action in reduction of methylene blue dye. *Sustain. Environ. Res., 27*, 245–250. https://doi.org/10.1016/j.serj.2017.04.003.
41. Ngoepe, N. M., (2018). Mbita, Z., Mathipa, M., Mketo, N., Ntsendwana, B., & Hintsho-Mbita, N. C., (2018). Biogenic synthesis of ZnO nanoparticles using *Monsonia burkeana* for use in photocatalytic, antibacterial, and anticancer applications. *Ceram. Int., 44*, 16999–17006. https://doi.org/10.1016/j.ceramint.2018.06.142.
42. Edison, T. N. J. I., Atchudan, R., Kamal, C., & Lee, Y. R., (2016). *Caulerpa racemosa*: A marine green alga for eco-friendly synthesis of silver nanoparticles and its catalytic degradation of methylene blue. *Bioprocess Biosyst. Eng., 39*, 1401–1408. https://doi.org/10.1007/s00449-016-1616-7.
43. Abid, M. A., Abid, D. A., Aziz, W. J., & Rashid, T. M., (2021). Iron oxide nanoparticles synthesized using garlic and onion peel extracts rapidly degrade methylene blue dye. *Phys. B Condens. Matter., 622*, 413277. https://doi.org/10.1016/j.physb.2021.413277.
44. Kumar, B., Smita, K., Cumbal, L., & Debut, A., (2015). Ultrasound agitated phytofabrication of palladium nanoparticles using Andean blackberry leaf and its photocatalytic activity. *J. Saudi Chem. Soc., 19*, 574–580. https://doi.org/10.1016/j.jscs.2015.05.008.

45. Suresh, K. C., Surendhiran, S., Manoj, K. P., Ranjth, K. E., Khadar, Y. A. S., & Balamurugan, A., (2020). Green synthesis of SnO_2 nanoparticles using *Delonix elata* leaf extract: Evaluation of its structural, optical, morphological, and photocatalytic properties. *SN Appl. Sci., 2*, 1735. https://doi.org/10.1007/s42452-020-03534-z.
46. Luque, P. A., Chinchillas-Chinchillas, M. J., Nava, O., Lugo-Medina, E., Martínez-Rosas, M. E., Carrillo-Castillo, A., Vilchis-Nestor, A. R., et al., (2021). Green synthesis of tin dioxide nanoparticles using Camellia sinensis and its application in photocatalytic degradation of textile dyes. *Optik., 229,* 166259. https://doi.org/10.1016/j.ijleo.2021.166259.
47. Bhargava, A., Jain, N., Khan, M. A., Pareek, V., Dilip, R. V., & Panwar, J., (2016). Utilizing metal tolerance potential of soil fungus for efficient synthesis of gold nanoparticles with superior catalytic activity for degradation of rhodamine B. *J. Environ. Manage., 183*, 22–32. https://doi.org/10.1016/j.jenvman.2016.08.021.
48. Ebrahimian, J., Mohsennia, M., & Khayatkashani, M., (2021). Green synthesis, and characterization of Ud-SnO_2-ZnO using *Urtica dioica* leaf extract: A nanocomposite photocatalyst for degradation of rhodamine B dye. *Res. Chem. Intermed., 47*, 4789–4802. https://doi.org/10.1007/s11164-021-04546-z.
49. Fatimah, I., Sahroni, I., Muraza, O., & Doong, R., (2020). One-pot biosynthesis of SnO_2 quantum dots mediated by *Clitoria ternatea* flower extract for photocatalytic degradation of rhodamine B. *J. Environ. Chem. Eng., 8*, 103879. https://doi.org/10.1016/j.jece.2020.103879.
50. Basnet, P., Samanta, D., Chanu, T. I., Jha, S., & Chatterjee, S., (2020). Glycine-A bio-capping agent for the bioinspired synthesis of nano-zinc oxide photocatalyst. *J. Mater. Sci. Mater. Electron., 31*, 2949–2966. https://doi.org/10.1007/s10854-019-02839-z.
51. Kamran, U., Bhatti, H. N., Iqbal, M., Jamil, S., & Zahid, M., (2019). Biogenic synthesis, characterization, and investigation of photocatalytic and antimicrobial activity of manganese nanoparticles synthesized from *Cinnamomum verum* bark extract. *J. Mol. Struct., 1179*, 532–539. https://doi.org/10.1016/j.molstruc.2018.11.006.
52. Ahmed, A., Usman, M., Yu, B., Shen, Y., & Cong, H., (2021). Sustainable fabrication of hematite (α-Fe_2O_3) nanoparticles using biomolecules of *Punica granatum* seed extract for unconventional solar-light-driven photocatalytic remediation of organic dyes. *J. Mol. Liq., 339*, 116729. https://doi.org/10.1016/j.molliq.2021.116729.
53. Kolya, H., Maiti, P., Pandey, A., & Tripathy, T., (2015). Green synthesis of silver nanoparticles with antimicrobial and azo dye (Congo red) degradation properties using *Amaranthus gangeticus* Linn leaf extract. *J. Anal. Sci. Technol., 6*, 33. https://doi.org/10.1186/s40543-015-0074-1.
54. Singhal, A., Singhal, N., Bhattacharya, A., & Gupta, A., (2017). Synthesis of silver nanoparticles (AgNPs) using *Ficus retusa* leaf extract for potential application as antibacterial and dye decolorizing agents. *Inorg. Nano-Met. Chem., 47*, 1520–1529. https://doi.org/10.1080/24701556.2017.1357604.
55. Najjar, M., Hosseini, H. A., Masoudi, A., Hashemzadeh, A., & Darroudi, M., (2020). Preparation of tin oxide (IV) nanoparticles by a green chemistry method and investigation of its role in the removal of organic dyes in water purification. *Res. Chem. Intermed., 46*, 2155–2168. https://doi.org/10.1007/s11164-020-04084-0.
56. Najjar, M., Hosseini, H. A., Masoudi, A., Sabouri, Z., Mostafapour, A., Khatami, M., & Darroudi, M., (2021). Green chemical approach for the synthesis of SnO_2 nanoparticles

and its application in photocatalytic degradation of Eriochrome black T dye. *Optik., 242*, 167152. https://doi.org/10.1016/j.ijleo.2021.167152.

57. Zinatloo-Ajabshir, S., Salehi, Z., & Salavati-Niasari, M., (2018). Green synthesis and characterization of $Dy_2Ce_2O_7$ nanostructures using *Ananas comosus* with high visible-light photocatalytic activity of organic contaminants. *J. Alloys Compd., 763*, 314–321. https://doi.org/10.1016/j.jallcom.2018.05.311.
58. Kumar, P., Govindaraju, M., Senthamilselvi, S., & Premkumar, K., (2013). Photocatalytic degradation of methyl orange dye using silver (Ag) nanoparticles synthesized from *Ulva lactuca. Colloids Surf. B Biointerfaces., 103*, 658–661. https://doi.org/10.1016/j.colsurfb.2012.11.022.
59. Francis, S., Joseph, S., Koshy, E. P., & Mathew, B., (2017). Green synthesis and characterization of gold and silver nanoparticles using *Mussaenda glabrata* leaf extract and their environmental applications to dye degradation. *Environ. Sci. Pollut. Res., 24*, 17347–17357. https://doi.org/10.1007/s11356-017-9329-2.
60. Varadavenkatesan, T., Selvaraj, R., & Vinayagam, R., (2016). Phyto-synthesis of silver nanoparticles from *Mussaenda erythrophylla* leaf extract and their application in catalytic degradation of methyl orange dye. *J. Mol. Liq., 221*, 1063–1070. https://doi.org/10.1016/j.molliq.2016.06.064.
61. Vidhu, V. K., & Philip, D., (2014). Catalytic degradation of organic dyes using biosynthesized silver nanoparticles. *Micron., 56*, 54–62. https://doi.org/10.1016/j.micron.2013.10.006.
62. Huang, L., Weng, X., Chen, Z., Megharaj, M., & Naidu, R., (2014). Synthesis of iron-based nanoparticles using oolong tea extract for the degradation of malachite green. *Spectrochim. Acta. A. Mol. Biomol. Spectrosc., 117*, 801–804. https://doi.org/10.1016/j.saa.2013.09.054.
63. Huang, L., Luo, F., Chen, Z., Megharaj, M., & Naidu, R., (2015). Green synthesized conditions impacting on the reactivity of Fe NPs for the degradation of malachite green. *Spectrochim. Acta. A. Mol. Biomol. Spectrosc., 137*, 154–159. https://doi.org/10.1016/j.saa.2014.08.116.
64. Weng, X., Huang, L., Chen, Z., Megharaj, M., & Naidu, R., (2013). Synthesis of iron-based nanoparticles by green tea extract and their degradation of malachite. *Ind. Crops Prod., 51*, 342–347. https://doi.org/10.1016/j.indcrop.2013.09.024.
65. Sorbiun, M., Shayegan, M. E., Ramazani, A., & Taghavi, F. S., (2018). Green synthesis of zinc oxide and copper oxide nanoparticles using aqueous extract of oak fruit hull (Jaft) and comparing their photocatalytic degradation of basic violet 3. *Int. J. Environ. Res., 12*, 29–37. https://doi.org/10.1007/s41742-018-0064-4.
66. Hassan, A. K., Al-Kindi, G. Y., & Ghanim, D., (2020). Green synthesis of bentonite-supported iron nanoparticles as a heterogeneous fenton-like catalyst: Kinetics of decolorization of reactive blue 238 dye. *Water Sci. Eng., 13*, 286–298. https://doi.org/10.1016/j.wse.2020.12.001.
67. Wicaksono, W. P., Sahroni, I., Saba, A. K., Rahman, R., & Fatimah, I., (2020). Biofabricated SnO_2 nanoparticles using red spinach (*Amaranthus tricolor* L.) extract and the study on photocatalytic and electrochemical sensing activity. *Mater. Res. Express., 7*, 075009. https://doi.org/10.1088/2053-1591/aba55b.
68. Fatimah, I., Purwiandono, G., Hidayat, H., Sagadevan, S., Ghazali, S. A. I. S. M., Oh, W. C., & Doong, R. A., (2021). Flower-like SnO_2 nanoparticle biofabrication using

Pometia pinnata leaf extract and study on its photocatalytic and antibacterial activities. *Nanomaterials, 11*, 3012. https://doi.org/10.3390/nano11113012.

69. Ameri, A., Shakibaie, M., Ameri, A., Faramarzi, M. A., Amir-Heidari, B., & Forootanfar, H., (2016). Photocatalytic decolorization of bromothymol blue using biogenic selenium nanoparticles synthesized by terrestrial actinomycete *Streptomyces griseobrunneus* strain FSHH12. *Desalination Water Treat., 57*, 21552–21563. https://doi.org/10.1080/19443994.2015.1124349.
70. Hoag, G. E., Collins, J. B., Holcomb, J. L., Hoag, J. R., Nadagouda, M. N., & Varma, R. S., (2009). Degradation of bromothymol blue by 'greener' nano-scale zero-valent iron synthesized using tea polyphenols. *J. Mater. Chem., 19*, 8671. https://doi.org/10.1039/b909148c.
71. Fu, L., & Fu, Z., (2015). *Plectranthus amboinicus* leaf extract–assisted biosynthesis of ZnO nanoparticles and their photocatalytic activity. *Ceram. Int., 41*, 2492–2496. https://doi.org/10.1016/j.ceramint.2014.10.069.
72. Mata, R., Bhaskaran, A., & Sadras, S. R., (2016). Green-synthesized gold nanoparticles from *Plumeria alba* flower extract to augment catalytic degradation of organic dyes and inhibit bacterial growth. *Particuology., 24*, 78–86. https://doi.org/10.1016/j.partic.2014.12.014.
73. Quan, X., Zhang, X., & Xu, H., (2015). In-situ formation and immobilization of biogenic nanopalladium into anaerobic granular sludge enhances azo dyes degradation. *Water Res., 78*, 74–83. https://doi.org/10.1016/j.watres.2015.03.024.
74. Th. Devi, B., Begum, S., & Ahmaruzzaman, M., (2016). Photo-catalytic activity of plasmonic Ag@AgCl nanoparticles (synthesized via a green route) for the effective degradation of Victoria blue B from aqueous phase. *J. Photochem. Photobiol. B., 160*, 260–270. https://doi.org/10.1016/j.jphotobiol.2016.03.033.

CHAPTER 14

Nanoremediation: A Sustainable Reclamation Method for Future Deployment

KHAIR UL NISA,[1,2] NAJEEBUL TARFEEN,[2] BURHAN HAMID,[2] QADRUL NISA,[3] HUMAIRA,[2] SABA WANI,[4] ZAFFAR BASHIR,[2] ALI MOHD. YATOO,[2] and SHABIR H. WANI[5]

[1]*Department of Environmental Science, University of Kashmir, Srinagar, Jammu and Kashmir, India*

[2]*Center of Research for Development (CORD), University of Kashmir, Srinagar, Jammu and Kashmir, India*

[3]*Division of Plant Pathology, SKUAST-K, Shalimar, Srinagar, Jammu and Kashmir, India*

[4]*Department of Biochemistry, University of Kashmir, Srinagar, Jammu and Kashmir, India*

[5]*Mountain Research Center for Field Crops, Khudwani, Sher-e-Kashmir University of Agricultural Sciences and Technology, Srinagar, Jammu and Kashmir, India*

ABSTRACT

Nanotechnology has fascinated scientists and researchers for exploitation of unparalleled biological, physical, and chemical characteristics of nanoparticles. Nano-formed compounds are developed for utilization in a diverse number of fields from medicine to the space exploration. Because of high surface area to volume ratio, size-dependent attributes and high reactivity,

Nano-Bioremediation for Water and Soil Treatment: An Eco-Friendly Approach.
Vishnu D. Rajput, Arpna Kumari, and Tatiana M. Minkina (Eds.)

nano-compounds are used for the reclamation of soil and wastewater. Nanoremediation technology has proven to be more effective than the conventional remediation technologies due of its high speed of degradation and its ability to be effective across wide range of environmental conditions. Nanoremediation approaches like nanosensors, nanoadsorbent sand nano-material-based photocatalysts have shown promising results for remediation of wide range of persistent environmental toxicants. Nanoremediation has also potential for source term treatment as well as the enhancement effect by synergistic action that were very less observed in conventional technologies. These notable advantages make nanotechnology a suitable technology for upcoming future for the reclamation of resources.

14.1 INTRODUCTION

Environmental sustainability primarily focuses on the balanced interaction between humans and environmental resources for the stability of present and future generations. The mutual harmony between the ecology and environment is very crucial for the proper management of natural resources [1]. Among the UN sustainable development goals (SDGs), the Environment Sustainable and economic security goals were of prime importance [2]. The Sustainable Remediation Forum for UK (SuRF-UK) came up with sustainability criteria or checklist for the remediation of contaminated land along with the complete guidance assessment strategies [3]. As per the data furnished by the Global Sustainable Development Report [4], by 2030 no country in the world will be in the capacity to achieve UNSDG, indicating the risk the world is heading [4]. Climate change, global warming, resource depletion, and biodiversity loss are all key concerns facing the environment today. For example, a recent report by World Water Development (2019), estimated an overall increase in water consumption by 1% from 1980 and it may reach up to 3% by 2050 [5]. Likewise, indiscriminate use of fossil fuels has hastened global warming, with a 3–5°C increase in temperature predicted by 2050. The problem has been exacerbated by the poisoning of land and water by a range of toxicants. Globally, more than 20 million hectors of land are contaminated by heavy metals (Cd, As, Hg, Ni, Zn, Cu, Cr) [6]. Therefore, it is the need of the hour to develop an effective and sustainable method for sustaining life on this planet [7].

Nanotechnology has emerged as an effective and innovative field to improve overall environment remediation methods. Broadly, speaking nanotechnology involves the application of matter at dimensions approx. between 1 to 100 nanometers [8]. It involves well engineering technology to manipulate and model matter at this length's scale for novel application purposes. It has been widely used in numerous fields like health, biotechnology, industrial sectors (semiconductors, memory storage) and optical technologies. However, a lot of efforts and inclination to use this novel technology for cleanup hazardous environmental wastes has been taken up. Scientists, all around the world suggest it as an emerging sustainable way-out and effective replacement for conventional remediation technologies [9]. Nanoremediation involves nanomaterials to tackle down the pollution crisis of 21st century like management of contaminated land, restoration of degraded environments [10]. Nanotechnology has provided a successful alternative approach for reducing the time and cost of existing cleanup technologies (physical and chemical stabilization) while also lowering pollution levels to near-zero levels [11]. Nanoremediation technology potentially involves the use of reactive nanomaterials like metaloxides, nanodots, carbon nanotubes, fibers, enzymes BNP's to transform and degrade the wide range of e-toxins [8, 12]. The most common nanomaterial used is zero-valent iron (n-ZVI) [8]. Nanomaterials, because of their minute size are able to pervade minute spaces and can travel farther and in-depth in comparison to larger molecules and thus offer highly desired properties and cleaning efficiencies then macro-sized properties. Their huge surface area and highly reactive nature make them versatile degrading agents for stubborn contaminants [13, 14]. Nanosensors have been used to detect and degrade a wide range of contaminants, including pesticides, pharmaceuticals, and aromatic hydrocarbons. From the recent few years, the nano-based technology has gained much interest in the reclamation processes as they primarily focus on sustainable remediation strategies. This review aims to present a broader outline of nanoremediation and its major application in detection of numerous toxicants using nanosensors (Figure 14.1). Nanophotocatalysts for the degradation of various persistent pollutants has been also discussed.

14.2 CONVENTIONAL REMEDIATION TECHNOLOGIES V/S NANOREMEDIATION

Undoubtedly, number of conventional techniques are being employed for the remediation of contaminated environment. The major constrains

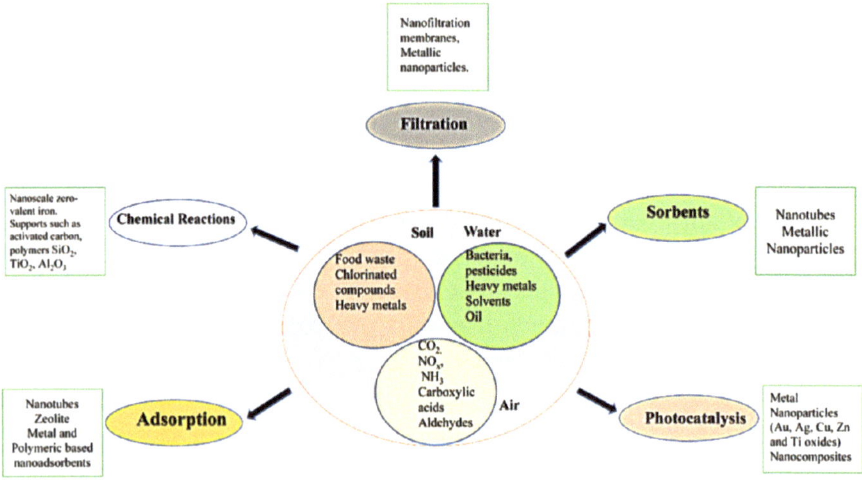

FIGURE 14.1 Depiction of various nanoremediation strategies take-up to clean different environmental media.

associated with these traditional technologies mainly include the cost and cleanup time. More than $1 billion has been provided to the US EPA for the reclamation of polluted sites under the American Recovery and Reinvestment Act (2009). Furthermore, using traditional processes, the US EPA was only able to complete a small portion of site rehabilitation [8]. Furthermore, it is projected that about $250 billion was spent on the cleanup of abandoned hazardous waste sites under the National Priorities List (NPL) of the United States (U.S. EPA, 2009). Pump and treat technique, for example, has been widely utilized to clean contaminated ground water. According to a US EPA study [15] the average cost incorporated for this technique was approximately 7,67,000 $/site representing huge cost. Similarly, several other in-situ techniques, such as permeable reactive barrier (PRB), thermal treatment, chemical oxidation, and surfactant cosolvent flushing, have high costs and lengthy operating procedures, making them difficult to use and sometimes infeasible in comparison to nanoremediation, which has significantly reduced both the financial and operational burden [11]. Nanoremediation has decreased operational costs and cleanup time in reclaiming contaminated sites by nearly 80%. The global commercial market of devices based on nanotechnology have also gained much share with net increase from $432 million to $4,100 million from 1997–2005 [16]. Similarly, the Asian nanotechnology market has grown at a breakneck pace, with an annual growth rate of 20.7%, showing a huge need for prospective nanomaterials [9].

14.3 NANOADSORBENTS AND NANOCATALYSTS FOR THE REMEDIATION OF WASTEWATER

Nanoremediation has proved to be efficient technique for the remediation of wastewater contaminated with heavy metals, organic solvents. Nanoparticle membrane permeability enables for full organic contaminant breakdown in wastewaters. For example, Co doped $BiVO_4$ nanocomposite have been effective in the elimination of malachite green tides caused by *Escherichia coli* and *Chlamydomonas plustilla* [17]. In another work, carbon nanotubes and nanoscale zero valent iron (nZVI) were employed to remove pesticides (DDT, lindane) and heavy metals from ground water [18]. For the treatment of organics in wastewater and activated sludge, carbon nanotubes have been used. Their remarkable ability to get attached with number of functional groups and consequently transforming them into environmental benign compounds thus increases their resilience [19, 20]. Owing to the excellent mechanical and electronic properties, researchers are now inclined to use these for the treatment of liquid or gaseous pollutants in wastewater [21]. Magnesium ferrite ($Mg_{0.27}Fe_{2.50}O_4$) nanocrystallites have shown positive results for the removal of As(III) and As(V) in the waste waters [22]. When compared to conventional bacterial aerobic granular sludge, algal-bacterial aerobic granular sludge effectively bioadsorbs chromium (CrVI) from synthetic wastewaters [23]. Similarly, the bioadsorption of Thorium (VI) was done using alginate-immobilized *Aspergillus niger* microsphere with about 303.95 mg/g bioadsorption potential at pH 6 [24]. Researchers are currently interested in graphene oxide (GO) as a strong adsorbent for wastewater treatment due to its simple design, low cost, and high efficiency in removing pollutants from effluents [25, 26]. Table 14.1 shows the most commonly used nanoadsorbents for the remediation of wastewater.

14.4 NANOADSORBENTS AND NANOCATALYSTS FOR THE RECLAMATION OF CONTAMINATED SOIL

Soil is a crucial natural resource and ever since the onset of industrialization, the condition of soil is degrading tremendously. Nanoremediation has effective potential to remediate the polluted soil chiefly via contaminant neutralization or by transforming them to non-toxic state [9]. For example, the nanoscale zero-valent iron (nZVI) have been widely utilized to remove organic pollutants as well as metal-polluted soils. Their high reactivity and tremendous adsorption capacity makes them more suitable [8]. Qaio et al. demonstrated that the iron-based nanoparticles supported on biochar were effective in degradation of Cd contaminated soil [27]. Similarly, nZnVI nanoparticles supported on biochar were used to treat Cr-contaminated soil [28]. In another study, remediation of soil contaminated with multiple metalloids (As, Pb, Cu, Cd, and Zn) was achieved by treating the soil with nZVI nanomaterials. The result showed that all metals were immobilized by nZVI except Cd [29]. In a recent study conducted by Arenas et al. [30] demonstrated three nanoparticles – hydroxapatite, maghemite, and hematite for the reclamation of mining site [30]. The three nanomaterials effectively immobilized As, Pb, and Sb elements. In addition, the findings found that hydroxapatite was effective against Pb, hematite was effective in removing As. Similarly, a wide range of nanomaterials has been investigated for the cleanup of organic polluted soils. For example, silica-based nanoparticles have been investigated for treating hydrocarbon polluted soil. The removal effectiveness of hydrophobic silica stabilized surfactant was around 95%, while hydrophilic stabilized silica surfactant had a removal efficiency of 75% [31]. Wang et al. [32] also communicated that the silica nanoparticles supported on zwitterionic lipid bilayer was efficient in removing benzo-a-pyrene from contaminated soils [32]. List of the various nanomaterials used for the reclamation pf contaminated soil is presented in Table 14.2.

14.5 NANOSENSORS IN NANOREMEDIATION PROCESS

Nanosensors are the devices that convert chemical signals to electrical signals, detect the nature of chemical substance, and quantify the same [9]. The monitoring of major contaminants that affect humans welfare (VOC's, persistent toxins, trace elements) is very essential [33]. Metal oxide nanosensors, for example, have sparked attention in recent years. These are typically tiny sensors in order to detect various hazardous toxic gases like VOC's, poisonous chemicals and biochemicals [34, 35] developed a nanosensor based on zinc doped copper oxide nanomaterials which shows good sensitivity and detection time. Likewise, Yuan et al. [36], developed a nanosensor for the detection of methanol with excellent sensitivity up to

500 ppm based on NiO fibers [36]. Barandun et al. [37] created a nanosensor to detect water soluble gases, especially ammonia. When compared to other ammonia sensors the efficiency was quite good. It has proven to be useful in the food packaging industry chiefly to verify the quality of packed foods [37].

1. **Nanosensors and Toxic Organic Pollutants:** Several techniques like voltammetry, amperometry, and linear sweep voltammetry has been used to identify organic solvents in various environmental media. As such a variety of nanoparticles can be integrated with the aforementioned techniques to speed up the remediation process. Rawat et al. [38] communicated that the nanocomposites-based electrochemical sensors can prove fruitful for the detection of number of hazardous toxic solvents [38]. Zinc-decorated graphene oxide based on glassy ascorbic acid showed better electrocatalytic activity towards the oxidation of ascorbic acid and was capable of sensing ascorbic acid over a wide range of concentrations (1 to 5,000 μM) [39]. In another study, Zircon-based carbon paste electrode was employed for the determination of phenol. This amperometric sensor showed remarkable efficiency [40]. Similarly, Zaibudeen et al. [41] utilized peptide-based non-enzymatic biosensor for the detection of ammonia and urea oxides [41].
2. **Nanosensors and Heavy Metal Detection:** A wide variety of nanoparticles has been used so far for the detection of heavy metals. For example, in the detection of copper (Cu^{2+}) ions in water, dopamine dithiocarbamate-functionalized gold nanoparticles were used [42]. In another study, Chitosan-poly-L-lysine nanocomposite were utilized for the detection of Cd, Cu, and Pb. The detection limits were found to be 0.01, 0.02 μg/L for Cu and 0.02 μg/l for Pb [43]. Yan et al. [44] employed gold nanoparticles modified with graphene and prepared by one pot redox for the colorimetric estimation of Hg^{2+}. The result confirmed that the nanomaterials possess efficient detection potential for Hg^{2+} in water sample analysis with overall detection limits to 0.16 nM [44]. DNA based MoS_2/AU hybrid FET sensors have been reported to be quiet efficient in monitoring Hg with detection limits up to 0.1 nm, i.e., lower than threshold levels of Hg in drinking waters [45].

TABLE 14.1 Application of Nanomaterials for Contaminated Water Remediation

Pollutant	Nanoadsorbent	Efficacy	References
MO	SiO_2/Fe_3O_4	53.19	[46]
MO	γ-Fe_2O_3/SiO_2/chitosan composite	34.29	[47]
Phenol	Magnetic nanopowder	13.5	[48]
Phenol	γ-Fe_2O_3/2C nanocomposite	42.34	[49]
MO	γ-Fe_2O_3/2C nanocomposite	72.68	[49]
Acridine orange	γ-Fe_2O_3	59	[50]
MB crystal violet	Polyacrylic acid-Fe_3O_4	1,99,116	[51]
Metoprolol (MTP)	Fe_3O_4@SiO_2/SiCRG	447	[52]
Congo red	α-Fe_2O_3	413.22	[53]

TABLE 14.2 Application of Nanomaterials for Contaminated Soil Remediation

Pollutants	Nanomaterial Used	Efficacy	References
As	Calcareous soil treated with nZVI	Increasing nZVI concentration (2.5–25 g/kg) and constant time treats As effectively.	[54]
Cl, Co, Cd, Cd, Mg	Soil leachate treated with DTPA functionalized maghemite NPs.	Efficient recovery of toxic metals and remediation of soil-derived contaminants.	[55]
Pb	Comparative treatment with phosphate, compost, nZVI.	nZVI was observed to be least effective in immobilizing Pb.	[56]
As, Cr, Cu, Pb, Zn	Micro- and nano-ZVI compared.	nZVI more effective than ZVI with highest immobilization for As.	[57, 58]
As	a-MnO_2 nanorods treated As in paddy soil.	Effective treatment of As with restricted As influx in rice plants aerial parts controlling As toxicity.	[58]
PCB (tri and tetrachlorobiphenyls)	nZVI combined thermal desorption (300–600°C).	nZVI enhances the treatment of PCB by thermal desorption with maximum treatment efficiency of 98.35%.	[59]

14.6 NANOPHOTOCATALYSTS AND ENVIRONMENTAL REMEDIATION

Photocatalysis is counted as an ecofriendly technique to treat the persistent organic pollutants and other environmental contaminants [60]. This technique speeds up the chemical degradation of pollutants by utilizing photocatalysts, thereby generating electron-hole pairs (e^-/h^+) consequently producing highly reactive species like hydroxyl (OH^-) and superoxide (O_2^-) radicals that ultimately oxidizes the target pollutant [9]. Table 14.3 enlists various types of photocatalysts employed for the remediation of contaminated environments. Pollutants of different types in various like media, air (NO_2, CO, NH_3), waste plastics, water toxicants and pathogens being reported to have been eliminated using this technique. A large number of photocatalysts in their nano-form have been employed to remediate contaminated soil, air, and wastewater [61]. It is noteworthy to mention here that the photocatalysis technique not only aid in degradation process but can be exploited to generate molecular hydrogen, an excellent alternative for fossil fuel [62]. For instance, the most extensively researched nanoform photocatalyst is TiO_2, for its fascinating properties of purifying air, water as well as producing photocatalytic hydrogen [63]. A study it was observed that almost 99% of rhodamine B-dye was degraded by the photocatalytic behavior of anatase TiO_2 [64]. Likewise, another photocatalyst, ZnO has received much attention because of its remarkable electronic properties and lower cost compared to TiO_2. In a comparative study conducted by Tian et al. [65], it was revealed that ZnO possesses better degradation potential for photodegradation of methylene orange compared to DegussaP25 TiO_2 [65]. Combination of a carbon source (GO, g-C_3N_4) with Ag, Zn, and other metals has a lot of potential for removing organic contaminants. Akir et al. [66]; He et al. [67], reported that the efficiency of the nano-photocatalyst could be enhanced by doping the photocatalyst with suitable dopant [66, 67]. It was recently conveyed that the positive interaction between photocatalytic system and microorganisms can help in the degradation of number of organic pollutants. For example, Marsolek et al. [68] proposed the photocatalysis system and microorganism interactions markedly accelerated the degradation process [68].

TABLE 14.3 List of Metal-based Nano-Photocatalysts and Their Use in Environmental Remediation

Nanomaterial Type	Light Used for Irradiation	Environmental Application	References
Ag_2S/Bi_2WO_6	Visible light	RhB degradation.	[69]
CuS/ZnS	Visible light	H_2 production, acid-blue degradation.	[70]
SnO_2 Nanospheres	Visible light	Photo-oxidation of RhB.	[71]
$TiO_2/ZnFe_2O_4$	UV and visible light	Photocatalytic and photo-electrochemical activity.	[72]
$BiOCl/Bi_4Ti_3O_{12}$	Visible light	Photodegradation of methyl orange and para-nitrophenol.	[73]
SnO_2/SrO	UV light	Degradation of azo-dye, drug, and pesticide.	[74]
$Ag_3PO_4/BiOBr$	UV light	Degradation of RhB and phenol.	[75]

14.7 NANOREMEDIATION AND AGRICULTURE

Nanotechnology has a considerable potential to significantly improve agriculture methods. The major challenges linked with the agriculture involves environmental changes, reduced crop yield, nutrient deficient soils, environmental implication related to pesticides and fertilizers and the contamination of agricultural soil with toxic heavy metal pollution [76]. The application of nanotechnology has significantly enhanced crop yields, improved soil nutrient concentrations while minimizing the inputs (chemical fertilizers, pesticides, insecticides). Nanoparticles (NPs) have gotten a lot of attention in recent years because of their potential applications in the environment and agriculture. Nanoremediation uses nanoparticles (NPs) to successfully reduce the levels of harmful heavy metals (HMs) in the soil-plant system. For example, a study conducted by Camilli et al. showed that carbon nanotube (CNT) nano-sponges removed the hazardous organic solvent dichlorobenzene along with Cd from the soil [77]. Another investigation found that the nZVI ion was employed to treat Cr contaminated soil with a 100% immobilization efficacy for Cr(VI). Furthermore, a considerable increase in plant growth has been seen in conjunction with a reduction in Cr(VI) phytotoxicity [28]. Another study found that FeO NPs reduced Cd stress in wheat plants by increasing plant growth, antioxidant levels, and chlorophyll content. Not only this it has transformed every prospect and field of agriculture like food science, fisheries, aquaculture, horticulture, vegetable science and animal husbandry [78]. For example, the sustainable agriculture targets the minimal use of fertilizers and pesticides besides improving crop yield [79]. The use of nanofertilizers and nanopesticides can be an excellent way to reduce contamination and toxicity caused by indiscriminate pesticide and insecticide application. Switching to such an environmentally friendly option will significantly lower the amount of hazardous substances in the soil-plant system. The use of nano-emulsions or nutrient encapsulated nanomaterials have improved nutrient supply and increased crop yield [80]. Liu and Lal [81] have concluded that the nano-fertilizers can potentially supply multiple nutrients in addition these can enhance the performance of chemical fertilizers [81]. This can directly pave way towards environmental protection by preventing soil pollution and accelerated eutrophication [82]. The use of nanopesticides has also reduced the demand of commercial pesticides thereby protecting the environmental pollution and decreased the risk of developing resistance. Furthermore, nanopesticides have improved the solubility and bioavailability of active chemicals in chemical pesticides to target species while avoiding deleterious effects on non-target organisms

[83]. Silver (Ag) based nanoparticles have shown effective results against variety of plant pathogens in dose dependent manner [84, 85]. In addition, a wide range of nanoparticles like silicon, TiO_2 and Cu have proven potential candidates against number of agricultural pests, biotic, and abiotic stresses [86, 87]. Substituting nano-fertilizers and nano-pesticides for the conventional chemical fertilizers can be regarded as a promising approach towards sustainable agriculture chiefly to tackle down the challenges related to environmental pollution and food security [81]. In addition to this nanotechnology is being employed in food packaging industries mainly to increase the shelf life and to improve food stability [88]. Nanopolymers, nanosensors, nanobarcodes are reported to increase the germination ability of wide variety of agricultural seeds [89] which otherwise show low germinating capacity.

14.8 CONCLUSION

Nanoremediation has evolved as an environmentally friendly and innovative solution for addressing the pollution challenge. This technology allows the manipulation of matter resulting in attractive and multitudinal applications. A vast number of nano-engineered materials like nanosensors, nanomaterial based photocatalysts, nanoadsorbents with remarkable properties have been efficiently used for reclamation of the deteriorated environments. Furthermore, nanoremediation has reduced total cost and cleanup time, making it a viable and long-term replacement for traditional remediation technologies. Although, nanoparticles can deliver effective and facile option for the remediation of contaminants, the ecological and health concerns associated with it needs proper consideration.

KEYWORDS

- **environmental remediation**
- **nanoadsorbents**
- **nanophotocatalysts**
- **nanoremediation**
- **nanosensors**
- **nanotechnology**
- **sustainability**

REFERENCES

1. Arora, N. K., et al., (2018). Environmental sustainability: Challenges and viable solutions. *Environmental Sustainability, 1*(4), 309–340.
2. Arora, N. K., & Mishra, I., (2019). *United Nations Sustainable Development Goals 2030 and Environmental Sustainability: Race Against Time* (pp. 339–342). Springer.
3. Bardos, R. P., et al., (2018). The development and use of sustainability criteria in SuRF-UK's sustainable remediation framework. *Sustainability, 10*(6), 1781.
4. Messerli, P., (2018). *Global Sustainable Development Report GSDR 2019*. https://sdgs.un.org/gsdr/gsdr2019.
5. Nations, U., (2018). *World Water Development Report 2019* (Vol. 12, pp. 12–22). UN-Water Publications, United Nations.
6. Liu, L., et al., (2018). Remediation techniques for heavy metal-contaminated soils: Principles and applicability. *Science of the Total Environment, 633*, 206–219.
7. Shields, M. W., et al., (2019). History, current situation, and challenges for conservation biological control. *Biological Control, 131*, 25–35.
8. Karn, B., Kuiken, T., & Otto, M., (2009). Nanotechnology and in situ remediation: A review of the benefits and potential risks. *Environmental Health Perspectives, 117*(12), 1813–1831.
9. Shafi, A., et al., (2021). Nanoremediation technologies for sustainable remediation of contaminated environments: Recent advances and challenges. *Chemosphere*, 130065.
10. Das, A., et al., (2019). Nanotechnology and it's applications in environmental remediation: An overview. *Vegetos, 32*(3), 227–237.
11. Corsi, I., et al., (2018). Ecofriendly nanotechnologies and nanomaterials for environmental applications: Key issue and consensus recommendations for sustainable and ecosafe nanoremediation. *Ecotoxicology and Environmental Safety, 154*, 237–244.
12. Guerra, F. D., et al., (2018). Nanotechnology for environmental remediation: Materials and applications. *Molecules, 23*(7), 1760.
13. Fajardo, C., et al., (2020). Evaluation of nanoremediation strategy in a Pb, Zn and Cd contaminated soil. *Science of the Total Environment, 706*, 136041.
14. Gil-Díaz, M., et al., (2019). Nanoremediation and long-term monitoring of brownfield soil highly polluted with As and Hg. *Science of the Total Environment, 675*, 165–175.
15. EPA, U., (2001). *Cost Analyses for Selected Groundwater Cleanup Projects: Pump and Treat Systems and Permeable Reactive Barriers*. in Tech. Rep. EPA 542-R-00-013, US EPA.
16. Roco, M. C., (2005). International perspective on government nanotechnology funding in 2005. *Journal of Nanoparticle Research, 7*(6), 707–712.
17. Regmi, C., et al., (2017). Cobalt-doped $BiVO_4$ (Co-$BiVO_4$) as a visible-light-driven photocatalyst for the degradation of malachite green and inactivation of harmful microorganisms in wastewater. *Research on Chemical Intermediates, 43*(9), 5203–5216.
18. Lu, H. J., et al., (2016). Mechanism, synthesis, and modification of nano zerovalent iron in water treatment. *Nanoscale, 8*(19), 9962–9975.
19. Tyagi, I., et al., (2017). Nanoparticles as adsorbent; a positive approach for removal of noxious metal ions: A review. *Science Technology and Development, 34*(3), 195–214.
20. Anjum, M., et al., (2019). Remediation of wastewater using various nanomaterials. *Arab J. Chem., 12*(8), 4897–4919.

21. Baby, R., Saifullah, B., & Hussein, M. Z., (2019). Carbon nanomaterials for the treatment of heavy metal-contaminated water and environmental remediation. *Nanoscale Research Letters, 14*(1), 1–17.
22. Tang, W., et al., (2013). Superparamagnetic magnesium ferrite nanoadsorbent for effective arsenic (III, V) removal and easy magnetic separation. *Water Research, 47*(11), 3624–3634.
23. Yang, X., et al., (2020). Enhanced biosorption of Cr (VI) from synthetic wastewater using algal-bacterial aerobic granular sludge: Batch experiments, kinetics, and mechanisms. *Separation and Purification Technology, 251*, 117323.
24. Ding, H., et al., (2019). Alginate-immobilized *Aspergillus Niger*: Characterization and biosorption removal of thorium ions from radioactive wastewater. *Colloids and Surfaces A: Physicochemical and Engineering Aspects, 562*, 186–195.
25. Jayakaran, P., Nirmala, G., & Govindarajan, L., (2019). Qualitative and quantitative analysis of graphene-based adsorbents in wastewater treatment. *International Journal of Chemical Engineering, 2019.*
26. Kyzas, G. Z., Deliyanni, E. A., & Matis, K. A., (2014). Graphene oxide and its application as an adsorbent for wastewater treatment. *Journal of Chemical Technology & Biotechnology, 89*(2), 196–205.
27. Qiao, Y., et al., (2017). Remediation of cadmium in soil by biochar-supported iron phosphate nanoparticles. *Ecological Engineering, 106*, 515–522.
28. Su, H., et al., (2016). Remediation of hexavalent chromium contaminated soil by biochar-supported zero-valent iron nanoparticles. *Journal of Hazardous Materials, 318*, 533–540.
29. Gil-Díaz, M., et al., (2016). A nanoremediation strategy for the recovery of an As-polluted soil. *Chemosphere, 149*, 137–145.
30. Arenas-Lago, D., et al., (2019). Is nanoremediation an effective tool to reduce the bioavailable As, Pb and Sb contents in mine soils from Iberian pyrite belt*? Catena, 176*, 362–371.
31. Chattopadhyay, P., & Karthick, R. A., (2017). *Characterization and Application of Surfactant Foams Produced from Ethanol-Sodium Lauryl Sulfate-Silica Nanoparticle Mixture for Soil Remediation. in Macromolecular Symposia.* Wiley Online Library.
32. Wang, H., Kim, B., & Wunder, S. L., (2015). Nanoparticle-supported lipid bilayers as an in-situ remediation strategy for hydrophobic organic contaminants in soils. *Environmental Science & Technology, 49*(1), 529–536.
33. Hernandez-Vargas, G., et al., (2018). Electrochemical biosensors: A solution to pollution detection with reference to environmental contaminants. *Biosensors, 8*(2), 29.
34. Solanka, P., Kaushik, A., & Agrawal, V., (2011). Nanostructured metal oxide–based sensors. *NPG Asia Mater, 3*, 17–24.
35. Cretu, V., et al., (2016). Synthesis, characterization and DFT studies of zinc-doped copper oxide nanocrystals for gas sensing applications. *Journal of Materials Chemistry A, 4*(17), 6527–6539.
36. Yuan, Z., et al., (2019). Approaches to enhancing gas sensing properties: A review. *Sensors, 19*(7), 1495.
37. Barandun, G., et al., (2019). Cellulose fibers enable near-zero-cost electrical sensing of water-soluble gases. *ACS Sensors, 4*(6), 1662–1669.

38. Rawat, S., & Singh, J., (2022). Green biocomposite materials for sustainable remediation application. In: *Sustainable Nanotechnology for Environmental Remediation* (pp. 273–297). Elsevier.
39. Nithya, M., (2015). Electrochemical sensing of ascorbic acid on ZnO-decorated reduced graphene oxide electrode. *Journal of Biosensors & Bioelectronics, 6*(1), 1.
40. Qin, W., et al., (2014). Amperometric sensors for detection of phenol in oilfield wastewater using electrochemical polymerization of zincon film. *Analytical Methods, 6*(15), 5734–5740.
41. Zaibudeen, A., & Philip, J., (2018). Magnetic nanofluid based non-enzymatic sensor for urea detection. *Sensors and Actuators B: Chemical, 255*, 720–728.
42. Mehta, V. N., Kumar, M. A., & Kailasa, S. K., (2013). Colorimetric detection of copper in water samples using dopamine dithiocarbamate-functionalized Au nanoparticles. *Industrial & Engineering Chemistry Research, 52*(12), 4414–4420.
43. Guo, H., et al., (2017). Highly selective detection of Pb^{2+} by a nanoscale Ni-based metal–organic framework fabricated through one-pot hydrothermal reaction. *Sensors and Actuators B: Chemical, 248,* 430–436.
44. Yan, Z., et al., (2014). Identification of multifunctional graphene–gold nanocomposite for environment-friendly enriching, separating, and detecting Hg^{2+} simultaneously. *ACS Applied Materials & Interfaces, 6*(24), 22761–22768.
45. Zhou, G., et al., (2016). Ultrasensitive mercury ion detection using DNA-functionalized molybdenum disulfide nanosheet/gold nanoparticle hybrid field-effect transistor device. *ACS Sensors, 1*(3), 295–302.
46. Shariati-Rad, M., et al., (2014). Magnetic solid phase adsorption, preconcentration and determination of methyl orange in water samples using silica coated magnetic nanoparticles and central composite design. *International Nano Letters, 4*(4), 91–101.
47. Zhu, H., et al., (2011). Preparation, characterization, and dye adsorption properties of γ-Fe_2O_3/SiO_2/chitosan composite. *Applied Surface Science, 258*(4), 1337–1344.
48. Mihoc, G., Ianoş, R., & Păcurariu, C., (2014). Adsorption of phenol and p-chlorophenol from aqueous solutions by magnetic nanopowder. *Water Science and Technology, 69*(2), 385–391.
49. Istratie, R., et al., (2019). Single and simultaneous adsorption of methyl orange and phenol onto magnetic iron oxide/carbon nanocomposites. *Arabian Journal of Chemistry, 12*(8), 3704–3722.
50. Qadri, S., Ganoe, A., & Haik, Y., (2009). Removal and recovery of acridine orange from solutions by use of magnetic nanoparticles. *Journal of Hazardous Materials, 169*(1–3), 318–323.
51. Mak, S. Y., & Chen, D. H., (2004). Fast adsorption of methylene blue on polyacrylic acid-bound iron oxide magnetic nanoparticles. *Dyes and Pigments, 61*(1), 93–98.
52. Soares, S. F., et al., (2016). Hybrid nanoadsorbents for the magnetically assisted removal of metoprolol from water. *Chemical Engineering Journal, 302*, 560–569.
53. Satheesh, R., et al., (2016). Removal of Congo red from water using quercetin modified α-Fe_2O_3 nanoparticles as effective nanoadsorbent. *Materials Chemistry and Physics, 180*, 53–65.
54. Azari, P., & Bostani, A. A., (2017). Reducing as availability in calcareous soils using nanoscale zero valent iron. *Environmental Science and Pollution Research, 24*(25), 20438–20445.

55. Hughes, D., et al., (2018). Metal removal from soil leachates using DTPA-functionalized maghemite nanoparticles, a potential soil washing technology. *Chemosphere, 209*, 480–488.
56. Gil-Díaz, M., et al., (2018). Comparison of nanoscale zero-valent iron, compost, and phosphate for Pb immobilization in an acidic soil. *Water, Air, & Soil Pollution, 229*(10), 1–11.
57. Danila, V., et al., (2020). Immobilization of metal (loid) s in two contaminated soils using micro and nano zerovalent iron particles: Evaluating the long-term stability. *Chemosphere, 248*, 126054.
58. Li, B., et al., (2019). Mitigating arsenic accumulation in rice (*Oryza sativa* L.) from typical arsenic contaminated paddy soil of southern China using nanostructured α-MnO_2: Pot experiment and field application. *Science of the Total Environment, 650*, 546–556.
59. Liu, J., et al., (2014). Thermal desorption of PCBs from contaminated soil using nano zerovalent iron. *Environmental Science and Pollution Research, 21*(22), 12739–12746.
60. Bates, M. E., et al., (2016). Emerging technologies for environmental remediation: Integrating data and judgment. *Environmental Science & Technology, 50*(1), 349–358.
61. Krishnan, S., et al., (2017). Comparison of various advanced oxidation processes used in remediation of industrial wastewater laden with recalcitrant pollutants. In: *IOP Conference Series: Materials Science and Engineering*. IOP Publishing.
62. Takata, T., & Domen, K., (2019). Particulate photocatalysts for water splitting: Recent advances and future prospects. *ACS Energy Letters, 4*(2), 542–549.
63. Kato, S. I., & Masuo, F., (1964). Titanium dioxide-photocatalyzed oxidation. I. Titanium dioxide photocatalyzed liquid phase oxidation of tetralin. *Kogyo Kagaku Zasshi, 67*, 1136–1140.
64. Yang, H., & Yang, J., (2018). Photocatalytic degradation of rhodamine B catalyzed by TiO_2 films on a capillary column. *RSC Advances, 8*(22), 11921–11929.
65. Tian, C., et al., (2012). Cost-effective large-scale synthesis of ZnO photocatalyst with excellent performance for dye photodegradation. *Chemical Communications, 48*(23), 2858–2860.
66. Akir, S., et al., (2017). Facile synthesis of carbon-ZnO nanocomposite with enhanced visible light photocatalytic performance. *Applied Surface Science, 400*, 461–470.
67. He, S., et al., (2018). High efficient visible-light photocatalytic performance of Cu/ZnO/rGO nanocomposite for decomposing of aqueous ammonia and treatment of domestic wastewater. *Frontiers in Chemistry, 6*, 219.
68. Marsolek, M. D., et al., (2008). Intimate coupling of photocatalysis and biodegradation in a photocatalytic circulating-bed biofilm reactor. *Biotechnology and Bioengineering, 101*(1), 83–92.
69. Mehraj, O., et al., (2015). Ag_2S sensitized mesoporous Bi_2WO_6 architectures with enhanced visible light photocatalytic activity and recycling properties. *RSC Advances, 5*(53), 42910–42921.
70. Harish, S., et al., (2017). Synergetic effect of CuS@ZnS nanostructures on photocatalytic degradation of organic pollutant under visible light irradiation. *RSC Advances, 7*(55), 34366–34375.
71. Parthibavarman, M., Sathishkumar, S., & Prabhakaran, S., (2018). Enhanced visible light photocatalytic activity of tin oxide nanoparticles synthesized by different microwave optimum conditions. *Journal of Materials Science: Materials in Electronics, 29*(3), 2341–2350.

72. Mugunthan, E., Saidutta, M., & Jagadeeshbabu, P., (2018). Visible light assisted photocatalytic degradation of diclofenac using TiO_2-WO_3 mixed oxide catalysts. *Environmental Nanotechnology, Monitoring & Management, 10*, 322–330.
73. Zhang, M., et al., (2015). BiOCl nanosheet/$Bi_4Ti_3O1_2$ nanofiber heterostructures with enhanced photocatalytic activity. *Catalysis Communications, 58*, 122–126.
74. Sultana, S., et al., (2015). SnO_2–SrO based nanocomposites and their photocatalytic activity for the treatment of organic pollutants. *Journal of Molecular Structure, 1098*, 393–399.
75. Mehraj, O., et al., (2015). Fabrication of novel Ag_3PO_4/BiOBr heterojunction with high stability and enhanced visible-light-driven photocatalytic activity. *Applied Surface Science, 332*, 419–429.
76. Parisi, C., Vigani, M., & Rodríguez-Cerezo, E., (2015). Agricultural nanotechnologies: What are the current possibilities? *Nano Today, 10*(2), 124–127.
77. Camilli, L., et al., (2014). A three-dimensional carbon nanotube network for water treatment. *Nanotechnology, 25*(6), 065701.
78. Prasad, R., Kumar, V., & Prasad, K. S., (2014). Nanotechnology in sustainable agriculture: Present concerns and future aspects. *African Journal of Biotechnology, 13*(6), 705–713.
79. Mukhopadhyay, S. S., (2014). Nanotechnology in agriculture: Prospects and constraints. *Nanotechnology, Science and Applications, 7*, 63.
80. DeRosa, M., Monreal, C., Schnitzer, M., Walsh, R., & Sultan, Y., (2010) Nanotechnology in fertilizers. *Nat. Nanotechnol. J., 5*, 91.
81. Liu, R., & Lal, R., (2015). Potentials of engineered nanoparticles as fertilizers for increasing agronomic productions. *Science of the Total Environment, 514,* 131–139.
82. Sekhon, B., (2014). Nanotechnology in agri-food production: An overview. *Nanotechnol. Sci., Appl., 7*, 31–53.
83. Kumar, R. S., et al., (2013). Distinctive effects of nano-sized permethrin in the environment. *Environmental Science and Pollution Research, 20*(4), 2593–2602.
84. Jung, J. H., et al., (2010). The effect of nano-silver liquid against the white rot of the green onion caused by *Sclerotium cepivorum*. *Mycobiology, 38*(1), 39–45.
85. Kim, S. W., et al., (2009). An in vitro study of the antifungal effect of silver nanoparticles on oak wilt pathogen *Raffaelea* sp. *Journal of Microbiology and Biotechnology, 19*(8), 760–764.
86. Paret, M. L., Palmateer, A. J., & Knox, G. W., (2013). Evaluation of a light-activated nanoparticle formulation of titanium dioxide with zinc for management of bacterial leaf spot on rosa 'noare.' *HortScience, 48*(2), 189–192.
87. Paret, M. L., et al., (2013). Photocatalysis: Effect of light-activated nanoscale formulations of TiO_2 on *Xanthomonas perforans* and control of bacterial spot of tomato. *Phytopathology, 103*(3), 228–236.
88. Sharma, C., et al., (2017). Nanotechnology: An untapped resource for food packaging. *Frontiers in Microbiology, 8*, 1735.
89. Mondal, A., et al., (2011). Beneficial role of carbon nanotubes on mustard plant growth: An agricultural prospect. *Journal of Nanoparticle Research, 13*(10), 4519–4528.

Index

F

G

N